Springer
Berlin
Heidelberg
New York
Hong Kong
London
Milan
Paris
Tokyo

Igor R. Shafarevich

Discourses
on Algebra

Translated from the Russian
by William B. Everett

 Springer

Igor R. Shafarevich
Mathematical Institute of the
Russian Academy of Sciences
Ul. Gubkina 8
117 966 Moscow, Russia
e-mail: shafar@mech.math.msu.su

Cataloging-in-Publication Data applied for

Die Deutsche Bibliothek - CIP-Einheitsaufnahme

Shafarevic, Igor R.:
Discourses on algebra / Igor R. Shafarevich. Transl. from the Russian by William B. Everett. - Berlin ;
Heidelberg ; New York ; Barcelona ; Hong Kong ; London ; Milan ; Paris ; Tokyo : Springer, 2002
(Universitext)
ISBN 3-540-42253-6

Originally published as Izbrannye glavy algebry
Matematicheskoe obrazovanie (zhurnal), Moscow 2000

ISBN 3-540-42253-6 Springer-Verlag Berlin Heidelberg New York

---

Mathematics Subject Classification (2000): 11-XX, 12-XX, 13-XX, 15-XX

Springer-Verlag Berlin Heidelberg New York
a member of BertelsmannSpringer Science+Business Media GmbH

http://www.springer.de

© Springer-Verlag Berlin Heidelberg 2003
Printed in Italy

Cover design: *design & production,* Heidelberg
Typesetting by the translator using a TEX macro package
Printed on acid-free paper   SPIN 10838057   46/3142db - 5 4 3 2 1 0

Abstract

In this book by a leading Russian mathematician and full member of the Russian Academy of Sciences, Igor Rostislavovich Shafarevich, the elements of algebra as a field of contemporary mathematics are laid out based on material bordering the school program as closely as possible.

The book can be used as enrichment materials for students in grades 9–12 in both ordinary schools and schools with a deeper study of mathematics and the sciences and also as a book for mathematics teachers.

Preface

I wish that algebra would be the Cinderella of our story. In the mathematics program in schools, geometry has often been the favorite daughter. The amount of geometric knowledge studied in schools is approximately equal to the level achieved in ancient Greece and summarized by Euclid in his *Elements* (third century B.C.). For a long time, geometry was taught according to Euclid; simplified variants have recently appeared. In spite of all the changes introduced in geometry courses, geometry retains the influence of Euclid and the inclination of the grandiose scientific revolution that occurred in Greece. More than once I have met a person who said, "I didn't choose math as my profession, but I'll never forget the beauty of the elegant edifice built in geometry with its strict deduction of more and more complicated propositions, all beginning from the very simplest, most obvious statements!"

Unfortunately, I have never heard a similar assessment concerning algebra. Algebra courses in schools comprise a strange mixture of useful rules, logical judgments, and exercises in using aids such as tables of logarithms and pocket calculators. Such a course is closer in spirit to the brand of mathematics developed in ancient Egypt and Babylon than to the line of development that appeared in ancient Greece and then continued from the Renaissance in western Europe. Nevertheless, algebra is just as fundamental, just as deep, and just as beautiful as geometry. Moreover, from the standpoint of the modern division of mathematics into branches, the algebra courses in schools include elements from *several* branches: algebra, number theory, combinatorics, and a bit of probability theory.

The task of this book is to show algebra as a branch of mathematics based on materials closely bordering the course in schools. The book does not claim to be a textbook, although it is addressed to students and teachers. The development presumes a rather small base of knowledge: operations with integers and fractions, square roots, opening parentheses and other operations on expressions involving letter symbols, the properties of inequalities. All these skills are learned by the 9th grade. The complexity of the mathematical considerations increases somewhat as we move through the book. To help the reader grasp the material, simple problems are given to be solved.

The material is grouped into three basic themes—**Numbers, Polynomials**, and **Sets**—each of which is developed in several chapters that alternate with the chapters devoted to the other themes.

Certain matters related to the basic text, although they do not use more ideas than are already present, are more complicated and require that the reader keep more facts and definitions in mind. These matters are placed in supplements to the chapters and are not used in subsequent chapters.

For the proofs of assertions given in the book, I chose not the shortest but the most "understandable." They are understandable in the sense that they connect the assertion to be proved with a larger number of concepts and other assertions; they thus clarify the position of the assertion to be proved within the structure of the presented area of mathematics. A shorter proof often appears later, sometimes as a problem to be solved.

At the first acquaintance with mathematics, the history of its development usually retreats into second place. Sometimes it even seems that mathematics was born in the form of a perfected textbook. In fact, mathematics has arisen as the result of the work of uncounted scholars throughout many milleniums. To give some attention to that aspect of mathematics, the dates of the lives of the mathematicians (and physicists) mentioned in the text are listed at the end of the book.

There are quite many formulas. For convenience in referring to them, they are numbered. If I only give the formula number when referring to it, then the formula is in the current chapter. For example, if "multiplying equality (16), we obtain ... " is said in Chap. 2, then the formula with the number (16) in Chap. 2 is meant. If a formula in a different chapter is intended, then the number of the chapter is also given, for example, "using formula (12) in Chap. 1." To help find the necessary chapter, the chapter numbers are printed at the top of every left-hand page. Theorems and lemmas are numbered in order throughout the entire book.

The Foundation for Mathematical Education and Enlightenment and especially S. I. Komarov and V. M. Imaikin helped me greatly in preparing the manuscript. S. P. Demushkin took upon himself the labor of reading the manuscript and made many important comments. I convey my heartfelt gratitude to all of them.

I. R. Shafarevich   Moscow, 2000

Added to the English edition:

Finally, I express my cordial gratitude to Bill Everett, who translated this book into English. As far as I can judge, this is beautiful English. However, I am not an expert in this. But certainly, he greatly improved the text as he showed me several mistakes and urged me by his questions to clarify the exposition in some places.

I. R. S.   Moscow, 2002

Contents

1
Integers

Topic: Numbers

1. $\sqrt{2}$ Is Not Rational

Natural numbers arose as a result of *counting*. An important step in mankind's development of logic was the recognition that two eyes, two persons walking together, and the two oars of a boat have something in common, something expressed in the abstract concept *two*. The next step was taken with difficulty, which is evident because the word *three* in many languages is equivalent to *many* or *too much*. But the concept of an endless series of natural numbers was gradually worked out.

After that it was natural to use numbers not only for *counting* but also for *measuring* lengths, areas, weights, and so on. For definiteness, we discuss measuring the lengths of line segments. We first choose a unit of length: millimeter (mm), centimeter (cm), kilometer (km), light year.... Let the line segment U define the unit of length. When the unit of length is chosen, we can try to use it to measure other line segments. If U is completely placed on a line segment A exactly n times, we say that the length of the line segment A is n (Fig. 1a). But as a rule, some small bit, smaller than the line segment U, remains uncovered (Fig. 1b).

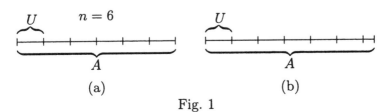

(a)   (b)

Fig. 1

Then we can reduce the unit of length, dividing U into m identical smaller segments U'. If U' is completely placed on the line segment A exactly n' times, then the length of A is equal to n'/m (in the original unit of length U).

People in different lands in the course of milleniums used this procedure in different situations until finally the question arose: **Is such a division of the unit of length always possible?** This totally novel question is related to a specific historical epoch; the question came up in the school of Pythagoras in ancient Greece in the sixth or fifth century B.C. The segments A and U are said to be *commensurable* if there exists a segment U' that can be completely placed on the segment U exactly m times and on the segment A exactly n times. Thus, the question is: **Are any two given line segments commensurable?** Or (yet another form of the question), is the length of any given line segment always a rational number n/m (in terms of a specified unit of length)? The answer is **NO!** And it is very easy to give an example of noncommensurable line segments. We consider a square whose side is the unit of length U. Then we take the diagonal of the square to be the line segment A.

Theorem 1. *The side and the diagonal of a square are noncommensurable.*

Before beginning the proof, we state the theorem in another form. We compare the side and diagonal of a square using the famous Pythagorean theorem: the area of the square constructed on the hypotenuse of a right triangle is equal to the sum of the areas of the two squares constructed on the other two sides of the triangle. Or, in other words, the hypotenuse squared is equal to the sum of the two legs squared.

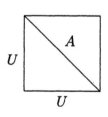

Fig. 2

But the diagonal A of our square is the hypotenuse of an isosceles right triangle whose two legs coincide with the sides U of the square (Fig. 2). Therefore, in our case, $A^2 = 2U^2$, and if A and U are commensurable, that is, if there exists U' such that $A = nU'$ and $U = mU'$, then we would have $(n/m)^2 = 2$, which is to say $n/m = \sqrt{2}$. Thus Theorem 1 can be stated differently.

Theorem 2. $\sqrt{2}$ *is not a rational number.*

We prove the theorem in this form. But first we make one observation. Although we refer to the Pythagorean theorem, we use it in the very special case of an isosceles right triangle. In this case, it is completely obvious.

It immediately follows from the known conditions for the equality of triangles that all five small isosceles right triangles in Fig. 3 are equal. Therefore they have the same area S. But the square constructed on the line segment U consists of two such triangles, and its area is U^2. Therefore $U^2 = 2S$. Analogously, $A^2 = 4S$. Therefore $A^2 = 2U^2$ and $(A/U)^2 = 2$.

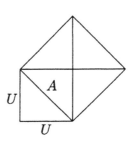

Fig. 3

Now we can turn to the proof of Theorem 2. Because we want to prove the *impossibility* of representing $\sqrt{2}$ in the form $\sqrt{2} = n/m$, it is natural to use *proof by contradiction*. We suppose that $\sqrt{2} = n/m$, where n and m are natural numbers. We take them to be relatively prime, that is, if they had a common divisor, then it was canceled without changing the value of the fraction n/m. By the definition of a square root, the equality $\sqrt{2} = n/m$ means that $2 = (n/m)^2 = n^2/m^2$. Multiplying both sides by m^2, we obtain the equality

$$2m^2 = n^2, \tag{1}$$

where n and m are two relatively prime natural numbers.

Because we have the factor 2 in the left-hand side, naturally, the question is tied to the divisibility of natural numbers by 2. Numbers that are divisible by 2 are called *even numbers*, and those not divisible by 2 are called *odd numbers*. By this definition, each even number k can be represented in the form $k = 2l$, where l is a natural number. That is, we have a certain obvious expression for even numbers, while odd numbers are so far defined purely negatively—this expression is impossible for them. But we can easily obtain an obvious expression for odd numbers, too.

Lemma 1. *Each odd number r can be represented in the form $r = 2s + 1$, where s is a natural number or 0. Conversely, each such number is odd.*

The converse assertion is totally obvious: if $r = 2s + 1$ were even, then it would have a representation $r = 2l$, whence

$$2l = 2s + 1, \qquad 2(l - s) = 1.$$

And this assertion is obviously contradictory.

To prove the primary assertion, we note that if the odd number $r \leq 2$, then $r = 1$ and the expression to be found has the form with $s =$

0. And if the odd number $r > 1$, then already $r \geq 3$. Subtracting 2 from it, we obtain the number $r_1 = r - 2 \geq 1$, where r_1 is again odd. If r_1 is still greater than 1, then we again subtract 2 and obtain $r_2 = r_1 - 2$. With this procedure, we obtain a decreasing series of odd numbers $r, r_1, r_2, \ldots$ in which each successive number is 2 less than the preceding number. We can continue this as long as $r_i > 1$. Because natural numbers cannot decrease without limit, we eventually arrive at the case where subtracting 2 is no longer possible, that is, $r_i = 1$. We obtain

$$r_i = r_{i-1} - 2 = r_{i-2} - 2 - 2 = \cdots = r - 2 - 2 - \cdots - 2 = r - 2i = 1.$$

This means that $r = 2i + 1$, which was to be proved.   □

We can now introduce a fundamental property of even and odd numbers.

Lemma 2. *The product of two even numbers is even, the product of an even number and an odd number is even, and the product of two odd numbers is odd.*

The first two assertions are obvious from the definition of an even number: if $k = 2l$, then no matter what the second factor m is, even or odd, always $km = 2lm$, which means it is even. But to prove the final assertion, we need Lemma 1. Let k_1 and k_2 be odd numbers. According to Lemma 1, we can represent them in the form

$$k_1 = 2s_1 + 1, \qquad k_2 = 2s_2 + 1,$$

where s_1 and s_2 are natural numbers or 0. Then

$$k_1 k_2 = (2s_1 + 1)(2s_2 + 1) = 4s_1 s_2 + 2s_1 + 2s_2 + 1 = 2s + 1,$$

where $s = 2s_1 s_2 + s_1 + s_2$. We saw that any number of the form $2s + 1$ is odd, and this means that $k_1 k_2$ is odd.   □

We note a special case of Lemma 2: *the square of an odd number is odd.*

Now it is completely easy to finish the proof of Theorem 2. We suppose that equality (1) is satisfied, where m and n are relatively prime natural numbers. If n were odd, then by Lemma 2, n^2 would be odd, but it is even according to equality (1). Therefore, n is even and can be represented in the form $n = 2s$. But m and n are relatively prime, which means that m must be odd (otherwise m and n would have the common divisor 2). Substituting the expression for n in equality (1) and dividing by 2, we obtain

$$m^2 = 2s^2,$$

that is, the square of an odd number is even, which contradicts Lemma 2. Theorem 2, and that means Theorem 1 also, is proved. □

We can look at Theorems 1 and 2 from a different point of view, presuming that the result of measuring the length of a line segment (with a given unit of length) is a certain number and that the square root of any positive number is a certain number. Then Theorems 1 and 2 assert that in the case of the diagonal of a square or in the case of $\sqrt{2}$, this certain number is not rational, in other words, it is *irrational*. This is the simplest example of an irrational number. Many irrational numbers exist. We meet some of them later. Probably, the most "well-known" irrational number (after $\sqrt{2}$) is π, the ratio of the circumference of a circle to its diameter. But to prove that π is irrational requires more complex means than we use here, and so we do not prove it in this book.

All numbers both rational and irrational together comprise the real numbers. In one of the following chapters, we try to state more precisely the logical meaning of the concept of a real number. Until then, we continue to use real numbers in the form that they are usually taught in school, not thinking especially about their logical basis.

Why did it take mankind so long to recognize the existence of such a simple, and at the same time important, circumstance as the existence of irrational numbers? The answer is simple—because for any practical purpose, $\sqrt{2}$, for example, can be considered a rational number. We state this assertion in the form of a theorem.

Theorem 3. *No matter how small a number ε is given, it is possible to find a rational number $a = m/n$ such that $a < \sqrt{2}$ and $\sqrt{2} - a < \varepsilon$.*

Because all practical measurements by necessity are taken with a certain accuracy, we can consider that $\sqrt{2}$ is rational with that degree of accuracy; we can say that our *measurement* gives us $\sqrt{2}$ as a rational number.

To prove Theorem 3, we take our "no matter how small" number ε in the form $1/10^n$ with sufficiently large n. We then find a natural number k such that

$$\frac{k}{10^n} \leq \sqrt{2} < \frac{k+1}{10^n}. \tag{2}$$

Then we can let $a = k/10^n$ because $\sqrt{2} - k/10^n < 1/10^n$.

Inequality (2) is equivalent to the inequality

$$\frac{k^2}{10^{2n}} \leq 2 < \frac{(k+1)^2}{10^{2n}}$$

or
$$k^2 \leq 2 \cdot 10^{2n} < (k+1)^2.$$

Because the number n (and therefore the number $2 \cdot 10^{2n}$ also) is a given specific number, there exists the greatest natural number k whose square is not greater than $2 \cdot 10^{2n}$. This k gives the value of a that was needed.

Obviously, the conclusion in Theorem 3 holds not only for the number $\sqrt{2}$ but also for any positive real number x (we restrict ourselves here to positive numbers for simplicity). This becomes obvious if we represent x as a point on the number line, divide the unit of length U into the small segments $U/10^n$, and then cover the number line with these small segments (Fig. 4).

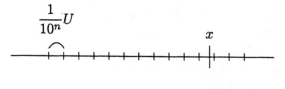

Fig. 4

Then the last mark that is not to the right of x gives the needed rational number. If this mark is the kth mark, then

$$a = \frac{k}{10^n} \leq x \quad \text{and} \quad x - a < \frac{1}{10^n}.$$

Theorem 3 is proved.   □

But now estimate, please, the depth of the assertion contained in Theorems 1 and 2. This assertion could never be confirmed by any kind of experiment, because an experiment is always conducted with a certain accuracy. And with *any* specified accuracy, $\sqrt{2}$ can be expressed as a rational number! This achievement of pure reason could not appear, even as a result of the accumulation of many milleniums of human experience, until the revolution in mathematics that was accomplished in ancient Greece during the 7th–5th centuries B.C. It is not surprising that in the school of Pythagoras, this knowledge was considered a sacred secret that must be withheld from the uninitiated. But one of the Pythagoreans, Hippas, revealed the secret. According to one legend, the gods punished him for this with death resulting from a shipwreck. A hundred years later, Plato related in the book *Laws* how he, no longer young, was smitten when he learned that it is not always possible "to measure a length with a length." He told of his "disgraceful ignorance."

"It seemed to me that this was characteristic not of a human but of some sort of swine. And I was embarrassed not only for myself but also for all Greeks."

Proved Theorems 1 and 2 can shed light on a question mathematicians often pose: Why are theorems proved? The first answer that comes to mind is to become convinced of the truth of some assertion. But it sometimes happens that the accumulation of particular cases of such verification is so large that the truth of the assertion no longer elicits any doubt (and often elicits the ridicule of physicists who say that mathematicians prove truths that were already beyond doubt). But we saw that sometimes the proof introduces mathematicians to a totally new world of mathematical ideas that would never have been known without the proof.

Problems:
1. Prove that the numbers $\sqrt{6}$ and $\sqrt[3]{2}$ are irrational.
2. Prove that the number $\sqrt{2} + \sqrt{3}$ is irrational.
3. Prove that the number $\sqrt[3]{3} + \sqrt{2}$ is irrational.
4. Find $\sqrt{2}$ to the accuracy of $1/100$.
5. Prove that every natural number can be represented as a sum of terms of the form 2^k such that a given term occurs no more than once. Prove that there is only one such representation for each natural number.

2.   The Irrationality of Other Square Roots

It would be interesting now to try generalizing the results we obtained in the previous section. For example, can we use the same method to prove that $\sqrt{3}$ is irrational? It is obviously natural to attempt adapting our previous arguments to the new situation. We must now prove the impossibility of the equality $3 = (n/m)^2$ or

$$3m^2 = n^2, \tag{3}$$

where, as in Sec. 1, we can consider the fraction n/m to be reduced, that is, the natural numbers m and n are relatively prime. Because the factor 3 appears in equality (3), we naturally call upon the property of divisibility by 3. We see how Lemmas 1 and 2 can be adapted to this new case.

Lemma 3. *Each natural number r either is divisible by 3 or can be represented in one of the two forms $r = 3s + 1$ or $r = 3s + 2$, where s is a natural number or 0. Conversely, numbers of the form $3s + 1$ or $3s + 2$ are not divisible by 3.*

The converse assertion is obvious. If, for example, $n = 3s + 1$ were divisible by 3, then we would have $3s+1 = 3m$, that is, $3(m-s) = 1$. But such an assertion is a contradiction. And if $n = 3s + 2$ were divisible by 3, then we would have $3s + 2 = 3m$ and $3(m - s) = 2$, which is again a contradiction. The primary assertion in Lemma 3 is proved by a literal repetition of the arguments in the proof of Lemma 1. If r is not divisible by 3 and is less than 3, then either $r = 1$ or $r = 2$, and the needed representation is obtained with $s = 0$. And if $r > 3$, then subtracting 3 from it, we obtain $r_1 = r - 3 > 0$, where r_1 is again not divisible by 3. Continuing succesive subtractions of 3, we obtain a decreasing series of numbers r, $r_1 = r - 3$, $r_2 = r - 3 - 3$, $\ldots$, $r_s = r - 3 - 3 - \cdots - 3$, where, as previously noted, we cannot continue subtracting 3 indefinitely; therefore, either $r_s = 1$ or $r_s = 2$. As a result, we have two possibilities: $r - 3s = 1$, that is, $r = 3s + 1$, or $r - 3s = 2$, that is, $r = 3s + 2$, as was asserted in the lemma.   □

The statement of the next lemma only includes the part in Lemma 2 that we actually use in what follows.

Lemma 4. *The product of two natural numbers that are not divisible by 3 is itself not divisible by 3.*

Let r_1 and r_2 be two natural numbers that are not divisible by 3. According to Lemma 3, we have two possibilities for each of them: representation in either the form $2s + 1$ or the form $2s + 2$. We thus have four possible cases:

1. $r_1 = 3s_1 + 1$ and $r_2 = 3s_2 + 1$,
2. $r_1 = 3s_1 + 1$ and $r_2 = 3s_2 + 2$,
3. $r_1 = 3s_1 + 2$ and $r_2 = 3s_2 + 1$, or
4. $r_1 = 3s_1 + 2$ and $r_2 = 3s_2 + 2$.

Cases 2 and 3 only differ in the labeling of r_1 and r_2, and we can therefore examine just one of these cases, for example, case 2. In the remaining three cases, we multiply and obtain

1. $r_1 r_2 = 9s_1 s_2 + 3s_1 + 3s_2 + 1 = 3t_1 + 1$, where $t_1 = 3s_1 s_2 + s_1 + s_2$;
2. $r_1 r_2 = 9s_1 s_2 + 6s_1 + 3s_2 + 2 = 3t_2 + 2$, where $t_2 = 3s_1 s_2 + 2s_1 + s_2$; and
4. $r_1 r_2 = 9s_1 s_2 + 6s_1 + 6s_2 + 4 = 3t_4 + 1$, where $t_4 = 3s_1 s_2 + 2s_1 + 2s_2 + 1$.

In the last case, we separated the term 4 into the sum of 3 and 1 and combined the 3 with the other terms divisible by 3. As a result, we obtain numbers of the form $3t + 1$ or $3t + 2$, which are not divisible by 3 (see Lemma 3).   □

Now, transferring Theorem 2 to our case does not present any difficulties.

Theorem 4. $\sqrt{3}$ *is not a rational number.*

The proof repeats the proof of Theorem 2 almost literally. We must show that equality (3), $3m^2 = n^2$, where m and n are relatively prime natural numbers, leads to a contradiction. If the number n were not divisible by 3, then its square would not be divisible by 3 according to Lemma 4. But n is equal to $3m^2$, which means that n is divisible by 3: $n = 3s$. Substituting this in equality (3) and dividing the equality by 3, we obtain $m^2 = 3s^2$. But because m and n are relatively prime and n is divisible by 3, m is not divisible by 3. And according to Lemma 4, the square of m is not divisible by 3; but it is equal to $3s$. This contradiction proves the theorem. □

Such parallelism of all arguments in the two cases makes us think that this path can be continued farther. Of course, there is no sense in examining $\sqrt{4}$, because $\sqrt{4} = 2$, but we can apply our arguments to $\sqrt{5}$. The number of products that must be listed in proving the analogues of Lemmas 2 and 4 obviously increases noticeably. But all the same, we can verify that $\sqrt{5}$ is irrational by using such an approach. It is entirely possible to continue the argument for other cases, examining $\sqrt{6}$, $\sqrt{7}$, and so on. However, the number of verifications that must be accomplished becomes increasingly large. Going through all integers up to 20, for example, we can be convinced that $\sqrt{n}$ is irrational excluding the obvious exceptions where n is the square of an integer ($n = 4$, 9, or 16). In this way, at the cost of ever more complicated calculations, we can be convinced that $\sqrt{2}$, $\sqrt{3}$, $\sqrt{5}$, $\sqrt{6}$, $\sqrt{7}$, $\sqrt{8}$, $\sqrt{10}$, $\sqrt{11}$, $\sqrt{12}$, $\sqrt{13}$, $\sqrt{14}$, $\sqrt{15}$, $\sqrt{17}$, $\sqrt{18}$, and $\sqrt{19}$ are irrational. This suggest a general hypothesis: for any natural number n that is not the square of a natural number, $\sqrt{n}$ is irrational. But our arguments are insufficient for proving this general hypothesis, because there is a place in our arguments where we list all possible cases and verify them directly.

It is interesting that the path we have followed in our arguments was once actually followed by humanity. As we already mentioned, the irrationality of $\sqrt{2}$ was proved in the school of Pythagoras. Later, the irrationality of $\sqrt{n}$ was proved for several different small n before the general problem, which we just formulated, was recognized. We learn of its solution from Plato's dialogue *Theaetetus*.

The events take place about 400 B.C. The author relates how the famous philosopher Socrates meets with the mathematician Theodorus of Cyrene, who acquaints him with his very talented young pupil by the name of Theaetetus. Theaetetus was then about the age of contemporary students: he was 14 or 15 years old. Theodorus describes his abilities thus: "He approaches his studies or any investigation lightly, smoothly, and accurately, as calmly as oil flows from a bowl—and I am surprised at how much can be achieved at such an age." Further on, Theaetetus himself is already telling Socrates about his investigations undertaken with his friend, who is coincidentally also named Socrates, the same as the philosopher. Theaetetus explains that Theodorus had told them about the noncommensurability (if we translate into our modern terminology) of the side of a square and the unit of length if the area of the square is an integer, but not the square of an integer. If the area is n, then this means that $\sqrt{n}$ is irrational. Theodorus had proved this for $n = 2, 3, 5$, "and thus going through the cases one after the other, he came to 17." Theaetetus became interested in the general problem and solved it together with his friend Socrates, as he informs us at the end of the dialogue. We will not try to resurrect those arguments that Theodorus used (there exist several hypotheses), but lay out the proof of the general case, essentially following the approach of Euclid, which quite likely is analogous to the proof by Theaetetus (with one simplication found by Gauss 2000 years later).

We first prove an analogue of Lemmas 1 and 3.

Theorem 5. *For any natural numbers n and m, there exist numbers t and r such that they are either natural or equal to zero, $r < m$, and*

$$n = mt + r. \tag{4}$$

For a given n and m, such a representation is unique.

Representation (4) is called the *division of n by m*, the number t is called the *quotient*, and the number r is called the *remainder*.

The proof follows our known principle. If $m > n$, then the equality is obviously satisfied with $t = 0$ and $r = n$. And if $n \geq m$, then we take $n_1 = m - n$. Obviously, $n_1 \geq 0$. If still $n_1 \geq m$, then we take $n_2 = n_1 - m$. We continue to thus subtract m until we obtain the number $n_t = n - m - m - \cdots - m = r$, where still $r \geq 0$ but already $r < m$. We obtain the needed representation $n - mt = r$, $n = mt + r$.

We prove the uniqueness of representation (4) for a given n and m. Let

$$n = mt_1 + r_1, \qquad n = mt_2 + r_2.$$

We suppose that $t_1 \neq t_2$ and, for example, $t_1 > t_2$. Subtracting the second representation from the first, we obtain $m(t_1 - t_2) + r_1 - r_2 = 0$, that is, $m(t_1 - t_2) = r_2 - r_1$. Because by our supposition $t_1 > t_2$ (and consequently $r_2 > r_1$, but $r_2 < m$), we have a positive number less than m on the right and a number divisible by m on the left. This is impossible.   □

Before turning to the proof of the analogue of Lemmas 2 and 4, we should introduce (more precisely, recall) an important concept. A natural number different from 1 is said to be *prime* if it has no divisors except itself and 1. For example, among the first 20 numbers, we have the prime numbers 2, 3, 5, 7, 11, 13, 17, and 19.

Although obvious, it is an important property that *every natural number different from 1 has at least one prime divisor*. Indeed, if a number n has no divisors except itself and 1, then it is its own prime divisor. And if n has other divisors, then $n = ab$, where $a < n$ and $b < n$. We now consider a, which again is either prime (which means it is a prime divisor of n) or has two factors, $a = a_1 b_1$. Then $n = a_1(b_1 b)$, where $a_1 < a$, that is, a_1 is a divisor of n. Applying the same arguments to a_1 and so on, we obtain decreasing divisors of n: $a_r < \cdots < a_1 < a < n$. We must stop somewhere. If we cannot obtain a smaller divisor than a_r, then a_r is a prime divisor of n.

Now we can prove the analogue of Lemmas 2 and 4.

Theorem 6. *If the product of two integers is divisible by a certain prime number, then at least one of those two numbers is divisible by that prime number.*

We suppose that we want to prove the theorem for the prime number p. We will prove it for all prime numbers in order of increasing size (as we in essence have done, proving it in Lemma 2 for $p = 2$ and in Lemma 4 for $p = 3$). Therefore, when we reach the prime number p, we can consider that the lemma is already proved for smaller prime numbers $q < p$. Let $n_1 n_2$ be divisible by p, but neither n_1 nor n_2 be divisible by p. Then

$$n_1 n_2 = pa. \qquad (5)$$

We apply Theorem 4 to the numbers n_1 and p and to the numbers n_2 and p. We obtain the representations

$$n_1 = pt_1 + r_1, \qquad n_2 = pt_2 + r_2,$$

where r_1 and r_2 are natural numbers less than p (they are not equal to 0, because otherwise n_1 or n_2 would be divisible by p). Substituting these representations in relation (5) and gathering the numbers that are divisible by p, we obtain

$$r_1 r_2 = p(a - t_1 r_2 - t_2 r_1 - p t_1 t_2),$$

that is,

$$r_1 r_2 = pb, \qquad b = a - t_1 r_2 - t_2 r_1 - p t_1 t_2,$$

where now, in contrast to (5), $r_1 < p$ and $r_2 < p$. If $r_1 = 1$ and $r_2 = 1$, then we obtain the contradiction $1 = pb$. Let $r_1 > 1$. We know that r_1 has a prime factor q, which is not greater than r_1 and therefore is less than p. Let $r_1 = q a_1$. Then equality (6) yields

$$q(a_1 r_2) = pb. \tag{7}$$

As we already said, we can consider that the theorem is already proved for prime numbers smaller than p, in particular, for this q. Because pb is divisible by q, one of the two factors must be divisible by q. It cannot be p, because p is a prime number. Therefore, b is divisible by q: $b = q b_1$. Substituting in equality (7) and simplifying, we obtain

$$a_1 r_2 = p b_1$$

with $a_1 < r_1$ and $b_1 < b$. If $a_1 \neq 1$, then we apply the same arguments to it and reduce the equality by yet another prime number. Because the numbers a, a_1, and so on constantly decrease, we must reach a stopping point, that is, we come to the number 1. We finally obtain $r_2 = p b'$, and this is impossible because $r_2 < p$ (and, as we saw, $r_2 > 0$). This proves the theorem. □

You see that the argument is similar to the proof of Lemmas 2 and 4: the assertion is reduced to the case where n_1 and n_2 in equality (5) (that is, r_1 and r_2) are less than p. But here the listing of all cases and the direct verification is replaced by an elegant argument using the fact that the theorem can be considered to hold for all smaller values of p. (Euclid proved Theorem 5 somewhat differently. The argument by moving to smaller numbers presented here apparently belongs to Gauss.)

Now, the proof of irrationality in the general case does not demand any new considerations.

Theorem 7. *If c is a natural number and is not the square of a natural number, then c is not the square of any rational number, that is, $\sqrt{c}$ is irrational.*

We can again verify our assertion, moving from one natural number to the next larger one, and we can therefore consider that the theorem is proved for all smaller values of c. Then we can assume that c is not divisible by any square of a natural number greater than 1. Indeed, if $c = d^2 f$ with $d > 1$, then $f < c$ and f is not the square of a natural number: $f = g^2$ would give us $c = (dg)^2$ and thus contradict the hypothesis of the theorem. We can therefore consider that the theorem is already proved for f, that is, $\sqrt{f}$ is an irrational number. But then $\sqrt{c}$ cannot be a rational number. Indeed, if $\sqrt{c} = n/m$ and we consider that $\sqrt{c} = d\sqrt{f}$, then $n/m = d\sqrt{f}$ and $\sqrt{f} = n/(dm)$, that is, $\sqrt{f}$ would be rational.

We now turn to the basic part of our proof. We suppose that $\sqrt{c}$ is rational and $\sqrt{c} = n/m$, where, as before, we can take n and m to be relatively prime. Then $m^2 c = n^2$. We take an arbitrary prime divisor of the number c, for example, p. Let $c = pd$, where d is not divisible by p, because otherwise c would be divisible by p^2 and we already assumed that c is not divisible by the square of any natural number. From the equality $m^2 c = n^2$, we see that n^2 is divisible by p. It follows from Theorem 6 that n is then divisible by p. Let $n = pn_1$. Substituting the equalities $n = pn_1$ and $c = pd$ in the relation $m^2 c = n^2$, we obtain $m^2 d = pn_1^2$. Because m and n are relatively prime and n is divisible by p, m is not divisible by p. According to Theorem 6, m^2 is not divisible by p. And as we saw, d is also not divisible by p, otherwise c would be divisible by p^2. The equality $m^2 d = pn_1$ leads us to a contradiction of Theorem 6. □

We note that more than once in this section, we constructed a proof of some assertion about the natural number n by assuming that we can take the numbers one after another. We verify that the assertion holds for $n = 1$. Then we prove it for an arbitrary n, supposing that it is already proved for all smaller values.

Here we rely upon an assertion that should be considered one of the axioms of arithmetic: *If a certain property of natural numbers n is true for $n = 1$ (or $n = 2$) and if its truth for all natural numbers less than n implies that it is true for n, then it is true for all natural numbers.*

This property is called the *principle of mathematical induction*. Sometimes it is assumed that the property holds not for all numbers less than n but just for $n - 1$. The assertion for the case where $n = 1$ or $n = 2$ with which we begin the argument (sometimes it is convenient to take $n = 0$) is called the *basis of induction*. The assertion for $n - 1$, which we assume to be already proved, is called the *induction assumption*.

The principle of mathematical induction is also used in definitions where a certain concept depending on a number or index n, which is a natural number, is defined on the assumption that it is already defined for $n - 1$. For example, when we define an arithmetic progression by stating that each member of the progression is obtained by adding the same number d (d is called the difference of the progression) to the preceding member, then we have defined the progression by induction. We can write this definition as a formula,

$$a_n = a_{n-1} + d.$$

To establish the entire progression based on this definition, we need only know its first (or zeroth) member: a_1 or a_0.

The principle of mathematical induction describes a fundamental property of the series of natural numbers. The work of the French mathematician, physicist, and philosopher Henri Poincaré is dedicated to analyzing the role of the induction principle. He poses a more general question: Why is mathematics not trivial? After all, they teach us that mathematics consists of proofs. And every proof consists of syllogisms. The typical syllogism is: "All men are mortal. Socrates is a man. Therefore, Socrates is mortal." And so, at the beginning, we know that all men are mortal; at the end, we learn only that Socrates is mortal. Then why, asks Poincaré, doesn't mathematics turn into a big pile of trivialities? In his opinion, mathematics deals with infinite collections of objects such as the natural numbers and the totality of points on a line. (Thus Theorem 7 is true for an infinite number of values of c. But even Theorem 2 in essence asserts that $2n^2 \neq m^2$ for any natural numbers n and m, the number of which is infinite.) But how is this possible in general? After all, all mathematical arguments consist of a finite number of words: how can they lead to an assertion that is true for an infinite totality? Poincaré sees the explanation in the principle of mathematical induction, which, as he says, "contains an infinite number of syllogisms compressed, as it were, into one formula." And this is the ability to describe the infinite series of natural numbers with a finite number of words.

Problems:

1. Prove that $\sqrt{5}$ is irrational using the same method that was used to prove Theorems 2 and 3.
2. Prove that the number of natural numbers that are not greater than n and are divisible by m is equal to the quotient of dividing n by m.
3. Prove that if a natural number c is not the cube of a natural number, then $\sqrt[3]{c}$ is irrational.
4. Replace the argument connected with the consecutive subtraction of the number m in the proof of Theorem 5 with a reference to the principle of mathematical induction.
5. Use the principle of mathematical induction to prove the formula

$$1 + 2 + \cdots + n = \frac{n(n+1)}{2}.$$

6. Use the principle of mathematical induction to prove the inequality $n < 2^n$.

3. Decomposition into Prime Factors

We saw in the preceding section that any natural number has a prime divisor. We can draw other conclusions from this.

Theorem 8. *Every natural number greater than 1 is a product of prime numbers.*

If the number p is itself prime, then we consider the equality $p = p$ a representation of p as the product of a single factor. And if the number $n > 1$ is not prime, then, as we saw, it has a prime divisor different from n: $n = p_1 \cdot n_1$. Because the prime number $p_1 > 1$ by definition, we have $n_1 < n$. We can now repeat the same argument for n_1, and so on. We obtain a decomposition into factors $n = p_1 \cdots p_k n_k$, where $p_1, \ldots, p_k$ are prime numbers and the constants n_k are decreasing, $n > n_1 > n_2 > \ldots$. This process must terminate. We obtain $n_r = 1$ for some value r, and we then have the desired decomposition, the representation of n as the product of prime numbers: $n = p_1 \cdots p_r$.   □

Of course, the reader can easily reformulate our proof more "rigorously" using the principle of mathematical induction.

The procedure used to prove Theorem 8 is not unequivocal: if the number n has several prime factors, then we can single out any of them first. For example, in the spirit of this proof, 30 can be first represented as $2 \cdot 15$ and then as $2 \cdot 3 \cdot 5$, but we can represent it as $3 \cdot 10$ and then $3 \cdot 2 \cdot 5$. It is impossible to foresee that the results of the decomposition into prime factors would differ only in the order of the factors. And if it is easy to grasp all possibilities for the number 30, is it so simple to become convinced that the number $740037721 = 23623 \cdot 31327$ does not have another decomposition into prime factors?

In school, it is usually taken for granted that there exists only one decomposition of a natural number into prime factors. But this assertion requires proof, as the following example shows. Suppose that we only know even numbers and cannot use the odd ones. (Possibly, such a supposition corresponds in some sense to an actual historical situation in that the word *odd* also means *strange* or *peculiar*.) Literally repeating our known definition, we must say that even numbers that do not have two different even divisors are *prime*. For example, the "primes" will be 2, 6, 10, 14, 18, 22, 26, 30,.... Then we can give two different decompositions of an even number into "prime" factors, for example,

$$60 = 2 \cdot 30 = 6 \cdot 10.$$

We can find a number with a large number of different decompositions, for example,

$$420 = 2 \cdot 210 = 6 \cdot 70 = 10 \cdot 42 = 14 \cdot 30.$$

Thus, if the decomposition of a natural number into prime factors is unique, then the proof must somehow use the fact that we are dealing with the totality of all natural numbers and not, for example, just even numbers.

Having become convinced that the uniqueness of the decomposition into prime factors is not obvious, we turn to its proof.

Theorem 9. *Two decompositions of a natural number into prime factors differ only in the order of the factors.*

The proof of the theorem is actually not totally obvious, but we overcame all the difficulties when proving Theorem 6. Everything follows from it quite simply.

We first note an obvious generalization of Theorem 6.

If the product of any number of factors is divisible by a prime number p, then at least one of them is divisible by p.

Let

$$n_1 n_2 \cdots n_r = pa.$$

We prove our assertion by induction on the number of factors r. For $r = 2$, it coincides with Theorem 6. For $r > 2$, we rewrite the equality in the form

$$n_1(n_2 \cdots n_r) = pa.$$

According to Theorem 6, either p divides n_1 (and then our assertion is already proved) or p divides $n_2 \cdots n_r$ (and then our assertion is true according to the induction hypothesis). We now prove Theorem 8. Let a certain number n have two decompositions into prime factors:

$$n = p_1 \cdots p_r = q_1 \cdots q_s. \tag{8}$$

We see that p_1 divides the product $q_1 \cdots q_s$. By the proved generalization of Theorem 6, it divides one of the numbers $q_1, \ldots, q_s$. But q_i is a prime number, and this means that q_i has only one divisor greater than 1, itself. Therefore, p_1 coincides with one of the q_i. Changing their numbering, we can consider that $p_1 = q_1$. Dividing equality (8) by p_1, we obtain

$$n' = \frac{n}{p_1} = p_2 \cdots p_r = q_2 \cdots q_s. \tag{9}$$

This assertion relates to a smaller number n', and using the principle of mathematical induction, we can consider that it is already proved. Therefore, the number of factors in the two decompositions is one and the same, that is, $r - 1 = s - 1$, and this means that $r = s$. Moreover, the factors $q_2, \ldots, q_s$ can be rewritten in such an order that $r_2 = q_2$, $r_3 = q_3, \ldots, p_r = q_r$. Because we already know that $p_1 = q_1$, this proves the theorem.   □

The theorem we have just proved is contained in Euclid. It was always considered an abstract mathematical theorem, although simple. However, in the last few decades, it has had unexpected practical applications, which we will say a few words about. The matter has to do with *ciphers* or *codes*, that is, writing information in such a way that it cannot be understood by a person who does not have certain additional information (the cipher key). It turns out that decomposing large numbers into their prime factors requires an enormous number of operations. It is incomparably more complicated than inverse problem of multiplying prime numbers. For example, the multiplication of two prime numbers, each of which is written with several tens of digits (say, a 30-digit number by a 40-digit number), is possible with great diligence. Working by hand all day, you can write the answer (about 70 digits) in the evening. But decomposing this number into prime factors, even with a good modern computer, requires more time than the Earth has existed. Thus, a pair of large numbers p and q on one hand and their product $n = pq$ on the other hand contain exactly the same information according to Theorem 8, just written in two different ways. But the transition from the pair p, q to the number $n = pq$ presents no difficulties, and the reverse transition from n to the pair p, q is practically impossible. This is the basic idea of the "cipher." We omit the technical details here.[1]

Some prime factors can appear several times in the decomposition of a number into prime factors, for example, $90 = 2 \cdot 3 \cdot 3 \cdot 5$. We can unite all equal prime factors into one factor raised to a power: $90 = 2 \cdot 3^2 \cdot 5$. For every natural number, we thus obtain a decomposition

$$n = p_1^{\alpha_1} p_2^{\alpha_2} \cdots p_r^{\alpha_r}, \tag{10}$$

where all the prime numbers $p_1, \ldots, p_r$ are different and the exponents $\alpha_i \geq 1$. Such a decomposition is said to be *canonical*. It is, of course, also unique for each n.

Knowing the canonical decomposition of the number n, we can learn everything we want to know about the divisors of n. In the first place, if the canonical decomposition has form (10), then the numbers

$$m = p_1^{\beta_1} p_2^{\beta_2} \cdots p_r^{\beta_r}, \tag{11}$$

---

[1] A new idea has recently appeared, the construction of a so-called quantum computer, which could solve the problem of decomposing large numbers into their prime factors in a reasonable time.

where $\beta_1 \leq \alpha_1, \beta_2 \leq \alpha_2, \ldots, \beta_r \leq \alpha_r$ and $p_1, p_2, \ldots, p_r$ are the same prime numbers as in formula (10), is a divisor of n (here, the value 0 is permitted for β_i, that is, a p_i that divides n might not divide m). Conversely, every divisor of n has form (11). Indeed, every prime divisor of the number m must also be a divisor of the number n, that is, coincide with one of the numbers $p_1, p_2, \ldots, p_r$. Therefore, the canonical decomposition of the number m has form (11) with some $\beta_1, \ldots, \beta_r$. If $n = mk$, the k is also a divisor of n, that is, has form (11),

$$k = p_1^{\gamma_1} p_2^{\gamma_2} \cdots p_r^{\gamma_r}.$$

Multiplying the canonical decompositions of m and k and combining powers of identical prime numbers, we must obtain decomposition (10) because such a decomposition is unique by virtue of Theorem 8. Because exponents are added when we multiply two powers of one prime number, we have $\beta_1 + \gamma_1 = \alpha_1$, which means that $\beta_1 \leq \alpha_1$; in exactly the same way, we obtain $\beta_2 \leq \alpha_2, \ldots, \beta_r \leq \alpha_r$.

For example, we can now find the sum of the divisors of the number n. It is convenient to include the numbers 1 and n in the list of divisors. Thus the number $n = 30$ has the divisors 1, 2, 3, 5, 6, 10, 15, 30 and the sum of the divisors is 72. We begin by considering the simplest case where n is a power of a single prime number: $n = p^\alpha$. Then its divisors are the numbers p^β, where $0 \leq \beta \leq \alpha$, that is, the numbers $1, p, p^2, \ldots, p^\alpha$. Consequently, we must find the sum $1 + p + p^2 + \cdots + p^\alpha$. There is a general formula (perhaps you already know it) that expresses the sum of consecutive powers of a given number,

$$s = 1 + a + a^2 + \cdots + a^r.$$

Deducing the formula is very simple: we must multiply both sides of the equality by a and open the parentheses on the right. We obtain

$$sa = a + a^2 + a^3 + \cdots + a^{r+1}.$$

We see the expressions for s and sa have almost the same terms. Only we have 1 in s and not in sa, and we have a^{r+1} in sa and not in s. Therefore, when we subtract s from sa, all terms cancel except for those two:

$$sa - s = a^{r+1} - 1,$$

that is, $s(a - 1) = a^{r+1} - 1$ and

$$s = 1 + a + a^2 + \cdots + a^r = \frac{a^{r+1} - 1}{a - 1}. \tag{12}$$

Because we divide by $a - 1$, we must assume that $a \neq 1$. A series of numbers $1, a, \ldots, a^r$ is called a *geometric progression*, and formula (12) is the formula for the sum of a geometric progression.

Therefore, if $n = p^\alpha$, then the sum of its divisors is equal to

$$1 + p + p^2 + \cdots + p^\alpha = \frac{p^{\alpha+1} - 1}{p - 1}.$$

We consider the next case in order of complexity, the case where n has two prime divisors p_1 and p_2. Its canonical decomposition accordingly has the form $n = p_1^{\alpha_1} p_2^{\alpha_2}$. According to formula (11), the divisors of the number n can be written in the form $p_1^{\beta_1} p_2^{\beta_2}$, where $0 \leq \beta_1 \leq \alpha_1$ and $0 \leq \beta_2 \leq \alpha_2$. We separate them into groups, each group containing those that have the same value for β_2. For $\beta_2 = 0$, we thus obtain the divisors $1, p_1, p_1^2, \ldots, p_1^{\alpha_1}$, whose sum is equal to

$$\frac{p_1^{\alpha_1+1} - 1}{p_1 - 1}.$$

For $\beta_2 = 1$, we obtain the group $p_2, p_1 p_2, p_1^2 p_2, \ldots, p_1^{\alpha_1} p_2$. Calculating the sum of its members, we can move p_2 outside the parentheses; we obtain a sum inside the parentheses whose form we already know,

$$(1 + p_1 + p_1^2 + \cdots + p_1^{\alpha_1})p_2 = \frac{p_1^{\alpha_1+1} - 1}{p_1 - 1} p_2.$$

In exactly the same way for a group with the arbitrary common value β_2, we obtain the sum

$$(1 + p_1 + p_1^2 + \cdots + p_1^{\alpha_1})p_2^{\beta_2} = \frac{p_1^{\alpha_1+1} - 1}{p_1 - 1} p_2^{\beta_2}.$$

The complete sum is separated into such partial sums and is consequently equal to

$$\frac{p_1^{\alpha_1+1} - 1}{p_1 - 1} + \frac{p_1^{\alpha_1+1} - 1}{p_1 - 1} p_2 + \frac{p_1^{\alpha_1+1} - 1}{p_1 - 1} p_2^2 + \cdots + \frac{p_1^{\alpha_1+1} - 1}{p_1 - 1} p_2^{\alpha_2}$$
$$= \frac{p_1^{\alpha_1+1} - 1}{p_1 - 1}(1 + p_2 + p_2^2 + \cdots + p_2^{\alpha_2}).$$

We again use formula (12) and calculate the sum in the parentheses. As a result, we obtain complete sum of the divisors of the number $n = p_1^{\alpha_1} p_2^{\alpha_2}$ in the form of the expression

$$\frac{p_1^{\alpha_1+1} - 1}{p_1 - 1} \frac{p_2^{\alpha_2+1} - 1}{p_2 - 1}.$$

Now you have already guessed the answer for the general case. We consider the product

$$S' = (1+p_1+p_1^2+\cdots+p_1^{\alpha_1})(1+p_2+p_2^2+\cdots+p_2^{\alpha_2})\cdots(1+p_r+p_r^2+\cdots+p_r^{\alpha_r})$$

and open all parentheses.

When we evaluate expressions with parentheses, if there is one pair of parentheses and the product has the form

$$(a + b + \cdots)k,$$

we multiply each component a, b, ... by k and then add all ak, bk, ... If there are two pairs of parentheses,

$$(a_1 + b_1 + c_1 + \cdots)(a_2 + b_2 + c_2 + \cdots),$$

we multiply each member of one pair of parentheses by each member of the other and add all the expressions obtained: a_1a_2, a_1b_2, a_1c_2, b_1a_2, b_1b_2, and so on. Finally, if the number of pairs of parentheses is arbitrary,

$$(a_1 + b_1 + c_1 + \cdots)(a_2 + b_2 + c_2 + \cdots)\cdots(a_r + b_r + c_r + \cdots),$$

we must take some term from each pair of parentheses, multiply them together, and add all such possible products. We apply this rule to our sum S'. The terms in different pairs of parentheses have the general form $p_1^{\beta_1}, p_2^{\beta_2}, \ldots, p_r^{\beta_r}$, $0 \le \beta_i \le \alpha_i$. Multiplying them, we obtain

$$p_1^{\beta_1} p_2^{\beta_2} \cdots p_r^{\beta_r},$$

that is, in accordance with formula (11), exactly a divisor of the number n and, according to Theorem 8, each one exactly once. Therefore, the sum S' is equal to the sum of the divisors of the number n. On the other hand, according to formula (12), the ith pair of parentheses is equal to $(p_i^{\alpha_i+1} - 1)/(p_i - 1)$, and the entire product is equal to

$$S = \frac{p_1^{\alpha_1+1} - 1}{p_1 - 1} \frac{p_2^{\alpha_2+1} - 1}{p_2 - 1} \cdots \frac{p_r^{\alpha_r+1} - 1}{p_r - 1}.$$

Such is the formula for the sum of the divisors. At the same time, we also found the *number* of divisors. Indeed, to determine the number of divisors, we must change each component in the sum of the divisors to 1. Returning to the preceding proof, we see that it is enough to change each term in the parentheses in the product S' to 1. The first pair of parentheses then becomes equal to $\alpha_1 + 1$, the second to $\alpha_2 + 1$, and so on up to the rth pair of parentheses, which is equal to $\alpha_r + 1$. For the total number of divisors, we obtain the expression

$$(\alpha_1 + 1)(\alpha_2 + 1) \cdots (\alpha_r + 1).$$

For example, the number of divisors for a number with the canonical decomposition $p^\alpha q^\beta$ is equal to $(\alpha + 1)(\beta + 1)$.

Using exactly the same method, we can obtain the formula for the sum of the squares or cubes of the divisors, or even for the sum of arbitrary kth powers. The argument does not differ in any way from the argument that we used to find the sum of the divisors. You can convince yourself that using this method in the general case, we will obtain the following formula for the sum of the kth powers of all divisors of the number n with canonical decomposition (10):

$$S = \frac{p_1^{k(\alpha_1+1)} - 1}{p_1^k - 1} \frac{p_2^{k(\alpha_2+1)} - 1}{p_2^k - 1} \cdots \frac{p_r^{k(\alpha_r+1)} - 1}{p_r^k - 1}. \tag{13}$$

We can also investigate *common divisors* of two natural numbers m and n. We write their canonical decompositions in the form

$$n = p_1^{\alpha_1} \cdots p_r^{\alpha_r}, \qquad m = p_1^{\beta_1} \cdots p_r^{\beta_r}, \tag{14}$$

where now one member of each pair (α_i, β_i) may take the value 0 (in order to consider prime numbers that divide one of the numbers m and n but not the other). Using what we know about divisors, we can say that a number k is a common divisor of m and n if and only if it has the form

$$k = p_1^{\gamma_1} \cdots p_r^{\gamma_r},$$

where simultaneously $\gamma_1 \leq \alpha_1$ and $\gamma_1 \leq \beta_1$, $\gamma_2 \leq \alpha_2$ and $\gamma_2 \leq \beta_2$, $\ldots$, $\gamma_r \leq \alpha_r$ and $\gamma_r \leq \beta_r$. In other words, if σ_i denotes the minimum of the two numbers α_i and β_i, then the conditions $\gamma_1 \leq \sigma_1$, $\gamma_2 \leq \sigma_2$, $\ldots$, $\gamma_r \leq \sigma_r$ must be satisfied. Let

$$d = p_1^{\sigma_1} \cdots p_r^{\sigma_r}. \tag{15}$$

Then the preceding arguments demonstrate that the following theorem is true.

Theorem 10. *For any two numbers that have canonical decompositions (14), the number d defined by formula (15) divides both m and n, and any common divisor of m and n is a divisor of d.*

The number d is called the *greatest common divisor* of m and n and is denoted by $\text{GCD}(m, n)$. We note that among the common divisors of the numbers m and n, there must, of course, be the greatest in absolute value. But it is not obvious that all other divisors will divide it. This follows only from Theorem 8 (concerning the uniqueness of the decomposition into prime factors). That is why we brought up these properties, which are usually given in school without proof.

As was mentioned above, decomposing a number into its prime factors is a very laborious task. We therefore recall another way to find the greatest common divisor, a way that does not require decomposition into factors. This method is often given in school. It is based on Theorem 5. Let n and m be two natural numbers and $n = mt + r$, $0 \leq r < m$ (the representation established in Theorem 5).

Lemma 5. If $r \neq 0$, then $\text{GCD}(n, m) = \text{GCD}(m, r)$.

It is easy to be convinced that even more is true: The two pairs (n, m) and (m, r) have exactly the same common divisors. In particular, they have the same common divisor that is the greatest, the one that all others divide. Indeed, any common divisor d of the numbers n and m is a divisor of both m and r because $r = n - mt$. Further, any common divisor d' of the numbers m and r is also a divisor of both m and n because $n = mt + r$. □

The advantage of moving from the pair (n, m) to the pair (m, r) is that $r < m$ by the definition of a remainder. Now we can apply the same argument to the pair (m, r) Let $m = rt_1 + r_1$, $0 \leq r_1 < r$. If $r_1 \neq 0$, then $\text{GCD}(m, r) = \text{GCD}(r, r_1)$. And thus we will continue until we stop. And stop we must when the next remainder is equal to 0, for example, $r_i = r_{i+1}t_{i+2} + 0$ (that is, $r_{i+2} = 0$). But then r_{i+1} divides r_i, and it is obvious that $\text{GCD}(r_i, r_{i+1}) = r_{i+1}$. Thus, the last nonzero remainder in the process of successive divisions of n by m, m by r, r by r_1, and so on will be equal to $\text{GCD}(n, m)$. This method for finding the GCD is called the *Euclidean algorithm*, and it can indeed be found in the *Elements*. For example, to find $\text{GCD}(8891, 2329)$, we do the divisions with remainders

$$8891 = 2329 \cdot 3 + 1904; \quad 2329 = 1904 \cdot 1 + 425;$$
$$1904 = 425 \cdot 4 + 204; \quad 425 = 204 \cdot 2 + 17; \quad 204 = 17 \cdot 12 + 0;$$
$$\text{GCD}(8891, 2329) = 17.$$

We recall that the numbers n and m are said to be *relatively prime* if they do not have any common divisors other than 1. This means that $\text{GCD}(n, m) = 1$. In particular, we can use the Euclidean algorithm to find out whether two numbers are relatively prime without decomposing them into prime factors.

We conclude this chapter by returning to the question with which we began, the question of irrationality. We prove a very wide generalization of our first assertion about the irrationality of $\sqrt{2}$. It is related to the concept to which the next chapter is devoted and is thus also an introduction to the next chapter.

An expression of the form ax^k, where a is a number, x is an unknown, and k is a natural number or 0 (then we simply write a), is called a *monomial*. The number k is called its *degree*, and a is its *coefficient*. In general, we can consider monomials that contain several unknowns, for example, ax^2y^8, but now we discuss only monomials with one unknown. A sum of monomials is called a *polynomial*. If a polynomial contains several monomials of the same degree, then we can replace them with one equivalent monomial (combining like terms). For example, if a polynomial contains the monomials ax^k and bx^k, then we can replace them with the monomial $(a+b)x^k$. In view of this, we assume in what follows that a polynomial contains only one term of a given degree, and we write that term as $a_k x^k$. For $k=0$, we have simply a_0. The highest degree of a monomial in a polynomial is called the *degree of the polynomial*. For example, the polynomial $2x^3 - 3x + 7$ has the degree 3. Its coefficients are $a_0 = 7$, $a_1 = -3$, $a_2 = 0$, and $a_3 = 2$. A polynomial of degree n thus has the general form

$$f(x) = a_0 + a_1 x + a_2 x^2 + \cdots + a_n x^n,$$

where some of the a_k can be 0, but $a_n \neq 0$ because otherwise the polynomial would have a degree less than n. The term a_0 is called the *free term*, and a_n is called the *coefficient of the leading term*. If $n=0$, the polynomial coincides with the number a_0. Such a polynomial is called a *constant* because if we substitute different values of x in it, we always obtain the same value a_0. The equality $f(x) = 0$ is called an *algebraic equation* with one unknown. A number α is called its *root* if $f(\alpha) = 0$. A root of the equation $f(x) = 0$ is also called a *root of the polynomial* $f(x)$. The degree of the polynomial $f(x)$ is called the *degree of the equation*. Obviously, the equations $f(x) = 0$ and $cf(x) = 0$, where c is a number different from 0, have exactly the same roots, that is, they are equivalent.

Now we study equations $f(x) = 0$ whose coefficients $a_0, a_1, \ldots, a_n$ are rational numbers, some of which may be equal to 0 or negative. Taking c to be a common denominator of all the nonzero coefficients, we can pass from the equation $f(x) = 0$ to the equation $cf(x) = 0$, whose coefficients are all integers. We deal with such equations with integer coefficients in what follows. We meet with properties of the divisibility of integers (not necessarily natural numbers). We recall that by definition, an integer a is divisible by an integer b if $a = bc$ for some integer c.

Theorem 11. *Let $f(x)$ be a polynomial with integer coefficients, and let the coefficient of the leading term be equal to 1. If the equation $f(x) = 0$ has a rational root α, then α is an integer, and α is a divisor of the free term of the polynomial $f(x)$.*

We represent α in the form $\alpha = \pm a/b$, where the fraction a/b is reduced, that is, the natural numbers a and b are relatively prime. By hypothesis, the polynomial $f(x)$ is equal to $a_0 + a_1 x + a_2 x^2 + \cdots + a_n x^n$ with integers a_i. We substitute α in the equation $f(x) = 0$. By hypothesis ($a_n = 1$), we have

$$a_0 + a_1 \left(\pm \frac{a}{b} \right) + \cdots + a_{n-1} \left(\pm \frac{a}{b} \right)^{n-1} + \left(\pm \frac{a}{b} \right)^n = 0. \qquad (16)$$

We multiply the equality by b^n and transfer $(\pm a)^n$ to the right-hand side. All terms that remain in the left-hand side are divisible by b:

$$\left[a_0 b^{n-1} + a_1 (\pm a) b^{n-2} + \cdots + a_{n-1} (\pm a)^{n-1} \right] b = (\pm 1)^{n-1} a^n.$$

We see that b divides a^n. If α were not an integer, then b would be greater than 1. In this case, let p be some prime divisor of $b > 1$. Then p must divide a^n, and p must then divide a according to Theorem 6. But by hypothesis, a and b are relatively prime, and we have obtained a contradiction. Therefore, $b = 1$ and $\alpha = \pm a$.

To verify the second assertion in the theorem, we leave only a_0 in the left-hand side of the equation and move all other terms to the right-hand side (remembering that $b = 1$). All the terms in the right-hand side are divisible by a:

$$a_0 = a \left(-(\pm a_1) - a_2 (\pm a) - \cdots - a_{n-1} (\pm a)^{n-2} - (\pm a)^{n-1} \right).$$

It is obvious from this equation that a, and therefore α, divides a_0. $\square$

The argument we used to prove Theorem 11 was found by the medieval mathematician Leonardo of Pisa (also known as Fibonacci). In a book that he wrote in about the year 1225 A.D., he applied this argument to the equation $x^3 + 2x^2 + 10x = 20$.

Theorem 11 gives us the possibility to find rational roots of equations with the given form: we must find all divisors of the free term (with $+$ and $-$ signs) and then check whether they are roots. For example, for the equation $x^5 - 13x + 6 = 0$, we must check the numbers ± 1, ± 2, ± 3, and ± 6. It turns out that only $x = -2$ is a root.

Thus, for a polynomial $f(x)$ whose coefficients are all integers and whose coefficient of the leading term is equal to 1, all roots are irrational except those roots that are integers dividing the free term. We established this fact at the beginning of the chapter for several special cases: first for $f(x) = x^2 - 2$ (Theorem 2); then for $f(x) = x^2 - 3$ (Theorem 3); and finally for $f(x) = x^2 - c$, where c is a natural number (Theorem 7). We have now obtained the widest generalization of all these assertions. It also has many other geometric applications in addition to Theorems 1, 2, 3, and 7.

For example, we consider the equation

$$x^3 - 7x^2 + 14x - 7 = 0.$$

According to Theorem 11, its rational roots can only be integers that divide the number -7, that is, one of the numbers 1, -1, 7, and -7. Substitution shows that not one of them satisfies the equation. We can say that the roots of this equations are irrational. True, we don't know that equation (17) has any roots at all. But we show later that it does have roots, and they are very interesting.

The square of the length of a side of a regular heptagon inscribed in a circle of radius 1 is a root. Moreover, equation (17) has three roots, one between 0 and 1, one between 2 and 3, and one between 3 and 4. They are equal to the squares of the lengths of the diagonals of a regular heptagon inscribed in a circle of radius 1. We call any line segment connecting two vertices of a polygon a diagonal; therefore, the sides are includ-

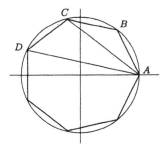

Fig. 5

ed among the diagonals. A regular heptagon has diagonals of three different lengths AB, AC, and AD (Fig. 5). All three lengths are thus irrational.

Problems:

1. Convince yourself that Theorem 6 is not true if the concept of a number and a "prime number" is understood only as applied to even numbers (as described in the beginning of this section). Which place in the proof of Theorem 6 turns out to be incorrect in this case?
2. Prove that if the numbers m and n are relatively prime, then divisors of the number mn are obtained by multiplying divisors of the number m by divisors of the number n, and each divisor of the number mn is obtained in this way exactly once. Deduce from this that if $S(N)$ denotes the sum of the kth powers of the

divisors of the number N, if m and n are relatively prime, and if $N = mn$, then $S(N) = S(m)S(n)$. Use this approach to deduce formula (13).

3. A natural number n is said to be *perfect* if it is equal to the sum of its *proper* divisors (that is, we exclude the number n itself from the divisors). For example, the numbers 6 and 28 are perfect. Prove that if the number $p = 2^r - 1$ is prime for a certain r, then the number $2^{r-1}p$ is perfect (but remember that the formula we derived for the sum S of the divisors includes *the number n itself*). This assertion was already contained in Euclid's *Elements*. About 2000 years later, Euler proved the converse assertion: any even perfect number has the form $2^{r-1}p$, where $p = 2^r - 1$ is a prime number. The proof does not use any facts that are not given already in this chapter, but the proof is not entirely simple. Try to establish it! To this time, it is unknown whether any *odd* perfect numbers exist.

4. Given two natural numbers m and n, if there exist integers a and b such that $ma + nb = 1$, then m and n are obviously relatively prime (any common divisor of m and n would divide 1). Prove the converse: If m and n are relatively prime natural numbers, then there exist integers a and b such that $ma + nb = 1$. Use division with remainder and mathematical induction.

5. Using the result in Problem 4, find a new proof of Theorem 6. As we saw, Theorem 9 follows easily from Theorem 6. This is the way Euclid proved it.

6. Prove formula (13) using induction on the number of prime divisors.

7. For what integers a does the polynomial $x^n + ax + 1$ have rational roots?

8. Let $f(x)$ be a polynomial whose coefficients are integers. Prove that if the reduced fraction $\alpha = \pm a/b$ is a root of $f(x) = 0$, then b is a divisor of the coefficient of the leading term, and a is a divisor of the free term. This assertion generalizes Theorem 11 to the case of a polynomial whose coefficients are integers and whose coefficient of the leading term is not necessarily equal to 1.

Simplest Properties of Polynomials

4.   Roots and the Divisibility of Polynomials

In this chapter, we consider equations of the form $f(x) = 0$, where f is a polynomial. We already met them at the end of the previous chapter. The equation $f(x) = 0$ should be understood as the problem "find all roots of the polynomial (or equation)." But it can happen that all coefficients of the polynomial $f(x)$ are zero and the equation $f(x) = 0$ becomes an identity. Then we write $f = 0$. In this case, we consider that the degree of the polynomial is undefined.

To add two polynomials, we must simply combine like terms. Polynomials are multiplied according to the rules for opening parentheses. If

$$f(x) = a_0 + a_1 x + \cdots + a_n x^n, \qquad g(x) = b_0 + b_1 g + \cdots + b_m x^m,$$

then

$$f(x)g(x) = (a_0 + a_1 x + \cdots + a_n x^n)(b_0 + b_1 g + \cdots + b_m x^m).$$

Opening parentheses, we obtain terms $a_k b_l x^{k+l}$, where $0 \le k \le n$ and $0 \le l \le m$. We must then combine like terms. As a result, we obtain the polynomial

$$c_0 + c_1 x + c_2 x^2 + \cdots$$

with the coefficients

$$c_0 = a_0 b_0, \quad c_1 = a_0 b_1 + a_1 b_0, \quad c_2 = a_0 b_2 + a_1 b_1 + a_2 b_0, \quad \ldots . \quad (1)$$

The coefficient c_m is equal to the sum of all the products $a_k b_l$ for which $k + l = m$.

Polynomials have many properties similar to those of integers. We can consider a polynomial written in the form

$$f(x) = a_0 + a_1 x + \cdots + a_n x^n$$

an analogue of a natural number written in the decimal system (or in a system with a different base). For example, when we write the number n in the form 1998, we assert that $n = 10^3 + 9 \cdot 10^2 + 9 \cdot 10 + 8$. It is of no importance that we usually write polynomials with increasing powers of x and numbers with decreasing powers of 10. The degree of a polynomial plays a role analogous to the role of the absolute value of an integer. If we use induction on the absolute value to prove a property of integers, then we usually use induction on the degree to prove the analogous property of polynomials.

We note an important property of polynomials: the degree of the product of two polynomials is equal to the sum of the degrees of the two polynomials. Indeed, we take two polynomials of degrees n and m

$$f(x) = a_0 + a_1 x + \cdots + a_n x^n$$

and

$$g(x) = b_0 + b_1 g + \cdots + b_m x^m,$$

that is,

$$a_n \neq 0 \quad \text{and} \quad b_m \neq 0.$$

Calculating the coefficients of the polynomial $f(x)g(x)$ accoding to formula (1), we obtain terms of the form

$$a_k b_l x^{k+l}, \quad \text{where } k + l \leq n + m.$$

It is evident that the highest degree that we obtain is $n + m$. Furthermore, there is exactly one term with that degree: $a_n b_m x^{n+m}$. This term is different from zero because $a_n b_m \neq 0$, and it cannot be combined with any other term, because it has the highest degree. This property is analogous to the property $|xy| = |x||y|$ for the absolute value $|x|$ of the number x.

The theorem on division with remainder for polynomials is formulated and proved almost the same as for natural numbers (see Theorem 5 in Chap. 1).

Theorem 12. *For any polynomials $f(x)$ and $g(x)$ with $g \neq 0$, there exist polynomials $h(x)$ and $r(x)$ such that*

$$f(x) = g(x)h(x) + r(x), \tag{2}$$

where either $r = 0$ or the degree of the polynomial r is less than the degree of the polynomial g. The polynomials h and r are uniquely determined for a given f and g.

If $f = 0$, then the needed representation (2) is obvious: $f = g \cdot 0 + 0$.

We suppose that $f \neq 0$ and apply the method of mathematical induction on the degree of the polynomial $f(x)$. Let $f(x)$ have the degree n and $g(x)$ have the degree m:

$$f(x) = a_0 + a_1 x + \cdots + a_n x^n,$$
$$g(x) = b_0 + b_1 g + \cdots + b_m x^m.$$

If $m > n$, then representation (2) has the form

$$f = g \cdot 0 + f \quad \text{with } h = 0, \ r = f.$$

If $m \leq n$, then we set

$$f_1 = f - \frac{a_n}{b_m} x^{n-m} g$$

(we remember that $b_m \neq 0$ by hypothesis because the degree of the polynomial $g(x)$ is equal to m). It is obvious that the term with x^n is canceled in the polynomial f_1 (that is why we chose the coefficient $-a_n/b_m$), that is, the degree of f_1 is less than n. We can therefore consider that the theorem is already proved for such a polynomial and that we have a representation of form (2),

$$f_1 = gh_1 + r,$$

where $r = 0$ or the degree of the r is less than m. From this,

$$f = f_1 + \frac{a_n}{b_m} x^{n-m} g = \left(h_1 + \frac{a_n}{b_m} x^{n-m} \right) g + r.$$

We have obtained representation (2) with $h = h_1 + (a_n/b_m)x^{n-m}$.

We now prove that this representation of form (2) is unique. If $f = gk + s$ is another such representation in which the polynomial s is either equal to zero or has a degree less than m, then we subtract one from the other and obtain

$$g(h - k) + r - s = 0,$$
$$g(h - k) = s - r.$$

If the polynomial $s - r$ is equal to zero, then $s = r$ and $h = k$. And if $s - r \neq 0$, then its degree is less than m by hypothesis, and we obtain a contradiction because it is equal to the polynomial $g(h - k)$, which is obtained by multiplying g by $h - k$, that is, it has a degree not less than the degree of g, which is equal to m. $\square$

Now, please read the proof of Theorem 5 in Chap. 1 and convince yourself that our proof here is completely parallel to it. At the same time, if we carry out all the operations taken for granted in applying the method of mathematical induction (that is, moving from f_1 to the polynomial f_2 of even lower degree and so on until we obtain a remainder r of degree less than m), we obtain the usual rule for dividing a polynomial into a polynomial by "sections" as taught in school. For example, if

$$f(x) = x^3 + 3x^2 - 2x + 5 \qquad \text{and} \qquad g(x) = x^2 + 2x - 1,$$

then division by "sections" occurs according to the scheme

$$
\begin{array}{rl}
 & x^3 + 3x^2 - 2x + 5 \quad \underline{\mid \ x^2 + 2x - 1} \\
- & \qquad\qquad\qquad\qquad\qquad x\ +1 \\
 & \underline{x^3 + 2x^2 - \ \ x} \\
 & \qquad x^2 - \ \ x + 5 \\
- & \qquad \underline{x^2 + 2x - 1} \\
 & \qquad\qquad - 3x + 6.
\end{array}
$$

This means that we choose the leading term of the polynomial $h(x)$ such that when it is multiplied by the leading term of the polynomial $g(x)$ (that is, x^2), it yields the leading term of the polynomial $f(x)$ (that is, x^3). We thus choose x as the leading term below the division line. In the first line, we have written $f(x)$, and in the second line, we have written the product $g(x)x$ (the result of multiplying $g(x)$ by the leading term of the quotient). The difference between these two expressions is given in the third line. We now choose the next term of the quotient such that its product with the leading term of the polynomial $g(x)$ (that is, x^2) yields the leading term obtained in the third line (that is, x^2). Therefore, the second term under the division line is 1. We now repeat the operation. Because the polynomial we obtain in the fifth line is of the first degree (less than the degree of $g(x)$, which is equal to two), the process is complete. We see that

$$x^3 + 3x^2 - 2x + 5 = (x^2 + 2x - 1)(x + 1) - 3x + 6.$$

As in the case of numbers, representation (2) is called *division with re-mainder* of the polynomial $f(x)$ by the polynomial $g(x)$, the polynomial

$h(x)$ is called the *quotient*, and $r(x)$ is called the *remainder* of dividing $f(x)$ by $g(x)$.

Division of a polynomial by a polynomial with remainder is analogous to division of a number by a number, and even simpler because when subtracting terms of a given degree, we obtain a term of the same degree. We do not have the process of borrowing from the next higher column that occurs with numbers.

Repeating the arguments used in Chap. 1 for the case of numbers, we can use Theorem 12 to find the greatest common divisor of two polynomials. In terms of Theorem 12, we have the analogue of Lemma 5

$$GCD(f,g) = GCD(g,r).$$

More precisely, the common divisors of the pair (f,g) and the pair (g,r) are the same. We can now use the same Euclidean algorithm as in Chap. 1: we divide g by r with remainder ($g = rh_1 + r_1$), divide r by r_1, and so on, obtaining a sequence of polynomials of decreasing degree $r, r_1, r_2, \ldots, r_k$. We must stop when we reach the polynomial $r_{k+1} = 0$, that is, when $r_{k-1} = r_k h_k$. We can see from the chain of equalities

$$GCD(f,g) = GCD(g,r) = GCD(r,r_1) = \cdots = GCD(r_{k-1}, r_k)$$

that

$$GCD(f,g) = GCD(r_{k-1}, r_k).$$

But because r_k is itself a divisor of r_{k-1}, we have $GCD(r_{k-1}, r_k) = r_k$, and this means that $GCD(f,g)$ is equal to r_k, that is, the last nonzero remainder in the Euclidean algorithm: $GCD(f,g) = r_k$. However, we must note that in contrast to the natural numbers, $GCD(f,g)$ is not uniquely determined: together with each polynomial $d(x)$ that is a common divisor of $f(x)$ and $g(x)$, the polynomial $cd(x)$, where c is a nonzero number, is also a comon divisor. Thus, $GCD(f,g)$ is determined up to multiplication by a numerical factor.

Theorem 12 has an especially simple and useful form in the case where $g(x)$ is a first-degree polynomial. We can then write

$$g(x) = ax + b, \quad \text{where } a \neq 0.$$

Because the property of divisibility by g does not change when it is multiplied by a number, we multiply it by a^{-1}. We thus obtain the coefficient of x equal to 1. We then write $g(x)$ as $g(x) = x - \alpha$ (why it is convenient to write α with a minus sign quickly becomes evident). In accordance with Theorem 12, we have the representation

$$f(x) = (x - \alpha)h(x) + r \tag{3}$$

for any polynomial $f(x)$. But in our case, the degree of r is less than 1, that is, equal to 0: r *is a number.* Can we learn which number without dividing with remainder? We can, and even very simply. For this, it is enough to substitute $x = \alpha$ in equality (3). We obtain the expression $r = f(\alpha)$ for the number r. We can rewrite equality (3) as

$$f(x) = (x - \alpha)h(x) + f(\alpha). \tag{4}$$

The polynomial $f(x)$ is divisible by $x - \alpha$ if and only if the remainder when $f(x)$ is divided by $x - \alpha$ is equal to 0. But according to equality (4), the remainder is equal to $f(\alpha)$. We obtain the assertion known as the Bézout theorem.

Theorem 13. *The polynomial $f(x)$ is divisible by $x - \alpha$ if and only if α is a root of $f(x)$.*

For example, the polynomial $x^n - 1$ has the root $x = 1$. Therefore, $x^n - 1$ is divisible by $x - 1$. We did this division previously: see formula (12) in Chap. 1 (and replace $r + 1$ with n and a with x).

Despite the simple proof, the Bézout theorem relates two totally different concepts—divisibility and root—and therefore has many important applications. For example, what can we say about the *common roots* of the polynomials f and g, that is, about the solutions of the system of equations $f(x) = 0$, $g(x) = 0$? According to the Bézout theorem, the number α is a common root if both f and g are divisible by $x - \alpha$. But then $x - \alpha$ divides $\mathrm{GCD}(f, g)$, which we can use the Euclidean algorithm to find. If $d(x) = \mathrm{GCD}(f, g)$, then $d(x)$ is divisible by $x - \alpha$, that is, $d(\alpha) = 0$. The question about the common roots of the polynomials f and g thus leads to the question about the roots of d, which generally has a lower degree. We illustrate how to find the greatest common divisor of polynomials in the case of two second-degree polynomials, which we write in the form

$$f(x) = x^2 + ax + b \quad \text{and} \quad g(x) = x^2 + px + q$$

(we can bring them to this form by multiplying by a number). In accordance with the general rule, we divide f by g with remainder:

$$
\begin{array}{rrrr|l}
x^2 & + \ ax & + \ b & & \underline{x^2 + px + q} \\
& & & & 1 \\
\underline{x^2} & + \ px & + \ q & & \\
(a - p)x & + (b - q). & & &
\end{array}
$$

The remainder is $r(x) = (a-p)x + (b-q)$. We know that $\mathrm{GCD}(f,g) = \mathrm{GCD}(g,r)$. Here, we must consider the case where $a = p$ separately. If $b = q$ in addition to that, then $f(x) = g(x)$, and the system of equations $f(x) = 0$, $g(x) = 0$ becomes the single equation $f(x) = 0$. And if $b \neq q$, then $r(x)$ is a nonzero number, and f and g do not have a common factor. Finally, if $a \neq p$, then we simply note that $r(x)$ has the single root $\alpha = (b-q)/(p-a)$. We know that $\mathrm{GCD}(f,g) = \mathrm{GCD}(g,r)$. It is enough to substitute α in $g(x)$ to learn whether $g(x)$ is divisible by $x - \alpha$. We obtain the relation

$$\left(\frac{b-q}{p-a}\right)^2 + p\left(\frac{b-q}{p-a}\right) + q = 0,$$

which when multiplied by the nonzero number $(p-a)^2$ is equivalent to the relation

$$(b-q)^2 + p(b-q)(p-a) + q(p-a)^2 = 0. \tag{5}$$

The second and third terms in this equality have the common factor $p-a$. Taking it outside the parentheses, opening parentheses, and combining like terms, we write relation (5) in the form

$$(b-q)^2 + (p-a)(pb-aq) = 0.$$

The expression $D = (b-q)^2 + (p-a)(pb-aq)$ is called the *resultant* of the polynomials f and g. We saw that the condition $D = 0$ is necessary and sufficient for the existence of a common variable factor of $f(x)$ and $g(x)$ given that $p \neq a$. But for $p = a$, the condition $D = 0$ becomes $q = b$, and that, as we saw, is also equivalent to the existence of a common variable factor of f and g. In all cases, the condition $D = 0$ is necessary and sufficient for the polynomials $f(x)$ and $g(x)$ to have a common variable factor. In principle (but technically more complicated, of course), we can similarly find an expression involving the coefficients of any polynomials $f(x)$ and $g(x)$ whose equality to zero is necessary and sufficient for the two polynomials to have a common variable factor.

Another important application of the Bézout theorem is connected with the number of roots of a polynomial. Let the polynomial $f(x)$ not be identical to zero, that is, $f \neq 0$. We suppose that the polynomial $f(x)$ has another root α_2 in addition to the root α_1, $\alpha_2 \neq \alpha_1$. According to the Bézout theorem, $f(x)$ is divisible by $x - \alpha_1$:

$$f(x) = (x - \alpha_1)f_1(x). \tag{6}$$

We substitute the value $x = \alpha_2$ in this equality. Because α_2 is also a root of the polynomial, we have $f(\alpha_2) = 0$. This means that $(\alpha_2 -$

$\alpha_1) f_1(\alpha_2) = 0$, and because $\alpha_2 \neq \alpha_1$, we have $f_1(\alpha_2) = 0$, that is, α_2 is a root of the polynomial $f_1(x)$. Now applying the Bézout theorem to the polynomial $f_1(x)$, we obtain the equality $f_1(x) = (x - \alpha_2) f_2(x)$. Substituting this in equality (6), we see that

$$f(x) = (x - \alpha_1)(x - \alpha_2) f_2(x).$$

We suppose that the polynomial $f(x)$ has k different roots $\alpha_1, \alpha_2, \ldots,$ α_k. Repeating our arguments k times, we find that $f(x)$ must be divisible by $(x - \alpha_1)(x - \alpha_2) \cdots (x - \alpha_k)$:

$$f(x) = (x - \alpha_1)(x - \alpha_2) \cdots (x - \alpha_k) f_k(x). \qquad (7)$$

Let the degree of the polynomial $f(x)$ be equal to n. We have a polynomial of degree not less than k in the right-hand side of equality (7) and a polynomial of degree n in the left-hand side. This means that $n \geq k$. This assertion can be formulated as a theorem.

Theorem 14. *The number of different roots of a polynomial that is not identical to zero does not exceed the degree of the polynomial.*

Of course, all numbers are roots of a polynomial that is identical to zero. Theorem 14 was proved by the philosopher and mathematician Descartes in the 17th century.

Theorem 14 gives us the possibility to answer a question that we have avoided up to now: What does *equality of polynomials* mean?

One meaning is as follows: we arrange both polynomials with like terms, that is, we write them in the form

$$f(x) = a_0 + a_1 x + \cdots + a_n x^n, \qquad g(x) = b_0 + b_1 g + \cdots + b_m x^m.$$

We consider them equal if all the coefficients are the same, that is, if $a_0 = b_0$, $a_1 = b_1$, and so on. For example, we understand that all the coefficients of f are zero for the identity $f = 0$.

Another meaning of the term "equality" is possible: The polynomials $f(x)$ and $g(x)$ are equal if they yield identical numbers for any substitution of a number c for x, that is, $f(c) = g(c)$ for all c.

We prove that these two meanings of "equality" in fact coincide. But first we must keep them apart. In the first case, we say that "all coefficients coincide for $f(x)$ and $g(x)$," and in the second case, we say that "$f(x)$ and $g(x)$ yield identical values for all values of x."

It is obvious that if all cofficients coincide for $f(x)$ and $g(x)$, then they also yield identical values for all values of x. We prove the converse assertion in a stronger version: it is sufficient to suppose that $f(x)$ and $g(x)$ yield identical values not for all values of x but only for $n+1$ values of x, where n is not less than the degrees of both polynomials.

Theorem 15. *Let the degrees of the polynomials $f(x)$ and $g(x)$ not exceed n, and let $f(x)$ and $g(x)$ yield identical values for $n+1$ different values of x. Then all coefficients coincide for $f(x)$ and $g(x)$.*

Proof. We suppose that the polynomials $f(x)$ and $g(x)$ yield identical values for $n+1$ different values of x, $x = \alpha_1, \alpha_2, \ldots, \alpha_{n+1}$, that is,

$$f(\alpha_1) = g(\alpha_1), \qquad f(\alpha_2) = g(\alpha_2), \qquad \ldots, \qquad f(\alpha_{n+1}) = g(\alpha_{n+1}).$$

We consider the polynomial

$$h(x) = f(x) - g(x)$$

(here the sign $=$ denotes equality of coefficients). We see that it follows from this that

$$h(\alpha) = f(\alpha) - g(\alpha)$$

for any α and, in particular,

$$h(\alpha_1) = 0, \qquad h(\alpha_2) = 0, \qquad \ldots, \qquad h(\alpha_{n+1}) = 0,$$

that is, $\alpha_1, \alpha_2, \ldots, \alpha_{n+1}$ are $n+1$ different roots of the polynomial $h(x)$. But the degrees of the polynomials f and g do not exceed n, and the degree of the polynomial h therefore cannot exceed n. We obtain a contradiction of Theorem 14, unless we suppose that $h = 0$, that is, all coefficients of the polynomial h are equal to zero. It follows from this that all coefficients of the polynomials f and g coincide. $\square$

We can now use the term "equality" in relation to polynomials without specifying in which sense we understand it.

Theorem 15 indicates the interesting property of "rigidity" of polynomials. Namely, if we know the values of a polynomial $f(x)$ of degree not exceeding n for some $n+1$ values of the unknown x, then that already uniquely determines the coefficients of $f(x)$, which means its values for *all other* values of the unknown x are determined. We note that the words "uniquely determines the coefficients" in this phrase only mean that there *cannot be* two different polynomials with the given property. The question of the *existence* of such a polynomial therefore naturally arises.

We consider this question in more detail. Let there be given $n+1$ numbers $x_1, x_2, \ldots, x_{n+1}$ and $n+1$ numbers $y_1, y_2, \ldots, y_{n+1}$. Does there exist a polynomial $f(x)$ of degree not exceeding n such that $f(x_1) = y_1$, $f(x_2) = y_2$, $\ldots$, $f(x_{n+1}) = y_{n+1}$? Theorem 15 only asserts that if such a polynomial exists, then only one exists. The problem of constructing such a polynomial is called the *interpolation problem*. It often arises in processing experimental results when a certain quantity is only measured for definite values $x = x_1$, $x = x_2$, ..., $x = x_{n+1}$ and we must form a plausible hypothesis of its values for the other values of x. The data are then written in a table:

$$
\begin{array}{c|ccccc}
x & x_1 & x_2 & \cdots & x_{n+1} \\
\hline
f(x) & y_1 & y_2 & \cdots & y_{n+1}
\end{array}
\tag{8}
$$

One possible plausible hypothesis is the following: Construct a polynomial $f(x)$ of degree not exceeding n for which $f(x_1) = y_1$, $f(x_2) = y_2$, ..., $f(x_{n+1}) = y_{n+1}$ and let our quantity be equal to $f(x)$ for all other values of x. But does such a polynomial exist? We now prove that it exists and find a formula for it. It is called the *interpolation polynomial* corresponding to table (8).

To deduce the formula for the interpolation polynomial in the general case, we first do it for the *simplest interpolation problem*, where all values $y_1, y_2, \ldots, y_{n+1}$ in table (8) are equal to zero except one value. Let $y_1 = y_2 = \cdots = y_{k-1} = y_{k+1} = \cdots = y_{n+1} = 0$. Then the table becomes

$$
\begin{array}{c|ccccccccc}
x & x_1 & x_2 & \cdots & x_{k-1} & x_k & x_{k+1} & \cdots & x_{n+1} \\
\hline
f(x) & 0 & 0 & \cdots & 0 & y_k & 0 & \cdots & 0
\end{array}
$$

We are given that n numbers $x_1, x_2, \ldots, x_{k-1}, x_{k+1}, \ldots, x_{n=1}$ (that is, all the numbers except x_k) are roots of the interpolation polynomial $f_k(x)$ that is the solution of this interpolation problem. But then $f_k(x)$ must be divisible by the product of the corresponding differences $x - x_i$. Because there are n such differences and the degree of the polynomial cannot exceed n, the polynomial can only differ from that product by a constant factor, namely, we *must assume* that

$$
f_k(x) = c_k(x - x_1)(x - x_2) \cdots (x - x_{k-1})(x - x_{k+1}) \cdots (x - x_{n+1}). \tag{9}
$$

Conversely, any polynomial of this form satisfies the given conditions for all $x_1, \ldots, x_{n+1}$ except possibly $x = x_k$. For the polynomial to also satisfy the condition for x_k, we must substitute $x = x_k$ in equality (9) and find the value of c_k from the resulting equality. Because $f_k(x_k)$ must be equal to y_k, we obtain

$$
c_k = \frac{y_k}{(x_k - x_1) \cdots (x_k - x_{k-1})(x_k - x_{k+1}) \cdots (x_k - x_{n+1})},
$$
$$
f_k(x) = c_k(x - x_1)(x - x_2) \cdots (x - x_{k-1})(x - x_{k+1}) \cdots (x - x_{n+1}).
$$

We can rewrite this formula in a shorter form by introducing an auxiliary polynomial of degree $n + 1$: $F(x) = (x - x_1) \cdots (x - x_{n+1})$. Then the product $(x - x_1) \cdots (x - x_{k-1})(x - x_{k+1}) \cdots (x - x_{n+1})$ is equal to $F(x)/(x - x_k)$. Setting $F(x)/(x - x_k) = F_k(x)$, we obtain

$$c_k = \frac{y_k}{F_k(x_k)},$$

$$f_k(x) = \frac{y_k}{F_k(x_k)} F_k(x). \tag{10}$$

We note that $F_k(x_k) \neq 0$.

Preceding to the case of an arbitrary interpolation problem with table (8), we need only note that its solution is the sum of all the polynomials $f_k(x)$ corresponding to all the simplest interpolation problems:

$$f(x) = f_1(x) + f_2(x) + \cdots + f_{n+1}(x).$$

Indeed, if we take $x = x_k$, then the right-hand side becomes zero for all terms except $f_k(x_k)$, and because $f_k(x)$ corresponds to the kth simplest interpolation problem, $f_k(x_k) = y_k$. Finally, the degrees of the polynomials $f_1(x), \ldots, f_{n+1}(x)$ do not exceed n, and this means that the same is true for their sum. We can rewrite the derived formula as

$$f(x) = \frac{y_1}{F_1(x_1)} F_1(x) + \frac{y_2}{F_2(x_2)} F_2(x)$$

$$+ \cdots + \frac{y_{n+1}}{F_{n+1}(x_{n+1})} F_{n+1}(x), \tag{11}$$

where

$$F_k(x) = \frac{F(x)}{x - x_k},$$

$$F(x) = (x - x_1)(x - x_2) \cdots (x - x_{n+1}).$$

We draw attention to an unexpectable identity that follows from the formula for the interpolation polynomial. We consider the interpolation problem corresponding to the table

x	x_1	x_2	$\cdots$	$x_{n=1}$
$f(x)$	x_1^k	x_2^k	$\cdots$	x_{n+1}^k

where k is a natural number not exceeding n or $k = 0$. On one hand, it is obvious that this interpolation problem is satisfied by the polynomial $f(x) = x^k$. On the other hand, the polynomial must be expressible in accordance with formula (11). We thus obtain

$$x^k = \frac{x_1^k}{F_1(x_1)}F_1(x) + \frac{x_2^k}{F_2(x_2)}F_2(x) + \cdots + \frac{x_{n+1}^k}{F_{n+1}(x_{n+1})}F_{n+1}(x),$$

where

$$F(x) = (x - x_1)(x - x_2) \cdots (x - x_{n+1})$$

and

$$F_i(x) = \frac{F(x)}{x - x_i}.$$

The polynomial $F_i(x)$ has the degree n, and the coefficient of x^n is equal to one. If $k < n$, then a polynomial of a degree less than n must be in the right-hand side, and all terms of degree n must consequently cancel. Namely, the equality

$$\frac{x_1^k}{F_1(x_1)} + \frac{x_2^k}{F_2(x_2)} + \cdots + \frac{x_{n+1}^k}{F_{n+1}(x_{n+1})} = 0$$

must hold for $k < n$. And if $k = n$, then the coefficient of x^n must be equal to one. Writing the equality for these coefficients, we obtain

$$\frac{x_1^n}{F_1(x_1)} + \frac{x_2^n}{F_2(x_2)} + \cdots + \frac{x_{n+1}^n}{F_{n+1}(x_{n+1})} = 1.$$

We note that

$$F(x) = (x - x_1)(x - x_2) \cdots (x - x_{n+1}),$$
$$F_k(x) = \frac{F(x)}{x - x_k},$$

and we thus have certain identities involving totally arbitrary numbers $x_1, \ldots, x_{n+1}$.

Problems:

1. Write the last identities for $n = 1$ and $n = 2$, that is, for $F(x) = (x - x_1)(x - x_2)$ and $F(x) = (x - x_1)(x - x_2)(x - x_3)$. Then verify them by direct calculation.
2. Divide $x^{n+1} - 1$ by $x - 1$ by "sections" and thus obtain a different deduction of formula (12) in Chap. 1.
3. Divide $x^n - a$ by $x^m - b$ with remainder (*Hint:* The answer depends on dividing n by m with remainder).
4. Explain why the following shorter argument is impossible in deducing formula (7): because $f(x)$ is divisible by each $x - \alpha_i$, $f(x)$ is divisible by their product. Become convinced that the assertion "if n is divisible by a and by b, then n is divisible by ab" is false for numbers. Show that it is also false for polynomials.
5. Prove that any polynomial can be represented as the product of binomials $x - \alpha_i$ and a polynomial with no roots. Prove that this representation is unique for each given polynomial.
6. Let $F(x) = (x - x_1) \cdots (x - x_n)$, where $x_1, \ldots, x_n$ are different numbers, and let $f(x)$ be a polynomial of a degree less than n. Prove that the fraction $f(x)/F(x)$ is equal to the sum of fractions of the form $a_k/(x - x_k)$, where $k = 1, \ldots, n$. Find a formula for a_k.

7. Prove that if $g(x)$ is a polynomial of a degree less than n and the numbers $x_1, \ldots, x_{n+1}$ and the polynomial $F_i(x)$ have the same meaning as at the end of this section, then

$$\frac{g(x_1)}{F_1(x_1)} + \cdots + \frac{g(x_{n+1})}{F_{n+1}(x_{n+1})} = 0.$$

8. With the same conditions as in Problem 7 except that $g(x)$ has the degree n and its coefficient of x^n is a, prove that

$$\frac{g(x_1)}{F_1(x_1)} + \cdots + \frac{g(x_{n+1})}{F_{n+1}(x_{n+1})} = a.$$

5. Multiple Roots and the Derivative

The equation $x^2 - a = 0$ for $a > 0$ has two roots, which are given by the formulas $x = \sqrt{a}$ and $x = -\sqrt{a}$, where $\sqrt{a}$ is the arithmetic value of the square root of a. For $a = 0$, the same formula gives two identical values. In exactly the same way, the formula for the solution of an arbitrary quadratic equation sometimes gives two identical roots. Does the same phenomenon occur for equations of an arbitrary degree? At first glance, posing the question itself seems nonsense. What does it mean that the equation $f(x) = 0$ has two *equal* roots? We can write any root of an equation on paper as many times as we like, and they will always be equal numbers! But in the case of a quadratic equation, we appealed to the formula for its solution for an answer. And in the general case, we want to use some additional considerations to give a reasonable *definition* of when we must think that the equation $f(x) = 0$ has two equal roots $x = \alpha$ and $x = \alpha$.

The Bézout theorem (Theorem 13) suggests such considerations. Let the polynomial $f(x)$ have the root $x = \alpha$. According to the Bézout theorem, the polynomial is divisible by $x - \alpha$ and can be represented in the form $f(x) = (x - \alpha)g(x)$, where $g(x)$ is a polynomial of one lower degree than $f(x)$. If the polynomial $g(x)$ also has the root $x = \alpha$, then we say that the polynomial $f(x)$ has *two roots, equal* to α. According to the Bézout theorem, $g(x)$ can be represented in the form $g(x) = (x-\alpha)h(x)$, and this means that

$$f(x) = (x - \alpha)^2 h(x). \tag{12}$$

We can say that we have a representation of form (7) which contains two factors $x - \alpha$. This agrees with the intuitive idea of what it means to have two identical roots.

If the polynomial $h(x)$ in representation (12) again has the root α, then we say that $f(x)$ has *three roots*, equal to α. In general, if $f(x)$

can be represented in the form $f(x) = (x - \alpha)^r u(x)$, where $u(x)$ is a polynomial that does not have α as a root, then we say that $f(x)$ has *exactly r identical roots*, equal to α. If $r \geq 2$, then α is called a *multiple root*. Therefore, α is a multiple root if $f(x)$ is divisible by $(x - \alpha)^2$. If the polynomial $f(x)$ has exactly r roots equal to α, then r is called the *multiplicity* of the root α.

For example, let the quadratic equation $x^2 + px + q = 0$ have the root $x = \alpha$. Dividing $x^2 + px + q$ by $x - \alpha$, we obtain

$$
\begin{array}{ll}
\begin{array}{r}
x^2 \quad\quad + px \ + q \\
\underline{-\ x^2 \quad\quad\quad - \alpha x} \\
(p + \alpha)x \ + q \\
\underline{-\ (p + \alpha)x - \alpha(p + \alpha)} \\
q + p\alpha + \alpha^2,
\end{array}
&
\begin{array}{|l}
\underline{\ x - \alpha} \\
x + p + \alpha
\end{array}
\end{array}
$$

that is,

$$x^2 + px + q = (x - \alpha)(x + p + \alpha) + (\alpha^2 + p\alpha + q).$$

Because α is a root of the equation $x^2 + px + q = 0$, we have

$$\alpha^2 + p\alpha + q = 0$$

and therefore

$$x^2 + px + q = (x - \alpha)(x + p + \alpha).$$

By our definition, the equation has two roots equal to α if $x+p+\alpha$ has the root α, that is, $2\alpha+p = 0$. Hence, $\alpha = -p/2$. Because $\alpha^2+p\alpha+q = 0$, we can substitute $\alpha = -p/2$ and obtain $-p^2/4 + q = 0$. This is the well-known condition for the equation $x^2 + px + q = 0$ to have equal roots.

The calculation is only a bit more complicated for the third-degree equation $x^3 + ax^2 + bx + c = 0$. Let the equation have the root α. We divide $x^3 + ax^2 + bx + c$ by $x - \alpha$:

$$
\begin{array}{ll}
\begin{array}{r}
x^3 \quad\quad + ax^2 \quad\quad\quad\quad + bx + c \\
\underline{-\ x^3 \quad\quad - \alpha x^2} \\
(a + \alpha)x^2 \quad\quad\quad + bx + c \\
\underline{-\ (a + \alpha)x^2 \quad - \alpha(a + \alpha)x} \\
(b + a\alpha + \alpha^2)x + c \\
\underline{-\ (b + a\alpha + \alpha^2)x - \alpha(b + a\alpha + \alpha^2)} \\
c + b\alpha + a\alpha^2 + \alpha^3
\end{array}
&
\begin{array}{|l}
\underline{\ x - \alpha} \\
x^2 + (a + \alpha)x + b + a\alpha + \alpha^2
\end{array}
\end{array}
$$

By hypothesis,
$$\alpha^3 + a\alpha^2 + b\alpha + c = 0,$$

and therefore
$$x^3 + ax^2 + bx + c = (x - \alpha)\big(x^2 + (a + \alpha)x + b + a\alpha + \alpha^2\big).$$

In accordance with our definition, the equation $x^3 + ax^2 + bx + c = 0$ has two identical roots equal to α if, first, α is a root of the equation and, second, α is a root of the polynomial
$$x^2 + (a + \alpha)x + b + a\alpha + \alpha^2.$$

In other words,
$$\alpha^2 + (a + \alpha)\alpha + b + a\alpha + \alpha^2 = 0,$$

that is,
$$3\alpha^2 + 2a\alpha + b = 0.$$

We see that multiple roots of the equation $x^3 + ax^2 + bx + c = 0$ are *common roots* of the polynomials
$$x^3 + ax^2 + bx + c \qquad \text{and} \qquad 3x^2 + 2ax + b.$$

As we saw in the preceding section, these are roots of the polynomial
$$\text{GCD}(x^3 + ax^2 + bx + c, \ 3x^2 + 2ax + b),$$

and we can use the Euclidean algorithm to find this greatest common divisor.

We now apply the same arguments to a polynomial of arbitrary degree,
$$f(x) = a_0 + a_1 x + \cdots + a_n x^n.$$

Not considering α a root of the polynomial, we divide the polynomial by $x - \alpha$ with remainder for an arbitrary α. We obtain a polynomial $g(x)$ of degree $n - 1$ as the quotient; because α enters the coefficients of g, we write it as $g(x, \alpha)$. We saw (see formulas (3) and (4)) that the remainder is equal to $f(\alpha)$:
$$f(x) = (x - \alpha)g(x, \alpha) + f(\alpha). \tag{13}$$

Substituting $x = \alpha$ in the polynomial $g(x, \alpha)$, we obtain a polynomial in α, which is called the *derivative* of the polynomial $f(x)$ at $x = \alpha$ and is denoted by $f'(\alpha)$. Thus, by definition,

$$f'(\alpha) = \frac{f(x) - f(\alpha)}{x - \alpha}(\alpha). \tag{14}$$

Such an expression can elicit doubt, because when we substitute $x = \alpha$, both the numerator and the denominator of the expression $(f(x) - f(\alpha))/(x - \alpha)$ become zero, and we obtain $0/0$. The formula demands an explanation: in fact, the numerator is divisible by the denominator (before substituting $x = \alpha$), and we substitute $x = \alpha$ in the quotient, which is a polynomial. We could thus make sense of the expression

$$\frac{x^2 - 1}{x - 1}(1)$$

by explaining that $(x^2 - 1)/(x - 1) = x + 1$ and we in fact consider $(x + 1)(1) = 2$. The passage from the polynomial $f(x)$ to its derivative $f'(\alpha)$ is called *differentiation* of the polynomial $f(x)$.

Those readers who will study mathematics further will meet other concepts of the derivative for other cases, for example, $f(x) = \sin(x)$ or $f(x) = 2^x$. Strictly speaking, the derivative is defined by the same formula (14), but it is more difficult to give an exact meaning to the right-hand side in the general case. In the case of polynomials, everything is clarified by applying the Bézout theorem to the polynomial $f(x) - f(\alpha)$.

If α is a root of the polynomial $f(x)$ in formula (13), that is, $f(\alpha) = 0$, then we obtain $f(x) = (x - \alpha)g(x, \alpha)$. By our definition, α is a multiple root of the polynomial $f(x)$ if α is a root of $g(x, \alpha)$, that is, $g(\alpha, \alpha) = 0$. But that means that $f'(\alpha) = 0$. We have proved an assertion, which is formulated in Theorem 16.

Theorem 16. *A root of the polynomial $f(x)$ is a multiple root if and only if it is a root of the derivative $f'(x)$.*

We see that a multiple root α of $f(x)$ is a *common root* of the polynomials $f(x)$ and $f'(x)$. In other words, α is a root of $\mathrm{GCD}(f(x), f'(x))$, and we can find this greatest common divisor by using the Euclidean algorithm. As a rule, the degree of the greatest common divisor is significantly lower than the degree of the polynomial $f(x)$.

We now clarify the division by $x - \alpha$ with remainder, find the polynomial $g(x, \alpha)$ in formula (13), and deduce an obvious formula for the derivative of a polynomial.

It is possible to divide $f(x)$ by $x - \alpha$ with remainder by "sections" and find the quotient $g(x, \alpha)$ and $f(\alpha)$ in explicit form. But another path is shorter. We recall that $f(x)$ is a sum of terms of the form $a_k x^k$. Therefore, $f(x) - f(\alpha)$ is a sum of terms of the form $a_k(x^k - \alpha^k)$. The polynomial $x^k - \alpha^k$ has the root $x = \alpha$ and is therefore divisible by $x - \alpha$ according to the Bézout theorem. We already noted (immediately after formulating the Bézout theorem) that we had previously produced the division by $x - \alpha$. True, we did it only for $\alpha = 1$, but passage to the general case is simple. We should use formula (12) in Chap. 1 (substituting k for $r + 1$),

$$x^k - 1 = (x - 1)(x^{k-1} + x^{k-2} + \cdots + x + 1).$$

Substituting the ratio x/α for x, we have

$$\frac{x^k}{\alpha^k} - 1 = \left(\frac{x}{\alpha} - 1\right)\left(\frac{x^{k-1}}{\alpha^{k-1}} + \frac{x^{k-2}}{\alpha^{k-2}} + \cdots + \frac{x}{\alpha} + 1\right).$$

Multiplying both sides of the equality by α^k, we obtain

$$x^k - \alpha^k = (x - \alpha)(x^{k-1} + \alpha x^{k-2} + \cdots + \alpha^{k-2}x + \alpha^{k-1}). \qquad (15)$$

We deduced this formula given that $\alpha \neq 0$ (because we used x/α), but it is obviously also true for $\alpha = 0$.

We consider the polynomial $f(x) = a_0 + a_1 x + \cdots + a_n x^n$ and the difference $f(x) - f(\alpha)$. As we saw, this difference is equal to the sum of the terms $a_k(x^k - \alpha^k)$. Using formula (15) to divide each such term, we obtain

$$\frac{a_k(x^k - \alpha^k)}{x - \alpha} = a_k(x^{k-1} + \alpha x^{k-2} + \cdots + \alpha^{k-2}x + \alpha^{k-1}).$$

If we substitute $x = \alpha$ (in the right-hand side!), then we obtain the term $k a_k \alpha^{k-1}$. For the polynomial $g(x, \alpha)$ in formula (13) with $x = \alpha$, we thus find that $g(x, \alpha)(\alpha) = g(\alpha, \alpha)$ is the sum of the terms $k a_k \alpha^{k-1}$, that is, $a_1 + 2a_2 x + 3a_3 x^2 + \cdots + n a_n x^{n-1}$. In other words, we have deduced the formula for the derivative $f'(x)$ of the polynomial $f(x) = a_0 + a_1 x + \cdots + a_n x^n$:

$$f'(x) = a_1 + 2a_2 x + 3a_3 x^2 + \cdots + n a_n x^{n-1}. \qquad (16)$$

Compare this with what we previously obtained for second- and third-degree polynomials, and convince yourself that we then deduced special cases of formula (16) for $n = 2$ and $n = 3$.

The concept of the derivative of a polynomial is important not only in connection with the question of multiple roots; it has many other applications. Therefore, we now prove some basic properties of the derivative. All the proofs follow directly from the definition, that is, from equality (13).

Derivative of a Constant. If $f(x) = \alpha_0$, then by definition, $f(x) = f(\alpha)$ and $g(x, \alpha) = 0$. Therefore, $f'(\alpha) = 0$, that is, $f'(x) = 0$.

Derivative of a Sum. Let f_1 and f_2 be two polynomials, and let $f = f_1 + f_2$. We have the equalities

$$f_1(x) = f_1(\alpha) + (x - \alpha)g_1(x, \alpha),$$
$$f_2(x) = f_2(\alpha) + (x - \alpha)g_2(x, \alpha) \tag{17}$$

and consequently $f_1'(\alpha) = g_1(\alpha, \alpha)$ and $f_2'(\alpha) = g_2(\alpha, \alpha)$. Adding formulas (17), we obtain

$$f(x) = f(\alpha) + (x - \alpha)g(x, \alpha),$$

where $g(x, \alpha) = g_1(x, \alpha) + g_2(x, \alpha)$. Therefore, $f'(\alpha) = g(\alpha, \alpha) = g_1(\alpha, \alpha) + g_2(\alpha, \alpha) = f_1'(\alpha) + f_2'(\alpha)$, that is

$$(f_1 + f_2)' = f_1' + f_2'.$$

From this, we can immediately obtain the formula for the sum of any number of polynomials by induction on the number of addends,

$$(f_1 + f_2 + \cdots + f_r)' = f_1' + f_2' + \cdots + f_r'.$$

Multiplication by a Number. Let $f_1(x) = af(x)$. Multiplying the equalities $f(x) = f(\alpha) + (x - \alpha)g(x, \alpha)$ and $g(\alpha, \alpha) = f'(\alpha)$ by a, we then obtain

$$f_1(\alpha) = af(x) = af(\alpha) + (x - \alpha)ag(x, \alpha),$$

that is, $f_1(x) = f_1(\alpha) + (x - \alpha)ag(x, \alpha)$ and $f_1'(\alpha) = af'(\alpha)$:

$$(af)' = af'.$$

Derivative of a Product. Let $f = f_1 f_2$. Multiplying equalities (17), we obtain

$$f_1(x)f_2(x) = f_1(\alpha)f_2(\alpha) + (x - \alpha)g(x, \alpha),$$

where

$$g(x, \alpha) = g_1(x, \alpha)f_2(\alpha) + g_2(x, \alpha)f_1(\alpha) + (x - \alpha)g_1(x, \alpha)g_2(x, \alpha).$$

Hence,
$$f(x) = f(\alpha) + (x - \alpha)g(x, \alpha),$$
where $g(x, \alpha)$ is the same as just given. Therefore,
$$f'(\alpha) = g(\alpha, \alpha) = g_1(\alpha, \alpha)f_2(\alpha) + g_2(\alpha, \alpha)f_1(\alpha)$$
$$= f_1'(\alpha)f_2(\alpha) + f_2'(\alpha)f_1(\alpha),$$
that is,
$$(f_1 f_2)' = f_1' f_2 + f_2' f_1. \tag{18}$$

If f_1 is a constant (a polynomial of degree zero), then considering that the derivative of a constant polynomial is zero, it follows from (18) that $(af)' = af'$.

By induction on the number of factors, we obtain
$$(f_1 f_2 \cdots f_r)' = f_1' f_2 \cdots f_r + f_1 f_2' \cdots f_r + \cdots + f_1 f_2 \cdots f_r' \tag{19}$$

(in the right-hand side, each factor in the product $f_1 \cdots f_r$ is in turn replaced with its derivative). Indeed, according to formula (18),
$$(f_1 \cdots f_r)' = ((f_1 \cdots f_{r-1})f_r)' = (f_1 \cdots f_{r-1})' f_r + (f_1 \cdots f_{r-1})f_r'.$$

Substituting expression (19) for $(f_1 \cdots f_{r-1})'$ (which we can assume to be already proved), we obtain the formula sought.

An important particular case is when all the factors in formula (19) are equal,
$$(f^r)' = r f^{r-1} f'. \tag{20}$$

It is easy to verify from the definition of the derivative that $x' = 1$. Therefore, $(x^r)' = rx^{r-1}$. Combining this rule, we can once again prove the explicit formula (16) for the derivative.

We return to the question of multiple roots of polynomials. Let a polynomial $f(x)$ have the root α with the multiplicity k. This means that it can be represented in the form
$$f(x) = (x - \alpha)^k g(x),$$
where $g(x)$ already does not have α as a root. By formula (18),
$$f'(x) = \left((x - \alpha)^k\right)' g(x) + (x - \alpha)^k g'(x),$$
and by formula (20),

$$\left((x-\alpha)^k\right)' = k(x-\alpha)^{k-1}$$

(we recall that $(x-\alpha)' = 1$ by formula (16)). Therefore,

$$f'(x) = k(x-\alpha)^{k-1}g(x) + (x-\alpha)^k g'(x) = (x-\alpha)^{k-1}p(x),$$

where $p(x) = kg(x) + (x-\alpha)g'(x)$.

Here, the polynomial $p(x)$ does not have α as a root: $p(\alpha) = kg(\alpha) \neq 0$. We consider the polynomials

$$d(x) = \mathrm{GCD}(f(x), f'(x))$$

and $\varphi(x) = f(x)/d(x)$ (because $d(x)$ is a divisor of $f(x)$, it must be that $f(x)$ is divisible by $d(x)$). The polynomial $d(x)$ is divisible by $(x-\alpha)^{k-1}$ because both $f(x)$ and $f'(x)$ are divisible by $(x-\alpha)^{k-1}$, but $d(x)$ is not divisible by $(x-\alpha)^k$, because $p(\alpha) \neq 0$, which means that $p(x)$ is not divisible by $x-\alpha$. As a result, we see that $\varphi(x)$ is divisible by $x-\alpha$ exactly to the first power. Because $\varphi(x)$ is defined independently of the root α ($\varphi(x) = f(x)/\mathrm{GCD}(f(x), f'(x))$), the given argument holds for all roots of the polynomial $f(x)$. We see that $\varphi(x)$ has exactly the same roots as $f(x)$ but does not have any multiple roots. Thanks to this, we can always reduce an investigation of the roots of a polynomial to a question about a polynomial without multiple roots.

We note that we already met the derivative implicitly in connection with the formula for the interpolation polynomial. Indeed, let $F(x) = (x-x_1)\cdots(x-x_{n+1})$. We see from formula (16) that $(x-x_j)' = 1$. Therefore, formula (()19) gives us

$$F'(x) = (x-x_2)\cdots(x-x_{n+1}) + (x-x_1)(x-x_3)\cdots(x-x_{n+1})$$
$$+ \cdots + (x-x_1)(x-x_2)\cdots(x-x_n).$$

If we use the notation $F_k(x) = F(x)/(x-x_k)$ as in the preceding section, then $F'(x) = F_1(x)+F_2(x)+\cdots+F_{n+1}(x)$. Substituting one of the values $x = x_k$ for x here, we see that because all the $F_i(x)$ for $i \neq k$ contain the factor $x-x_k$, we must have $F_i(x_k) = 0$. Hence, $F'(x_k) = F_k(x_k)$, and we can rewrite formula (11) as

$$f(x) = \frac{y_1}{F'(x_1)}F_1(x) + \frac{y_2}{F'(x_2)}F_2(x) + \cdots + \frac{y_{n+1}}{F'(x_{n+1})}F_{n+1}(x).$$

Problems:

1. The polynomial $x^{2n} - 2x^n + 1$ obviously has the root $x = 1$ and is divisible by $x - 1$ according to the Bézout theorem. Find its quotient when it is divided by $x - 1$.

2. For which values of a and b does the polynomial $x^n + ax^{n-1} + b$ have a multiple root? Find the value of the root.
3. For which values of a and b does the polynomial $x^3 + ax + b$ have a multiple root?
4. Prove that the polynomial $x^n + ax^m + b$, $n > m$, does not have nonzero roots of multiplicity three or more.
5. The derivative of the polynomial $f'(x)$ is called the *second derivative* of the polynomial $f(x)$ and is denoted by $f''(x)$. Find a formula for $(f_1 f_2)''$ analogous to formula (18) (but, of course, somewhat more complicated).
6. Prove that the derivative of a polynomial is identical to zero if and only if the polynomial is a constant (that is, it has the degree zero).
7. Prove that for any polynomial $f(x)$, there exists a polynomial $g(x)$ such that $g'(x) = f(x)$ and that all such polynomials $g(x)$ (for a given $f(x)$) differ only in their free term.
8. Prove that the number of roots of a polynomial does not exceed its degree, not even if each root is counted as many times as its multiplicity.

6. Binomial Formula

In this section, we study an important formula that expresses the polynomial $(1 + x)^n$ in our accustomed form $a_0 + a_1 x + \cdots + a_n x^n$. For this, we need only open parentheses in the expression $(1 + x)^n = (1 + x)(1 + x) \cdots (1 + x)$, multiplying each term by each other term in n factors $1 + x$. Obviously, we obtain terms of the form x^k after opening the parentheses, but each such term may occur several times. After combining like terms, we obtain the formula sought. For example, for $n = 2$, as is well known,

$$(1 + x)^2 = (1 + x)(1 + x) = 1(1 + x) + x(1 + x)$$
$$= 1 + x + x + x^2 = 1 + 2x + x^2.$$

The formula for $n = 3$ is probably also well known to you. But if not, it is very easy to deduce by multiplying the formula for $(1 + x)^2$ by $1 + x$:

$$(1 + x)^3 = (1 + x)^2(1 + x) = (1 + 2x + x^2)(1 + x)$$
$$= (1 + 2x + x^2) + (1 + 2x + x^2)x$$
$$= 1 + 3x + 3x^2 + x^3.$$

The coefficient a_k in the polynomial $(1 + x)^n$ depends on the index k and also on the power n of the expression $(1 + x)^n$ that we want to expand in the form of a polynomial. To include the dependence on both n and k, the symbol C_n^k is used for the coefficient. Thus, *by definition,* C_n^k are the coefficients in the formula

$$(1 + x)^n = C_n^0 + C_n^1 x + C_n^2 x^2 + \cdots + C_n^n x^n. \qquad (21)$$

For example, $C_2^0 = 1$, $C_2^1 = 2$, $C_2^2 = 1$, $C_3^0 = 1$, $C_3^1 = 3$, $C_3^2 = 3$, and $C_3^3 = 1$.

The coefficients C_n^k are called *binomial coefficients*. Our problem is to write them in an explicit form. We note that we can recognize some of them immediately. Because the leading terms are multiplied together once when polynomials are multiplied, the leading term in the polynomial $(1 + x)^n$ is equal to x^n, that is,

$$C_n^n = 1. \tag{22}$$

In exactly the same way, the free terms (the value of the polynomial for $x = 0$) are multiplied together when polynomials are multiplied. Therefore, the free term of the polynomial $(1 + x)^n$ is equal to one, that is,

$$C_n^0 = 1. \tag{23}$$

In the general case, we consider the derivative of both sides of equality (21). On the left, according to formula (20), we obtain $n(1 + x)^{n-1}$ because $(1 + x)' = 1$ by formula (16). The derivative of the right-hand side is written according to formula (16). We obtain

$$n(1 + x)^{n-1} = C_n^1 + 1C_n^2 x + \cdots + kC_n^k x^{k-1} + \cdots + nC_n^n x^{n-1}.$$

We can expand the left-hand side according to formula (21) applied for $n - 1$. The coefficient of x^{k-1} is nC_{n-1}^{k-1} in the left-hand side and kC_n^k in the right-hand side. We see that $kC_n^k = nC_{n-1}^{k-1}$ or

$$C_n^k = \frac{n}{k}C_{n-1}^{k-1},$$

that is, the coefficient C_n^k is simply expressed in terms of the coefficient C_{n-1}^{k-1} with lower indices. Applying the same formula to C_{n-1}^{k-1}, we obtain

$$C_n^k = \frac{n(n - 1)}{k(k - 1)}C_{n-2}^{k-2}.$$

Repeating this process r times, we obtain the formula

$$C_n^k = \frac{n(n - 1)\cdots(n - r + 1)}{k(k - 1)\cdots(k - r + 1)}C_{n-r}^{k-r}$$

(we subtract the values $0, 1, \ldots, r - 1$ from n in the numerator and from k in the denominator, which is r subtractions giving r factors for each). Finally, we take $r = k$. Because we know that $C_m^0 = 1$ for any m, we obtain the formula for C_n^k:

$$C_n^k = \frac{n(n-1)\cdots(n-k+1)}{k(k-1)\cdots 1}. \tag{24}$$

This is the expression we sought.

Formula (21) with explicit expressions (24) for the binomial coefficients C_n^k substituted in it is called the *binomial formula* (or *Newton binomial*). The word "binomial" comes from a Latin word that means "having two names" and refers to an algebraic expression with two terms. This formula is understood as the formula for raising a binomial to a power.

The binomial formula has very many applications, and it is therefore useful to have various ways of writing the binomial coefficients given by (24). In the denominator, we have the product of all integers from 1 to k. A product of the form $1 \cdot 2 \cdots m$ is called m *factorial* and is denoted by $m!$. In the numerator, we have the product of integers from n to $n - k + 1$. If we multiply it by the product of integers from $n - k$ to 1 (that is, by $(n - k)!$), then we obtain $n!$. Therefore, multiplying the numerator and the denominator in formula (24) by $(n - k)!$, we convert the formula to the form

$$C_n^k = \frac{n!}{k!(n-k)!}. \tag{25}$$

It immediately follows that

$$C_n^k = C_n^{n-k}. \tag{26}$$

We note that it is not immediately obvious that the numerator is divisible by the denominator both in formula (24) and in formula (25), although we know that this must be so based on the sense of the coefficient C_n^k in formula (21). We can express the fact that the right-hand side of formula (24) is an integer most simply by stating that *the product of any k sequential integers is divisible by $k!$*. We see later how interesting properties of prime numbers follow from the fact that the right-hand sides of formulas (24) and (25) yield integers.

We now deduce important properties of the coefficients C_n^k. The first property follows from the obvious equality

$$(1+x)^n = (1+x)^{n-1}(1+x).$$

If we expand $(1+x)^n$ and $(1+x)^{n-1}$ according to formula (21), we obtain

$$C_n^0 + C_n^1 x + C_n^2 x^2 + \cdots + C_n^n x^n = \left(C_{n-1}^0 + C_{n-1}^1 x + \cdots + C_{n-1}^{n-1}\right)(1+x).$$

The coefficient of x^k is equal to C_n^k in the left-hand side and is obtained from the sum of the terms $C_{n-1}^k x^k$ and $C_{n-1}^{k-1} x^{k-1} x$ in the right-hand side, that is, it is equal to $C_{n-1}^k + C_{n-1}^{k-1}$. Therefore,

$$C_n^k = C_{n-1}^k + C_{n-1}^{k-1}. \tag{27}$$

This is a very convenient formula for calculating the coefficients C_n^k in terms of the coefficients with the index $n-1$. To present this formula more visually, we write the coefficients C_n^k in a triangular arrangement, writing C_n^k in the nth row. Taking formulas (22) and (23) into account (we have 1 at the beginning and end of each row), the triangle looks like

$$
\begin{array}{ccccccccccc}
&&&&& 1 \\
&&&& 1 && 1 \\
&&& 1 && 2 && 1 \\
&& \cdots && \cdots && \cdots && \cdots \\
& 1 && C_{n-1}^1 && \cdots && C_{n-1}^{n-2} && 1 \\
1 && C_n^1 && C_n^2 && \cdots && C_n^{n-1} && 1 \\
\cdots && \cdots && \cdots && \cdots && \cdots && \cdots && \cdots
\end{array}
$$

Formula (27) indicates that each binomial coefficient C_n^k is equal to the sum of the two coefficients immediately above to the left and right. Taking the first two rows as initial data, we can easily obtain the values for the coefficients

$$
\begin{array}{ccccccccccc}
&&&&& 1 \\
&&&& 1 && 1 \\
&&& 1 && 2 && 1 \\
&& 1 && 3 && 3 && 1 \\
& 1 && 4 && 6 && 4 && 1 \\
1 && 5 && 10 && 10 && 5 && 1
\end{array}
$$

and so on. This triangular arrangement is called the *Pascal triangle*.

We obtain the second property by substituting $x = 1$ in formula (21) (the definition of the binomial coefficients). We have 2^n on the left and the sum of all the binomial coefficients C_n^k for $k = 0, 1, \ldots, n$ on the right. Therefore, *the sum of all numbers in the nth row of the Pascal triangle is equal to 2^n.*

Finally, we consider two neighboring numbers in one row, C_n^{k-1} and C_n^k. According to formula (25),

$$C_n^k = \frac{n!}{k!(n-k)!}, \qquad C_n^{k-1} = \frac{n!}{(k-1)!(n-k+1)!}.$$

Because $k! = (k-1)!k$ and $(n-k+1)! = (n-k)!(n-k+1)$, we have

$$C_n^k = \frac{n-k+1}{k}C_n^{k-1}.$$

It is obvious that $(n-k+1)/k > 1$ if $n-k+1 > k$, that is, if $k < (n+1)/2$, and then we have $C_n^k > C_n^{k-1}$. Conversely, if $k > (n+1)/2$, then we similarly obtain $C_n^k < C_n^{k-1}$. Therefore, *the numbers in one row of the Pascal triangle increase to the middle of the row and then decrease.* If n is even, then we have one largest number $C_n^{n/2}$, and if n is odd, then we have two neighboring numbers both equal to the largest value, $C_n^{(n-1)/2}$ and $C_n^{(n+1)/2}$. In this case where $k = (n+1)/2$, we have

$$C_n^k = C_n^{k-1}.$$

Formula (21) (with the values of the binomial coefficients C_n^k determined by formula (24)) can be given in a somewhat more general form. For this, we substitute $x = b/a$ and multiply both sides of equality (21) by a^n. We obtain the formula

$$(a+b)^n = C_n^0 a^n + C_n^1 a^{n-1}b + C_n^2 a^{n-2}b^2 + \cdots + C_n^n b^n. \qquad (28)$$

We established this formula for $a \neq 0$ (because we divided by a), but it obviously also holds for $a = 0$. It is also called the *binomial formula*.

We now discuss some consequences of the binomial formula and their applications. As a rule, the simpler an assertion is, the more applications it has. We especially often have occasion to use the first coefficients in the binomial formula. We already noted that the very first coefficient C_n^0 is equal to one. The next coefficient C_n^1 is equal to n according to formula (24). We note that it hence follows from formula (26) that $C_n^n = 1$ (as we already mentioned) and $C_n^{n-1} = n$. Therefore,

$$(a+b)^n = a^n + na^{n-1}b + \cdots + nab^{n-1} + b^n.$$

This observation is useful for investigating equations. We wrote an equation of degree n in the form

$$a_0 + a_1 x + \cdots + a_{n-1}x^{n-1} + a_n x^n = 0.$$

That the degree of the equation is namely n means that $a_n \neq 0$. We can therefore divide the equation by a_n and obtain an equivalent equation in which we already have $a_n = 1$. In what follows, we assume that this has been done, and we write the equation in the form

$$f(x) = a_0 + a_1 x + \cdots + a_{n-1}x^{n-1} + x^n = 0.$$

We now develop another transformation for replacing an equation with an equivalent equation. For this, we set $x = y + c$, where y is a new unknown and c is some number. Substituting this value for x in our equation, we obtain a term of the form $a_m(y + c)^m$ from each term $a_m x^m$. We represent each such new term as a polynomial in y according to the binomial formula and then combine like terms. As a result, we obtain a new polynomial in y, which is denoted by $g(y) = f(y + c)$. Because y is expressed in terms of x ($y = x - c$), the equations $f(x) = 0$ and $g(y) = 0$ are equivalent: a root $y = \alpha - c$ of the equation $g(y) = 0$ corresponds to the root $x = \alpha$ of the equation $f(x) = 0$, and a root $x = \beta + c$ of the equation $f(x) = 0$ corresponds to the root $y = \beta$ of the equation $g(y) = 0$. We examine how the coefficients of the equation change under this substitution. First, the degree of the equation $g(y) = 0$ is n as before, and the coefficient of the leading term is 1 as before. This follows because the terms $a_m(y + c)^m$ yield terms of a degree less than or equal to m with respect to y when expanded in the binomial formula. Therefore, the term of degree n only comes from the term $(y + c)^n$ and is equal to y^n (again by the binomial formula). We now consider terms of degree $n - 1$. For the same reasons, terms of degree $n - 1$ with respect to y can arise from the term $(y + c)^n$ and the term $a_{n-1}(y + c)^{n-1}$. From the latter, we obtain the term $a_{n-1}y^{n-1}$. From the term $(y + c)^n$, we must take the *second* term in the expansion according to the binomial formula. As we saw, this term is equal to $ny^{n-1}c$. This means that the term of degree $n - 1$ in the polynomial $g(y) = f(y + c)$ has the form $(a_{n-1} + nc)y^{n-1}$. We can use this to simplify the equation, choosing the number c such that the term of degree $n - 1$ becomes zero: we need to set $a_{n-1} + nc = 0$, that is, $c = -a_{n-1}/n$. We have proved the assertion formulated in Theorem 17.

Theorem 17. *Substituting $x = y - a_{n-1}/n$ in the equation $f(x) = a_0 + a_1 x + \cdots + a_{n-1}x^{n-1} + x^n = 0$, we obtain an equivalent equation $g(y) = 0$ of degree n and with the coefficient of the leading term equal to one but without a term of degree $n - 1$ with respect to y.*

We note that for a second-degree equation, Theorem 17 gives the formula for its solution. Indeed, the polynomial $g(y)$ has the form $y^2 + b_2$, and we immediately find its roots: $y = \pm\sqrt{-b_2}$. Substitute as indicated in Theorem 17, calculate b_2 and hence the roots of the polynomial $f(x)$. Convince yourself that we thus obtain the usual formula for the solution of a quadratic equation. In the case of a polynomial of arbitrary degree, we obtain only a certain simplification, which is sometimes useful. For example, any third-degree equation is equivalent to an equation of the form $x^3 + ax + b = 0$.

In conclusion, we discuss the application of the binomial theorem to the calculation of the sum of powers of consecutive integers. We consider the sum

$$S_m(n) = 0^m + 1^m + 2^m + \cdots + n^m \tag{29}$$

of the mth powers of all natural numbers not exceeding n. You are probably familiar with the formula $S_1(n) = n(n+1)/2$ (see Problem 5 in Sec. 2). We begin with some remarks about calculating the sum in the general form. Let $a_0, a_1, a_2, \ldots, a_n, \ldots$ be an infinite sequence of numbers. We are interested in the sums of its consecutive members: $a_0, \ a_0 + a_1, \ a_0 + a_1 + a_2, \ \ldots, \ a_0 + a_1 + a_2 + \cdots + a_n, \ldots$. The first sequence is denoted by the letter a, and its $(n+1)$th member is then a_n (it is more convenient to denote the $(n+1)$th member thus, and not the nth, beginning the sequence with a_0). The associated sequence of sums is denoted by Sa, and its $(n+1)$th member is then equal to

$$(Sa)_n = a_0 + a_1 + a_2 + \cdots + a_n, \quad n = 0, 1, 2, \ldots .$$

For example, if $a_n = n^m$, $n = 0, 1, 2, \ldots$, then Sa is the sequence of sums $S_m(n)$. Clearly, knowing a sequence Sa, we can recover the sequence a. We must subtract the nth member of the sequence Sa from the $(n+1)$th member, and we obtain a_n. Indeed, if

$$b_n = (Sa)_n = a_0 + a_1 + \cdots + a_n, \tag{30}$$
$$b_{n-1} = (Sa)_{n-1} = a_0 + a_1 + \cdots + a_{n-1}, \tag{31}$$

then subtracting equality (31) from equality (30), we obtain

$$b_n - b_{n-1} = a_n.$$

We meet a new, important construction.

Along with an arbitrary sequence $b_0, b_1, b_2, \ldots, b_n, \ldots$, we consider the sequence $b_0, b_1 - b_0, b_2 - b_1, \ldots, b_{n+1} - b_n, \ldots$. If the first sequence is denoted by b, then the second is denoted by Δb. Its $(n+1)$th member is equal to

$$(\Delta b)_0 = b_0, \qquad (\Delta b)_n = b_n - b_{n-1}, \quad n = 1, 2, \ldots .$$

We can now write the relation we noted between the sequences a and Sa as the formula $\Delta Sa = a$. It turns out that the completely symmetric formula also holds, namely, both the following relations are valid:

$$\Delta Sa = a,$$
$$S\Delta b = b. \tag{32}$$

We can express this by saying that the operations on sequences S and Δ are inverse to each other.

We already verified the first formula. To prove the second, we write all equalities defining the numbers $a_k = (\Delta b)_k$ for $k = 0, 1, \ldots, n-1$:

$$a_0 = b_0,$$
$$a_1 = b_1 - b_0,$$
$$a_2 = b_2 - b_1,$$
$$\vdots$$
$$a_n = b_n - b_{n-1}.$$

We add them. On the left, we obtain $a_0 + \cdots + a_n$, that is, $(Sa)_n$; on the right, all terms cancel except b_n in the last formula. We thus obtain $(Sa)_n = b_n$, that is, the second formula in (32).

The advantage of the relations just proved is that it is often simpler not to calculate the sums forming Sa immediately but to work out the sequence b for which $\Delta b = a$. Then the second relation in (32) yields $Sa = b$.

We apply this idea to sum (29). We saw that $S_m(n) = (Sa)_n$, where $a_n = n^m$. How can we represent the sequence a, $a_n = n^m$, in the form $a = \Delta b$? The answer to this question follows from Theorem 18.

Theorem 18. *For each polynomial $f(x)$ of degree m, there exists a unique polynomial of degree $m+1$ such that*

$$g(x) - g(x-1) = f(x) \tag{33}$$

and the free term of the polynomial $g(x)$ is equal to zero.

It is very easy to prove that the polynomial $g(x)$ with the given conditions is unique. Let $g_1(x)$ be a different polynomial satisfying the relation $g_1(x) - g_1(x-1) = f(x)$ with the free term equal to zero. We subtract equality (33) from this relation. If we set $g_1(x) - g(x) = g_2(x)$, then we have $g_2(x) - g_2(x-1) = 0$, and the free term of $g_2(x)$ is equal to zero, that is, $g_2(0) = 0$. Substituting $x = 1$ in $g_2(x) - g_2(x-1) = 0$, we see that $g_2(1) = 0$. Substituting $x = 2$, we obtain $g_2(2) = g_2(1) = 0$. Thus, $g_2(n) = 0$ for all natural numbers n by induction, that is, all natural numbers are roots of the polynomial $g_2(x)$. By Theorem 14, this is possible only when $g_2 = 0$, and this means that $g = g_1$.

We prove the existence of the polynomial g by induction on the degree of the polynomial f, that is, on m. For $m = 0$, the polynomial f is a number a, and we see that $g(x) = ax$ satisfies condition (33). Let the assertion hold for polynomials f of a degree less than m. Let $a_m x^m$ be the leading term of the polynomial f. We choose a number a such that the leading term of the polynomial $ax^{m+1} - a(x-1)^{m+1}$ is equal to the leading term $a_m x^m$ of the polynomial $f(x)$. For this, we use the binomial formula to write

$$(x-1)^{m+1} = x^{m+1} - (m+1)x^m + \dots,$$

where the dots denote terms of a degree less than m. Hence,

$$x^{m+1} - (x-1)^{m+1} = (m+1)x^m + \dots.$$

Clearly, we must set

$$a = \frac{a_m}{m+1}. \tag{34}$$

Then terms of degree m cancel in the difference

$$f(x) - \frac{a_m}{m+1}\left[x^{m+1} - (x-1)^{m+1}\right],$$

and this difference has a degree less than m. Letting $h(x)$ denote this polynomial, we can already assume by induction that for it, there exists a polynomial $g_1(x)$ of a degree less than $m+1$ with the free term equal to zero such that

$$h(x) = g_1(x) - g_1(x-1),$$

that is,

$$f(x) - \frac{a_m}{m+1}\left[x^{m+1} - (x-1)^{m+1}\right] = g_1(x) - g_1(x-1).$$

Adding the term with square brackets to both sides, we obtain $f(x) = g(x) - g(x-1)$, where

$$g(x) = \frac{a_m}{m+1}x^{m+1} + g_1(x).$$

This proves the theorem.   □

For practical purposes, naturally, we must use not induction but successive subtractions applied to $h(x)$ and so on until we obtain a polynomial of degree zero.

We now return to calculating the sum $S_m(n)$. We saw that this sum is equal to b_n, where b is a sequence such that $\Delta b = a$, $a_n = n^m$. We apply Theorem 18 to the polynomial x^m. We obtain a polynomial $g(x)$ of degree $m+1$ such that

$$g(x) - g(x-1) = x^m$$

and the free term of the polynomial $g(x)$ is equal to zero. Substituting $x = n$ in this equality, we see that the sequence $b_n = g(n)$ for $n \geq 1$ and $b_0 = g(0) = 0$ satisfies the condition $\Delta b = a$, that is, $a = Sb$. We have thus proved the assertion formulated in Theorem 19.

Theorem 19. *The sums $S_m(n)$ are expressed in the form $g_m(n)$, where g_m is a polynomial of degree $m+1$ such that $g_m(x) - g_m(x-1) = x^m$ and the free term of $g_m(x)$ is equal to zero.*

We note that the proof of Theorem 18 provides a way to find the polynomial $g_m(x)$ for each m. We consider the case where $m = 2$ as an example. By analogy with sequences, we let Δg denote the polynomial $g(x) - g(x-1)$, that is, we set $(\Delta g)(x) = g(x) - g(x-1)$. For a beginning, we must take a monomial ax^3 such that $\Delta(ax^3)$ has the leading term equal to x^2. According to formula (34), we must set $a = 1/3$ (here $m = 2$ and $a_2 = 1$). According to the binomial formula,

$$\Delta\left(\frac{1}{3}x^3\right) = \frac{1}{3}x^3 - \frac{1}{3}(x-1)^3 = x^2 - x + \frac{1}{3}$$

and

$$x^2 - \Delta\left(\frac{1}{3}x^3\right) = x - \frac{1}{3}.$$

We must now find a monomial bx^2 such that $\Delta(bx^2)$ has the leading term equal to x. According to formula (34), we must set $b = 1/2$ (here $m = 1$ and $a_1 = 1$). According to the binomial formula,

$$\Delta\left(\frac{1}{2}x^2\right) = \frac{1}{2}x^2 - \frac{1}{2}(x-1)^2 = x - \frac{1}{2}$$

and

$$x^2 - \Delta\left(\frac{1}{3}x^3\right) - \Delta\left(\frac{1}{2}x^2\right) = -\frac{1}{3} + \frac{1}{2} = \frac{1}{6}.$$

We finally obtain

$$x^2 = \Delta\left(\frac{1}{3}x^3 + \frac{1}{2}x^2 + \frac{1}{6}\right),$$

and we must take the polynomial

$$g(x) = \frac{1}{3}x^3 + \frac{1}{2}x^2 + \frac{1}{6}x = \frac{(2x^2 + 3x + 1)x}{6} = \frac{(2x+1)(x+1)x}{6}.$$

We obtain

$$S_2(n) = \frac{(2n+1)(n+1)n}{6}.$$

We make two remarks concerning this result.

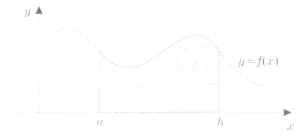

Fig. 6

Remark 1. We can summarize the formula obtained for the sum $S_m(n)$ as follows: For each m, there exists a unique polynomial $g_m(x)$ with the free term equal to zero such that $g_m(x) - g_m(x-1) = x^m$. The method for calculating it is contained in the proof of Theorem 18. Its degree is equal to $m+1$. The formula for $S_m(n)$ has the form $S_m(n) = g_m(n)$. The question is thus reduced to investigating the remarkable polynomials $g_m(x)$. The polynomials $g_m(x)$ are called the *Bernoulli polynomials*. In the supplement, we give a much more explicit expression for these polynomials using a remarkable sequence of rational numbers called the *Bernoulli numbers*.

Remark 2. (Historical.) The operations S and Δ we introduced above, which construct the sequence Sa from the sequence a and the sequence Δb from the sequence b, are very similar to the basic operations in analysis, which determine the *indefinite integral* $\int f\,dx$ of the function $f(x)$ (although not for every function!) and the *derivative* g' of the function g. Our operations S and Δ are elementary analogues of the operations $\int f\,dx$ and g'. Sums and differences similarly participate in the definitions of the integral and derivative, but in a more complicated way (differences also participate in our definition of the derivative for the case where the function is a polynomial; see formula (14)). As in the case of the operations S and Δ, the operations of taking the integral and the derivative are inverse to each other. As in our case, calculating the derivative is simpler than calculating the integral. It is therefore easiest to calculate the integral by choosing another function whose derivative is equal to the initial function.

There is not only an analogy but also a more direct connection between operations on sequences and operations on functions. For example, calculating the integral of the function $f(x)$ is equivalent to calculating the area bounded by the graph of the function, the x axis, and the two perpendiculars to the x axis at the points $x = a$ and $x = b$ (Fig. 6).

We do not prove this here, of course, because we have not even defined
the integral. But we use a simple example to illustrate the calculation
of such an area and its connection with the problem we investigated.
We attempt to find the area bounded by the parabola that is the graph
of the function $y = x^2$, the x axis, and the segment of the line $x = 1$
(Fig. 7).

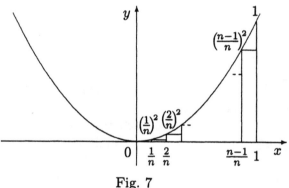

Fig. 7

For this, we divide the segment of the x axis between 0 and 1 into a
large number n of equal parts with the coordinates 0, $1/n$, $2/n$, ...,
$(n-1)/n$, 1 and calculate the corresponding values 0, $(1/n)^2$, $(2/n)^2$,
..., $((n-1)/n)^2$, 1 of the function $y = x^2$. We construct rectangles on
the base equal to the segment between i/n and $(i+1)/n$ with the height
$(i/n)^2$. The polygon composed of these rectangles is contained in the
piece of the parabola whose area we want to determine. Our eyes tell
us that for very large n, the area S_n of the polygon will differ from the
area of the piece of the parabola by very little (we cannot argue more
precisely, because we have not explicated the concept of area; however,
see Problem 1). The area of the polygon is equal to the sum of the areas
of the rectangles composing the polygon. The area of the ith rectangle
is equal to the product of its base $1/n$ times its height $(i/n)^2$, that is,
i^2/n^3. Therefore, the area S_n of the polygon is equal to

$$S_n = \frac{0^2}{n^3} + \frac{1^2}{n^3} + \frac{2^2}{n^3} + \cdots + \frac{(n-1)^2}{n^3} = \frac{S_2(n-1)}{n^3}.$$

Substituting the value $S_2(n) = n^3/3 + n^2/2 + n/6$, which we found above,
we obtain the expression (substituting $n - 1$ for n in the expression for
$S_2(n)$)

$$S_n = \frac{1}{3} - \frac{1}{2}\frac{1}{n} + \frac{1}{6}\frac{1}{n^2}$$

for the area of the polygon. Clearly, as the value of n increases, the
absolute value of the terms $-1/(2n)$ and $1/(6n^2)$ constantly decrease,

and the area of the polygon approaches $1/3$. This means that such is the area of the figure bounded by the parabola.

We have laid out the train of thought that, in principle, Archimedes (third century B.C.) followed in first solving this problem. (Archimedes invented an artificial approach allowing the use of the sum of a certain geometric progression instead calculating the sum $S_2(n)$ in this case. But he knew the formula for $S_2(n)$ and used it to calculate other areas and volumes.)

The mathematicians of the Enlightenment were obsessed with the dream of "surpassing the ancients" (that is, the ancient Greek mathematicians), of whom Archimedes was the most brilliant representative in their eyes. They were therefore much occupied with solving the problem we just considered for functions $y = x^n$ with n greater than two. The solution was apparently first found by the French mathematician Fermat (17th century), initially using practically the same method we described (and then somewhat more simply). The connection between the integral and the derivative, which we mentioned, was not yet known at that time, and the integral (that is, the area) was calculated directly from the definition. It was noticed only later that (using our modern terminology) the operations of taking the derivative and the integral are mutually inverse. This was established by Newton's teacher, Barrow. (Newton collaborated with Barrow when he studied at the university and then occupied Barrow's chair when it was vacated.) The systematic calculation of the integral of the function f by selecting a function g such that the derivative of g is equal to f was begun by Newton. After that, calculating integrals and areas by the method we studied became unnecessary. Now, a student in the upper classes can easily find the integral of x^m for any m without resorting to calculating the sum $S_m(n)$.

Thus, if we returned to the circle of ideas of ancient Greek mathematicians (Pythagoras, Theaetetus, Euclid) in Chap. 1, then we met the ideas of the Enlightenment mathematicians (17th and 18th centuries) in this chapter.

Problems:

1. Note that the area S_n of the polygon calculated at the end of the section is *less than* the area we sought for the figure bounded by the parabola $y = x^2$ because the polygon is included inside that figure. Construct a polygon divided into rectangles based on the segments from i/n to $(i + 1)/n$ with the height $((i + 1)/n)^2$ that *contains* the same figure. Its area S'_n is therefore *greater than* the area of the figure. Calculate the area S'_n, and prove that as n increases, this area also approaches $1/3$ without limit. This provides a more convincing (that is, more "rigorous") proof that the area sought for the figure is $1/3$.
2. Try to solve the analogous problem for a "parabola of degree m" given by the equation $y = x^m$. Convince yourself that for the answer, we do not need to know the entire Bernoulli polynomial $g_m(x)$ but need to calculate only the coefficient

a_{m+1} of its leading term $a_{m+1}x^{m+1}$. Prove that $a_{m+1} = 1/(m+1)$. From this, find the area of the figure bounded by the parabola with $y = x^m$, the x axis, and the line $x = 1$.

3. Prove that the area of the figure bounded by the parabola $y = x^m$, the x axis, and the line $x = a$ is equal to $a^{m+1}/(m+1)$. Note that the derivative of the polynomial $x^{m+1}/(m+1)$ is equal to x^m. This is an instance of Barrow's theorem that integrating and finding the derivative are mutually inverse operations.

4. Prove that the sums of binomial coefficients with even superscripts $C_n^0 + C_n^2 + \cdots$ and with odd superscipts $C_n^1 + C_n^3 + \cdots$ are equal to each other, and find the value of these sums.

5. Find a relation between binomial coefficients expressing the fact that

$$(1+x)^n(1+x)^m = (1+x)^{n+m}.$$

For $n = m$, deduce the formula for the sum of squares of binomial coefficients from this.

6. Prove that if p is a prime number, then all binomial coefficients C_p^k for $k \neq 0$ and $k \neq p$ are divisible by p. From this, deduce that $2^p - 2$ is divisible by p. Prove that for any integer n, $n^p - n$ is divisible by p. Fermat first proved this theorem.

7. What can be said about the sequence a if the sequence Δa consists of identical numbers? What does formula (32) yield in this case?

8. Find the sum $S_3(n)$, and verify that $S_3(n) = (S_1(n))^2$.

9. Let a be an arbitrary sequence $a_0, a_1, a_2, \ldots$. Apply the operation Δ to the sequence Δa. Let $\Delta^2 a$ denote the result. Define $\Delta^k a$ by induction as $\Delta(\Delta^{k-1}a)$. When is the "infinite interpolation problem" solvable for the sequence a, that is, when does there exist a polynomial $f(x)$ of a degree not exceeding m such that $f(n) = a_n$ for $n = 0, 1, 2, \ldots$? Prove that the condition $(\Delta^{m+1}a)_n = 0$ for $n \geq m$ is necessary and sufficient. This condition means that if we write the sequence a and the sequence of differences of two consecutive terms under it and so on,

$$a_0 \qquad a_1 \qquad a_2 \qquad a_3 \quad \cdots \quad a_n \qquad\qquad a_{n+1}$$
$$a_1 - a_0 \quad a_2 - a_1 \quad a_3 - a_2 \quad \cdots \quad \cdots \quad a_{n+1} - a_n$$

then there are only zeros in the $(m+1)$th row. Does there exist a polynomial $f(x)$ such that $f(n) = 2^n$ for all integers n?

10. Prove that if $a_n = q^n$, then $(\Delta a)_n = (q-1)a_{n-1}$. Using this, deduce formula (12) in Chap. 1 one more time.

11. Let $m_1 < m_2 < \cdots < m_{n+1}$ be integers, and let $f(x)$ be a polynomial of degree n with the coefficient of x^n equal to one. Prove that at least one of the values of $f(m_k)$ has an absolute value not less than $n!/2^n$. *Hint:* Use the result in Problem 8 in Sec. 4. Note that (in the notation in Sec. 4) the absolute value of the number $F_k(m_k) \geq k!(n-k)!$, and use the relations between binomial coefficients that we know.

12. Apply formula (12) in Chap. 1 to the sum $1 + (1+x) + (1+x)^2 + \cdots + (1+x)^n$. Equating coefficients of terms of the same degrees on the left and on the right, deduce the formula for the sum

$$C_k^k + C_{k+1}^k + \cdots + C_n^k.$$

Supplement: Polynomials and Bernoulli Numbers

We proved in Sec. 6 that the values of sums of powers of consecutive natural numbers, that is, the sums $S_m(n)$, coincide with the values $g_m(n)$ of the Bernoulli polynomials $g_m(x)$, which have two properties:

1. $g_m(x) - g_m(x-1) = x^m.$ $\hphantom{xxxxxxxxxxxxxxxxxxxx}$ (35)

2. The free term of the polynomial $g_m(x)$ is equal to zero.

There is only one such polynomial for each m, and its degree is equal to $m+1$.

We gave a method for constructing the polynomials $g_m(x)$. But we would like to have a more explicit formula. For this, we first try the way that we used to deduce the binomial formula. Namely, we calculate the derivative of the left-hand and right-hand sides of equality (35). For this, we must study how to calculate the derivative $f(x-1)'$ of the polynomial $f(x-1)$.

Lemma 6. $f(x-1)' = f'(x-1).$

The equality seems obvious at first glance, but it is not necessarily so. The idea is that if we substitute $x-1$ for x in $f(x)$, expand each term in powers of x, and then take the derivative, then the result is the same as if we substituted $x-1$ for x in the derivative $f'(x)$.

We obtain the proof directly from the definition, that is, formula (12). Let

$$f(x) - f(\alpha) = (x - \alpha)g(x, \alpha).$$

Replacing x with $x-1$ and α with $\alpha - 1$ in this identity, we obtain

$$f(x-1) - f(\alpha - 1) = (x - \alpha)g(x - 1, \alpha - 1).$$

By definition, $f(\alpha - 1)' = g(\alpha - 1, \alpha - 1)$ and $f'(\alpha) = g(\alpha, \alpha)$. Therefore, $f(\alpha - 1)' = f'(\alpha - 1)$ as was required. $\qquad\square$

Lemma 6 could be deduced from formulas (17)–(20) by reduction to the case of a monomial. Convince yourself of this.

We can now calculate the derivative of both sides of equality (35). Taking Lemma 6 and rule (16) for derivatives into account, we obtain

$$g'_m(x) - g'_m(x-1) = mx^{m-1}.$$

On the other hand, replacing m with $m-1$ in equality (35), we obtain

$$g_{m-1}(x) - g_{m-1}(x-1) = x^{m-1}.$$

We multiply the second equality by m and subtract it from the first. Setting $h_m = mg_{m-1} - g'_m$, we obtain

$$h_m(x) = h_m(x-1).$$

But it follows from this that the polynomial h_m is a constant (has the degree zero). Indeed, substituting $x = 1, 2, \ldots$ in this equality, we obtain

$h_m(0) = h_m(1) = h_m(2) = \ldots$, the polynomial $h_m(x)$ and the constant $h_m(0)$ yield identical values for all natural numbers x. It hence follows by Theorem 15 that they coincide: $h_m(x) = h_m(0)$. (We already met this argument at the beginning of the proof of Theorem 18.) Therefore, the polynomial $h_m(x)$ is equal to a constant number. We let α_m denote this number. Recalling the definition of the polynomial $h_m(x)$, we obtain the relation

$$g'_m = mg_{m-1} + \alpha_m. \tag{36}$$

This formula is meaningless for $m = 1$, but the same reasoning results in $g'_0(x) = \alpha_0$, where α_0 is the same number.

Now, as we did when deducing the binomial formula, we write the polynomial $g_m(x)$ in powers of x. As in that case, we indicate which polynomial the coefficient occurs in by the subscript, and the superscript denotes the power of x to which it applies. The coefficients are denoted by A_m^k. The polynomial $g_m(x)$ thus takes the form

$$g_m(x) = A_m^1 x + A_m^2 x^2 + \cdots + A_m^k x^k + \cdots + A_m^{m+1} x^{m+1}.$$

(We recall that the free term of the polynomial $g_m(x)$ is zero.) We write the derivative of the polynomial $g_m(x)$ according to formula (16):

$$g_m(x)' = A_m^1 + 2A_m^2 x + \cdots + kA_m^k x^{k-1} + \cdots + (m+1)A_m^{m+1} x^{m+1}.$$

We write the analogous formula for $g_{m-1}(x)$ (substituting $m-1$ for m):

$$g_{m-1}(x) = A_{m-1}^1 x + A_{m-1}^2 x^2 + \cdots + A_{m-1}^k x^k + \cdots + A_{m-1}^m x^m.$$

We substitute both formulas in relation (36) and equate the coefficients of x^{k-1}:

$$kA_m^k = mA_{m-1}^{k-1} \quad \text{for } k \geq 2, \tag{37}$$

$$A_m^1 = \alpha_m \qquad \text{for } k = 1 \text{ and } m \geq 0. \tag{38}$$

We obtain almost the same formulas as for the binomial coefficients C_m^k except that formula (37) is only suitable for $k \geq 2$ and we must use formula (38) for $k = 1$.

We proceed further as in the case of the binomial coefficients. We have $A_m^k = (m/k)A_{m-1}^{k-1}$. Applying the same formula to A_{m-1}^{k-1}, we obtain

$$A_m^k = \frac{m(m-1)}{k(k-1)} A_{m-2}^{k-2}.$$

Continuing this process for $k-1$ steps, we obtain

$$A_m^k = \frac{m(m-1)\cdots(m-k+2)}{k(k-1)\cdots 2}A_{m-k+1}^1$$
$$= \frac{m(m-1)\cdots(m-k+2)}{k(k-1)\cdots 2}\alpha_{m-k+1}$$

(substituting for A_{m-k+1}^1 according to formula (38)).

The coefficient of α_{m-k+1} is very similar to a binomial coefficient. It differs from C_m^k (see formula (22)) in that the numerator lacks the last factor $m-k+1$ (and the denominator lacks the factor 1, but this does not change the product). However, in the formula for C_{m+1}^k, the product in the numerator ends with exactly our last factor but begins with the factor $m+1$, which is now lacking in our formula. Therefore, we can write the coefficient of α_{m-k+1} in the form $C_{m+1}^k/(m+1)$, and the formula for A_m^k becomes

$$A_m^k = \frac{1}{m+1}C_{m+1}^k\alpha_{m+1-k}.$$

(We write α_{m-k+1} as α_{m+1-k} so that the factors appear more similar.) For the polynomial $g_m(x)$, we thus obtain the formula

$$g_m(x) = \frac{1}{m+1}\big(C_{m+1}^1\alpha_m x + C_{m+1}^2\alpha_{m-1}x^2$$
$$+\cdots+C_{m+1}^k\alpha_{m+1-k}x^k+\cdots+C_{m+1}^{m+1}\alpha_0 x^{m+1}\big). \tag{39}$$

The formula obtained is similar to the binomial formula. In the rest of this supplement, we use a notation that significantly shortens the formulas but requires getting used to. Namely, let a sequence $a = \alpha_0, \alpha_1, \alpha_2, \ldots, \alpha_n, \ldots$ and a polynomial $f(t) = a_0 + a_1 t + \cdots + a_k t^k$ be given. We let $f(a)$ and $a_0 + a_1 a + \cdots + a_k a^k$ denote the number obtained by replacing each t^i in $f(t)$ with α_i. Of course, this is only a new notation. Earlier, we considered numbers $f(c)$, where c is a number; now, by virtue of the new definition, we give a definite meaning to the numbers $f(a)$, where a is a sequence. According to this definition, to determine the value of $f(a)$, where $f(t)$ is a polynomial of degree k, it is sufficient to know the first $k+1$ members of the sequence a. In what follows, we use this notation for a given sequence a and polynomials $f(t)$ of different degrees. For example, we find that, by definition, $a^m = \alpha_m$ for any $m = 1, 2, \ldots$.

We now further expand our notation. If, in addition to t, the polynomial depends on another unknown, for example, x, then substituting a for t yields a polynomial in which t^i is replaced with α_i while the power of x does not change. For example, what is $(x+a)^m$, where a is the given sequence? We must first expand the polynomial $(x + t)^m$ according to the binomial formula:

$$(x + t)^m = x^m + C_m^1 x^{m-1} t + \cdots + C_m^m t^m.$$

Then we must replace t^i everywhere in this formula with α_i:

$$(x + a)^m = x^m \alpha_0 + C_m^1 x^{m-1} \alpha_1 + \cdots + C_m^m \alpha_m.$$

This almost gives us the right-hand side of formula (39). To obtain it exactly, we must, first, replace m with $m + 1$, second, introduce the multiplier $1/(m+1)$ before the parentheses, and, finally, note that $(x + a)^{m+1}$ will contain the term $C_{m+1}^{m+1} \alpha_{m+1} = \alpha_{m+1}$, which is absent from formula (39). We know that $\alpha_{m+1} = a^{m+1}$. We can therefore write formula (39) in the rather shorter form

$$g_m(x) = \frac{1}{m+1} \left((a + x)^{m+1} - a^{m+1} \right). \tag{40}$$

But we note that we cannot assert that the polynomial given by formula (40) satisfies relation (35). We have found a general form of a polynomial satisfying relation (36), which is only a *consequence* of relation (35). And indeed, the answer depends on the sequence α_m, which can be any sequence in formula (40), while Theorem 18 asserts that the polynomial $g_m(x)$ is unique for each m. Therefore, we have not yet solved our problem.

Among the polynomials $g_m(x)$ given by formula (40), we must select those that satisfy relation (35). We already know that such a polynomial $g_m(x)$ exists and that it is unique. Therefore, we must find the unique sequence a that gives it. This is very simple to do. It is sufficient to substitute $x = 1$ in relation (35). Because $g_m(0) = 0$ (the free term of the polynomial $g_m(x)$ is equal to zero), we obtain $g_m(1) = 1$. Written in form (40), this yields $(a + 1)^{m+1} - \alpha_{m+1} = m + 1$ for $m = 0, 1, 2, \ldots$ or

$$(a + 1)^m - \alpha_m = m \quad \text{for } m = 1, 2, 3, \ldots.$$

Definition. The numbers $B_1, B_2, \ldots$ are called the *Bernoulli numbers* if the sequence B they compose satisfies the relations

$$(B + 1)^m - B_m = m \quad \text{for } m = 1, 2, 3, \ldots.$$

The sequence of Bernoulli numbers is uniquely determined by these relations. Indeed, expanding them in accordance with the definition, we obtain

$$1 + mB_1 + C_m^2 B_2 + \cdots + mB_{m-1} = m \quad \text{for } m = 1, 2, 3, \ldots \qquad (41)$$

(B_m cancels). From the relation for $m = 2$, we obtain $B_1 = 1/2$, and each successive relation gives the opportunity to find B_{m-1} if we already know all the B_r with the indices $r < m - 1$.

The polynomial

$$B_m(x) = \frac{1}{m+1} \left((B + x)^{m+1} - B_{m+1} \right),$$

where B is the sequence of Bernoulli numbers, is called a *Bernoulli polynomial*. We proved that if a polynomial $g_m(x)$ satisfying relation (35) is written in form (40), then the sequence a corresponding to it must coincide with the sequence of Bernoulli numbers B. But we know that such a polynomial exists according to Theorem 18. Therefore, that polynomial must coincide with the Bernoulli polynomial $B_m(x)$, that is, $B_m(x) - B_m(x - 1) = x^m$, and this means that $S_m(n) = B_m(n)$. Our problem is solved.

We used an unusual, although quite logical, approach. We needed to find polynomials satisfying relation (35) (for all values of x, of course). From this relation, we deduced relation (36) and found all polynomials $g_m(x)$ that satisfy it (formula (40)). There are many of them: their general form depends on the arbitrary sequence a. It would be necessary to substitute them in relation (35) and find out which of them satisfies it. But we already previously proved that there is only one such polynomial $g_m(x)$ for any m. It turns out that substituting just the value $x = 1$ in relation (35) already uniquely determines the sequence a: it must coincide with the sequence of Bernoulli numbers. Therefore, the polynomial $g_m(x)$ must coincide with the Bernoulli polynomial $B_m(x)$.

The Bernoulli polynomials and numbers were discovered by Jakob Bernoulli (there was a large family of mathematicians with this surname). His basic work was done in the second half of the 17th century, but this discovery appeared in a book published after his death in the beginning of the 18th century. The name "Bernoulli" was given to the numbers B_n by Euler (18th century), who found many applications for them.

Substituting $k = 1, 2, 3, 4, 5, 6, 7, 8, 9, 10, 11, 12$ in formula (41), we obtain the following values for the numbers B_k (verify it yourself!):

$$B_1 = \frac{1}{2}, \qquad B_2 = \frac{1}{6}, \qquad B_3 = 0, \qquad B_4 = -\frac{1}{30},$$

$$B_5 = 0, \qquad B_6 = \frac{1}{42}, \qquad B_7 = 0, \qquad B_8 = -\frac{1}{30},$$

$$B_9 = 0, \qquad B_{10} = \frac{5}{66}, \qquad B_{11} = 0, \qquad B_{12} = -\frac{691}{2730},$$

and so on, from which it is easy to obtain

$$S_1(n) = \frac{n(n+1)}{2}, \qquad S_2(n) = \frac{n(n+1)(2n+1)}{6},$$

$$S_3 = \frac{n^2(n+1)^2}{4}, \qquad S_4(n) = \frac{n(n+1)(2n+1)(3n^2+3n-1)}{30},$$

$$S_5(n) = \frac{n^2(n+1)^2(2n^2+2n-1)}{12}.$$

Problems:

1. Find $B_m(-1)$.
2. Prove the formula $B_m = (B-1)^m$ for $m \geq 2$.
3. Deduce a relation analogous to (41) relating the Bernoulli numbers B_m with odd indices $m \geq 3$. Prove that all Bernoulli numbers B_m with an odd index m except B_1 are equal to zero.
4. Find $S_6(n)$.
5. Prove the formula for the derivative of a polynomial of a polynomial. If $f(x)$ and $g(x)$ are two polynomials, then

$$f\left(g(x)\right)' = f'\left(g(x)\right) g'(x).$$

This is a generalization of Lemma 6 (which is obtained for $g(x) = x - 1$).
6. Find $(a+x)^n$ if the sequence a has $a_n = q^n$ for some number q.

3
Finite Sets

Topic: Sets

7. Sets and Subsets

The concept *set* has a somewhat different sense in mathematics than in ordinary speech. The word *set* usually implies the presence of a relatively large number of objects ("a set or multitude of friends").[1] A mathematician understands a set as an absolutely arbitrary collection of objects distinguished by some exactly defined property. The objects composing a set are called its *elements*. It is entirely possible for a set to consist of only two (or fewer) elements. A set is usually denoted by a capital letter (for example, S), and its elements by lower-case letters (for example, $a, b, \ldots$; $\alpha, \beta, \ldots$). If a is an element of the set S, then we write $a \in S$, and we say that a belongs to S. If S comprises the elements $a_1, \ldots, a_n$, then we write $S = \{a_1, \ldots, a_n\}$.

A set comprising a finite number of elements is said to be *finite*; an *infinite* set comprises an infinite number of elements. The number of elements of a finite set S is denoted by $n(S)$. We mostly deal with finite sets in this chapter. Two finite sets S and S' are said to be *equipotent* if they consist of the same number of elements, that is, $n(S) = n(S')$. We describe how the equipotence of two sets is usually established. We consider a combination of elements into pairs (a, a') with $a \in S$ and $a' \in S'$ such that each element a of the set S is in a pair with one and only one element a' of the set S' and each element a' of the set S' is in a pair with one and only one element a of the set S. Such a combination into pairs is called a *one-to-one correspondence*. If we imagine that the

---

[1] [Unlike the English word *set*, which has a multitude of meanings, the Russian word *mnozhestvo* has exactly two meanings: 1. multitude, 2. (math.) set. Translator.]

elements of S and S' combined into one pair are connected by a line, then each element of S is connected to one and only one element of S', and vice versa (Fig. 8).

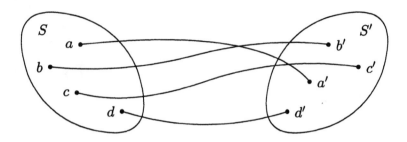

Fig. 8

For example, if $n(S) = n$, we number the elements of the set S in some way and let $S = \{a_1, \ldots, a_n\}$. We thus establish a one-to-one correspondence between the set S and the set N of numbers $1, 2, \ldots, n$.

Choosing an origin point O and a segment OU as the unit of measure on a straight line, we can associate a real number with a point A by taking the number $|OA|/|OU|$ with a plus sign if A is on the same side of O as U and with a minus sign in the opposite case. A one-to-one correspondence is thus established between the set of points on a straight line and the set of real numbers, usually denoted by $\mathbb{R}$. We discuss this in greater detail in a following chapter.

If the elements $a \in S$ and $a' \in S'$ are combined into a pair (a, a') by some one-to-one correspondence between the sets S and S', then we say that the element a' corresponds to the element a and that the element a corresponds to the element a'.

Two finite sets are equipotent if and only if a one-to-one correspondence can be established between them.

This assertion is so obvious that it is hard to call it a theorem. If $n(S) = n(S') = n$, then we can reindex our sets: $S = \{a_1, \ldots, a_n\}$, $S' = \{a'_1, \ldots, a'_n\}$. Combining elements with the same index into pairs (a_i, a'_i), we establish a one-to-one correspondence between the sets. Conversely, if there exists a one-to-one correspondence between S and S' and the elements are indexed in the set $S = \{a_1, \ldots, a_n\}$, then a_i is in a pair with one and only one element $a' \in S'$, which can be assigned the same index number, that is, $a' = a'_i$. By the definition of a one-to-one correspondence, we thus assign one index number $i = 1, \ldots, n$ to each element of the set S', that is, we obtain $S' = \{a'_1, \ldots, a'_n\}$.

R. Dedekind (second half of the 19th century) did much to explicate the role that the set concept plays in mathematics and even regarded the previous assertion as an implicit *definition* of a natural number. According to his view, we must first define the concept of a one-to-one correspondence, and then a natural number is that which is common to all finite sets between which a one-to-one correspondence can be established. The concept of a natural number probably arose historically in just this way (although without this terminology, of course). For example, the concept "two" arose from a process of abstraction (as we mentioned in Sec. 1), that is, by separating that which is common to two eyes, two persons walking together, the two oars of a boat, and generally any set that has a one-to-one correspondence with one of these.

The set concept is thus the most fundamental concept in mathematics. Even the concept of a natural number is based on the set concept.

In what follows, we often meet constructions defining a new set in terms of two given sets.

The set of all pairs (a, b), where a is an element of the set S_1 and b is an element of the set S_2, is called the *product* of the sets S_1 and S_2. The product of the sets S_1 and S_2 is denoted by $S_1 \times S_2$.

For example, if $S_1 = \{1, 2\}$ and $S_2 = \{3, 4\}$, then $S_1 \times S_2$ comprises the pairs $(1, 3)$, $(1, 4)$, $(2, 3)$, and $(2, 4)$.

If $S_1 = S_2$ and is the set $\mathbb{R}$ of all real numbers, then $S_1 \times S_2$ is the set of pairs (a, b) with real a and b. The coordinate system on the plane establishes a one-to-one correspondence between the set $S_1 \times S_2$ and the set of points in the plane (Fig. 9).

As another example, we suppose that the set S_1 comprises the numbers $1, 2, \ldots, n$ and the set S_2 comprises the numbers $1, 2, \ldots, m$. We introduce two new variables x and y and associate the monomial x^k with the number $k \in S_1$ and the monomial y^l with the number $l \in S_2$. An element of the set $S_1 \times S_2$ has the form (k, l), and the monomial $x^k y^l$ can be associated with it. We thus obtain a one-to-one correspondence between $S_1 \times S_2$ and the set of monomials of the form $x^k y^l$, where $k = 1, \ldots, n$ and $l = 1, \ldots, m$. In other words, this is the set of monomials in the right-hand side of the equality

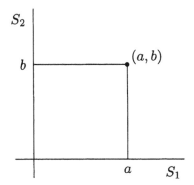

Fig. 9

$$(x + x^2 + \cdots + x^n)(y + y^2 + \cdots + y^m)$$
$$= xy + x^2y + xy^2 + \cdots + x^ny^m. \quad (1)$$

The set of these monomials is consequently equipotent to the set $S_1 \times S_2$.

Analogously, let $S_1, S_2, \ldots, S_r$ be arbitrary sets. Their product is the set comprising all sequences $(a_1, \ldots, a_r)$, where the ith member of the sequence is an element of the set S_i. The product of the sets $S_1, \ldots, S_r$ is denoted by $S_1 \times \cdots \times S_r$.

For example, if $S_1 = S_2 = S_3$ is the set $\mathbb{R}$ of all real numbers, then the coordinate system in space establishes a one-to-one correspondence between the points in space and the set $S_1 \times S_2 \times S_3$.

But we are interested in finite sets in this chapter.

Theorem 20. *If the sets* $S_1, \ldots, S_r$ *are finite, then the set* $S_1 \times \cdots \times S_r$ *is finite, and*
$$n(S_1 \times \cdots \times S_r) = n(S_1) \cdots n(S_r).$$

We first prove the theorem for the case of two sets, that is, for $r = 2$: it is the basis of induction. Let $S_1 = \{a_1, \ldots, a_n\}$ and $S_2 = \{b_1, \ldots, b_m\}$. Then all pairs (a_i, b_j) can be written in a rectangle

$$\begin{array}{ccc}
(a_1, b_1) & \cdots & (a_n, b_1) \\
(a_1, b_2) & \cdots & (a_n, b_2) \\
\vdots & \ddots & \vdots \\
(a_1, b_m) & \cdots & (a_n, b_m).
\end{array} \quad (2)$$

Here, pairs in the jth row are all those in which the second element b_j is the same. There are as many such pairs (a_i, b_j) as there are elements a_i, that is, n such pairs. The number of rows is equal to the number of different elements b_j, that is, m rows. Therefore, the number of pairs is equal to nm. We note that "rectangle" (2) is somewhat reminiscent of Fig. 9. (Arguing differently, we could say that the set S_1 is equipotent to the set $\{1, \ldots, n\}$ or to the set of monomials $\{x, x^2, \ldots, x^n\}$ and that the set S_2 is equipotent to the set of monomials $\{y, y^2, \ldots, y^m\}$. Then $n(S_1 \times S_2)$, as we saw, is the number of terms in the right-hand side of equality (1). Substituting $x = 1$ and $y = 1$ in it, we find that the number of terms is equal to nm.)

We consider the general case of r sets $S_1, \ldots, S_r$ using induction on the number r. We add a pair of parentheses in each sequence $(a_1, \ldots, a_r)$, writing it in the form $((a_1, \ldots, a_{r-1}), a_r)$. Clearly, this does not change the number of sequences. However, the composite sequence $((a_1, \ldots, a_{r-1}), a_r)$ is the pair (x, a_r), where $x = (a_1, \ldots, a_{r-1})$ can be considered an element of the set $S_1 \times \cdots \times S_{r-1}$. Therefore, the set

$S_1 \times \cdots \times S_r$ is equipotent to the set $P \times S_r$, where $P = S_1 \times \cdots \times S_{r-1}$. We have proved (in the case $r = 2$) that $n(P \times S_r) = n(P)n(S_r)$, and by the induction assumption, $n(P) = n(S_1) \cdots n(S_{r-1})$. Therefore, $n(S_1 \times \cdots \times S_r) = n(S_1) \cdots n(S_{r-1})n(S_r)$, as was to be proved. □

Theorem 20 allows again finding the expression for the number of divisors of a natural number n. Let n have the canonical decomposition

$$n = p_1^{\alpha_1} \cdots p_r^{\alpha_r}.$$

We saw in Sec. 3 that the divisors of n can be represented in the form

$$m = p_1^{\beta_1} \cdots p_r^{\beta_r},$$

where β_i can take any integer value from 0 to α_i (formula (11) in Chap. 1). In other words, the set of divisors is equipotent to the set of sequences $(\beta_1, \ldots, \beta_r)$ with the indicated values for β_i. This is nothing other than the product $S_1 \times \cdots \times S_r$, where S_i is the set $\{0, 1, \ldots, \alpha_i\}$. Because $n(S_i) = \alpha_i + 1$, the number of divisors is equal to $(\alpha_1 + 1)(\alpha_2 + 1) \cdots (\alpha_r + 1)$. We obtained this formula by a different route in Sec. 3.

If the sets $S_1, \ldots, S_r$ coincide $(S_1 = S_2 = \cdots = S_r = S)$, then the product $S_1 \times \cdots \times S_r$ is denoted by S^r. We consider the case where $S_1 = \cdots = S_r = I$ and the set I comprises the two elements a and b. An element of I^r is a sequence of r symbols, each of which is either a or b, like $aababbba$ (for simplicity, we omit the commas). We can consider it a text of r letters written in an alphabet containing two letters. Such a short alphabet is really used—the Morse code, where a and b correspond to dot and dash. Therefore, $n(I^r)$ is equal to the number of texts of length r written in the Morse code. As we saw, it is equal to 2^r (all $n_i = 2$).

The number 2^n increases very rapidly as n increases. For example, $2^{10} = 1024$, that is, already larger than 1000. (This observation is expressed in very ancient legends. For example, there is tale about a wise man who taught the king to play chess. As compensation, he asked the king to give him just one grain of wheat on the first square of the chess board, two grains on the next square, and so on, doubling the number of grains for each succesive square. The result was that all the king's granaries did not contain enough wheat. The king should give the wise man 2^{64} grains. Supposing that the weight of one grain is more than $1/100$ of a gram and using the fact that $2^{10} > 10^3$, we can easily calculate that 2^{64} grains weigh more than 10 billion metric tons.)

An enormous number of different texts can be written in a alphabet of two letters, while the length of the text is not very great. The value of $n(S^r)$ is even larger if $n(S)$ is greater than two. We use this in writing, creating an enormous number of different texts in the Russian alphabet of 32 letters or in the Latin alphabet of 26 letters. A rare book contains more than tens of millions of characters, and all the literature of humanity fits into these "texts." Moreover, the overwhelming majority of the possible texts remain unused because they make no sense. The same principle is used by Nature. All the hereditary information contained in a chromosome of a living organism can be compared to a text written in an alphabet of four letters.

Further in this chapter, we study sets contained in a given set S. They are called *subsets* of S. The subset N of the set S thus comprises only elements of the set S, but not necessarily all. We write $N \subset S$ to mean that N is a subset of the set S. The set S itself is included in the number of subsets of S. As becomes evident in what follows, it is very convenient to consider a subset of S that contains not a single element. It greatly simplifies formulating definitions and theorems. Such a subset is called the *empty subset* and is denoted by $\varnothing$. We assume that $n(\varnothing) = 0$ by definition.

If $N \subset S$, then the collection of elements of the set S that do not belong to N is called the *complement* of N and is denoted by $\overline{N}$. For example, if S is the set of natural numbers and N is the set of even numbers, then $\overline{N}$ is the set of odd numbers. If $N = S$, then $\overline{N} = \varnothing$.

If N_1 and N_2 are two subsets of the set S (that is, $N_1 \subset S$ and $N_2 \subset S$), then the set of elements that belong to both N_1 and N_2 is called their *intersection* and is denoted by $N_1 \cap N_2$. For example, if S is the set of natural numbers, N_1 is the subset of those that are divisible by two, and N_2 is the subset of those that are divisible by three, then $N_1 \cap N_2$ comprises those natural numbers that are divisible by six.

If N_1 and N_2 contain no elements in common, then $N_1 \cap N_2$ is equal to the empty set by definition. For example, if S and N_1 are as in the preceding example and N_2 is the set of odd numbers, then $N_1 \cap N_2 = \varnothing$.

The set comprising the elements that belong to the subset N_1 or N_2 is called their *union* and is denoted by $N_1 \cup N_2$. For example, if S is the set of natural numbers, N_1 is the set of even numbers, and N_2 is the set of odd numbers as in the preceding example, then $N_1 \cup N_2 = S$.

Intersections and unions of sets can be represented by figures like those in Fig. 10. In Fig. 10a, $S_1 \cup S_2$ is represented by the area that is shaded in some way, and $S_1 \cap S_2$ is only the most shaded area. In Fig. 10b, $(S_1 \cup S_2) \cap S_3$ is shaded.

In this chapter, we consider subsets of a certain finite set S that satisfy one or another condition, and we deduce formulas for the number of such subsets. The branch of mathematics dedicated to such questions is called

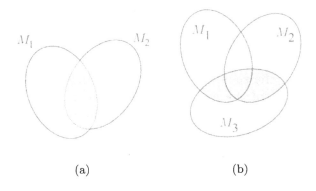

(a)   (b)

Fig. 10

combinatorics. Combinatorics is thus the theory of an arbitrary finite set. In this theory, we do not use concepts such as distance or the size of an angle, equation or its roots. Combinatorics uses only the concept of a subset and the number of its elements. It is all the more amazing that with such modest material, we can discover many nonobvious regularities and connections with other branches of mathematics.

Problems:

1. Let $S = S'$ be the set of natural numbers. Connect the numbers $a \in S$ and $b \in S'$ in a pair if $b = 2a$. Will this be a one-to-one correspondence between S and S'?
2. Let N be the set of natural numbers, $M = N \times N$, and M' be the set of positive rational numbers. Connect $(n_1, n_2) \in M$ and $a \in M'$ in a pair if $a = n_1/n_2$. Will this be a one-to-one corrrespondence?
3. How many different one-to-one correspondences exist between two sets S and S' if $n(S) = n(S') = 3$? Draw them analogously to the diagram in Fig. 8.
4. Each one-to-one correspondence between the sets S and S' defines a set of those pairs (a, a') of elements $a \in S$ and $a' \in S'$ that correspond to each other, that is, defines a certain subset $\tilde{A} \subset S \times S'$, called the *graph of the relation*. Let $\tilde{A}_1$ and $\tilde{A}_2$ be graphs of two one-to-one correspondences. Prove that $\tilde{A}_1 \cap \tilde{A}_2$ is a graph of a one-to-one correspondence if and only if $\tilde{A}_1 = \tilde{A}_2$ and the initial two correspondences coincide.
5. Let $n(S) = n(S') = n$ and Γ be the graph of a one-to-one correspondence between S and S' (see Problem 4). What is $n(\Gamma)$ equal to?
6. Let S be the set of all natural numbers, $N_1 \subset S$ be the subset of all numbers divisible by a given number a_1, and $N_2 \subset S$ be the subset of all numbers divisible by a given number a_2. Describe the subsets $N_1 \cup N_2$ and $N_1 \cap N_2$.
7. Prove that $\overline{(\overline{N})} = N$, that is, the complement of the complement of the subset N coincides with N.

8. Combinatorics

We begin with the simplest problem: the enumeration of all subsets of a finite set.

We enumerate the subsets of the set S for small values of $n(S)$. We list the subsets N of the set S, writing the subsets N with the same number of elements (that is, value of $n(N)$) in one row and arranging the rows in the order of increasing $n(N)$ (Table 1).

Table 1

1.	$n(S) = 1$	$S = \{a\}$
	$n(N) = 0$	$N = \varnothing$
	$n(N) = 1$	$N = S = \{a\}$
2.	$n(S) = 2$	$S = \{a, b\}$
	$n(N) = 0$	$N = \varnothing$
	$n(N) = 1$	$N = \{a\}, \quad N = \{b\}$
	$n(N) = 2$	$N = S = \{a, b\}$
3.	$n(S) = 3$	$S = \{a, b, c\}$
	$n(N) = 0$	$N = \varnothing$
	$n(N) = 1$	$N = \{a\}, \quad N = \{b\}, \quad N = \{c\}$
	$n(N) = 2$	$N = \{a, b\}, \quad N = \{a, c\}, \quad N = \{b, c\}$
	$n(N) = 3$	$N = S = \{a, b, c\}$

The total number subsets is equal to 2 for $n(S) = 1$, is equal to 4 for $n(S) = 2$, and is equal to 8 for $n(S) = 3$. This suggests a general statement.

Theorem 21. *The number of subsets of a finite set S is equal to $2^{n(S)}$.*

There is a general approach that reduces the study of an arbitrary finite set to the study of sets with a smaller number of elements. The set S is called the *sum* of two of its subsets $S_1 \subset S$ and $S_2 \subset S$ if $S_1 \cup S_2 = S$ and $S_1 \cap S_2 = \varnothing$. Obviously, this is equivalent to $S_2 = \overline{S_1}$ and $S_1 = \overline{S_2}$. Thus, each element of the set S belongs to one of the sets S_1 or S_2 (because $S_1 \cup S_2 = S$) and to only one (because $S_1 \cap S_2 = \varnothing$). Therefore, $n(S) = n(S_1) + n(S_2)$. We write $S = S_1 + S_2$ to mean that the set S is the sum of S_1 and S_2. Such a representation is also called a *partition* of S into S_1 and S_2.

Let $S = S_1 + S_2$, and let $N \subset S$ be an arbitrary subset. Then an element $a \in N$ belongs to either S_1 (and then $a \in N \cap S_1$) or S_2 (and then $a \in N \cap S_2$); moreover, only one of these cases holds (because $S_1 \cap S_2 = \varnothing$). Therefore, $N = (N \cap S_1) + (N \cap S_2)$. Conversely, if $N_1 \subset S_1$ and $N_2 \subset S_2$ are arbitrary subsets, then $N_1 \subset S$, $N_2 \subset S$, and $N = N_1 \cup N_2 \subset S$; moreover, $N \cap S_1 = N_1$ and $N \cap S_2 = N_2$. A one-to-one correspondence is thus established between subsets N of the set S and pairs (N_1, N_2), where N_1 is a subset of the set S_1 and N_2 is a subset of the set S_2.

We formulate this result in terms of sets. We let $U(S)$ denote the set of all subsets of the set S. There is no reason to be concerned that subsets are now viewed as elements of a new set. The Union of Metallugical Workers and the Sewing Industry Union enter as elements of the All-Russian Committee of Unions. We described the sets $U(S)$ for $n(S)$ equal to 1, 2, and 3 in Table 1. We can now formulate the result obtained above as follows: if there exists a partition of the set S, $S = S_1 + S_2$, then the set $U(S)$ is in one-to-one correspondence with the set $U(S_1) \times U(S_2)$. We let $v(S)$ denote the number $n(U(S))$; it is the number of all subsets that we seek. Applying Theorem 20 to the assertion we just found, we obtain

$$v(S_1 + S_2) = v(S_1)v(S_2). \tag{3}$$

Equality (3) already reduces calculating the quantity $v(S)$ to calculating the quantities $v(S_1)$ and $v(S_2)$ for the sets S_1 and S_2 with a smaller number of elements. To obtain the final result, we consider a partition of the set S not into two but into an arbitrary number of subsets. We can define this concept by induction, writing $S = S_1 + \cdots + S_r$ if $S = (S_1 + \cdots + S_{r-1}) + S_r$, where the meaning of the expression $S_1 + \cdots + S_{r-1}$ is assumed to be already defined. Simply speaking, the sense of this definition is that $S_1, \ldots, S_r$ are subsets of the set S and each element of S belongs to one and only one of the subsets $S_1, \ldots, S_r$. For example, if S is the set of natural numbers, S_1 is the subset of numbers divisible by three, S_2 is the subset of numbers representable as $3r + 1$, and S_3 is the subset of numbers representable as $3r + 2$, then $S = S_1 + S_2 + S_3$.

From equality (3), we immediately obtain

$$v(S_1 + \cdots + S_r) = v(S_1) \cdots v(S_r) \tag{4}$$

by induction, where the S_i are finite sets.

If $n(S) = n$, then the "finest" partition of the set S is into n subsets S_i, each of which contains exactly one element: $S = S_1 + \cdots + S_n$. If $S = \{a_1, \ldots, a_n\}$, then $S_i = \{a_i\}$. A set S_i with one element has two subsets: the empty set $\varnothing$ and the set S_i itself (see the first section in Table 1). Therefore, $v(S_i) = 2$, and applying formula (4) to the partition $S = S_1 + \cdots + S_n$, we obtain $v(S) = 2^n$, which is the assertion in Theorem 21. $\qquad\square$

The question of the number of all subsets of a given set is encountered in connection with certain problems about numbers. As an example, we find out how many ways a given natural number can be represented as the product of two relatively prime factors. Let $n = ab$, let a and b be relatively prime, and let $n = p_1^{\alpha_1} \cdots p_r^{\alpha_r}$ be the canonical decomposition of n into prime factors. Then a and b are divisors of n, and, as we saw in Sec. 3, each of them has a representation $p_1^{\beta_1} \cdots p_r^{\beta_r}$, where $0 \leq \beta_i \leq \alpha_i$. But because a and b are relatively prime, if a certain p_i divides a, then it cannot divide b and therefore must enter a with the power α_i. Therefore, to obtain the desired product $n = ab$, we must choose a subset N of the set $S = \{p_1, \ldots, p_r\}$ and set a equal to the product of the factors $p_i^{\alpha_i}$ for $p_i \in N$. Then a divides n, and $n = ab$ is the needed representation. According to Theorem 21, the number of representations of n as the product of two relatively prime factors is equal to 2^r, where r is the number of different prime divisors of n.

We must note, however, that we include in our count $n = ab$ and $n = ba$ as different representations. Indeed, if a (and therefore the representation $n = ab$) corresponds to the subset $N \subset \{p_1, \ldots, p_r\}$, then b corresponds to the subset containing those $p_i \in S$ that are not contained in N, that is, the complement $\overline{N}$ of the subset N. Consequently, two different subsets N and $\overline{N}$ correspond to the two representations $n = ab$ and $n = ba$ in our count. If we do not distinguish $n = ab$ and $n = ba$, then we must count the two subsets N and $\overline{N}$ as one. Then the number of different representations in this sense is two times less, that is, 2^{r-1}.

We now turn to a more subtle problem: find the number of subsets of a finite set S that contain a given number m of elements. For this, we again collect all subsets $N \subset S$ such that $n(N) = m$ into one set denoted by $U(S, m)$. We set $n(U(S, m)) = v(S, m)$, the number we want to find. In Table 1, we wrote the sets belonging to $U(S, m)$ in one line. From this table, we obtain the values of $v(S, m)$ for small values of $n(S)$:

Table 2

$n(S) = 1$	$v(S, 0) = 1,$	$v(S, 1) = 1$		
$n(S) = 2$	$v(S, 0) = 1,$	$v(S, 1) = 2,$	$v(S, 2) = 1$	
$n(S) = 3$	$v(S, 0) = 1,$	$v(S, 1) = 3,$	$v(S, 2) = 3,$	$v(S, 3) = 1$

Theorem 22. *If $n(S) = n$, then the number of subsets $N \subset S$ of the set S that contain m elements (that is, such that $n(N) = m$) is equal to the binomial coefficient C_n^m, that is, $v(S, m) = C_n^m$.*

Our proof is based on the same idea as in the proof of Theorem 21. Namely, we assume that the set S is the sum of two sets, $S = S_1 + S_2$, and we express the number $v(S, m)$ in terms of the numbers $v(S_1, m)$ and $v(S_2, m)$. If $S = S_1 + S_2$, then each subset $N \subset S$ can be represented as $N = N_1 + N_2$, where $N_1 = N \cap S_1$ and $N_2 = N \cap S_2$. If we apply the additional condition that $n(N) = m$, then we must have $n(N_1) + n(N_2) = m$. We specify certain natural numbers (or possibly zero) k and l such that $k + l = m$. We consider all subsets $N \subset S$ such that $n(N \cap S_1) = k$ and $n(N \cap S_2) = l$. We let $U(k, l)$ denote the set of all such subsets and set $n(U(k, l)) = v(S, k, l)$. Then, exactly as in the proof of Theorem 21, we see that

$$v(S, k, l) = v(S_1, k)v(S_2, l). \tag{5}$$

The set $U(S, m)$ itself can obviously be partitioned into the sets $U(S, k, l)$ for different pairs k, l for which $k + l = m$. Therefore, the number $v(S, m)$ of its elements is equal to the sum of all numbers $v(S, k, l)$ for which $k + l = m$, that is, the possible values $k = 0$ and $l = m$, $k = 1$ and $l = m - 1$, and so on to $k = m$ and $l = 0$. From relation (5), we obtain

$$v(S, m) = v(S_1, m)v(S_2, 0)$$
$$+ v(S_1, m - 1)v(S_2, 1) + \cdots + v(S_1, 0)v(S_2, m), \tag{6}$$

where if $k > n(S_1)$ in the product $v(S_1, k)v(S_2, l)$, then we naturally set $v(S_1, k) = 0$, and similarly for $v(S_2, l)$.

We have obtained a relation similar to relation (3), although it is more complicated. But we already met expression (6) in connection with a completely different question. This is the expression for the coefficient of x^m in the product of two polynomials $f(x)$ and $g(x)$ if the coefficient of x^k in $f(x)$ is equal to $v(S_1, k)$ and the coefficient of x^l in $g(x)$ is equal to $v(S_2, l)$ (see formula (1) in Chap. 2). To relate these two assertions, given a finite set S, we define a polynomial $f_S(x)$ whose coefficients coincide with the numbers $v(S, s)$:

$$f_S(x) = v(S, 0) + v(S, 1)x + \cdots + v(S, n)x^n, \tag{7}$$

where $n = n(S)$.

For example, in accordance with Table 2, if $n(S) = 1$, then $f(x) = 1 + x$; if $n(S) = 2$, then $f(x) = 1 + 2x + x^2$; and if $n(S) = 3$, then $f(x) = 1 + 3x + 3x^2 + x^3$. Comparing relation (6) and (1) in Chap. 2, we can now write

$$f_S(x) = f_{S_1}(x) \cdot f_{S_2}(x) \quad \text{if } S = S_1 + S_2. \tag{8}$$

At the cost of introducing the polynomial $f_S(x)$, we have thus achieved complete similarity to formula (3). We see that the polynomial $f_S(x)$

is the proper substitution for the number $v(S)$ in our more complicated problem. Such a manifestation is often met. If we deal with not a single number but a finite sequence of numbers $(a_0, \ldots, a_n)$, then a polynomial $a_0 + a_1 x + \cdots + a_n x^n$ often reflects its property. We see this later in other examples.

Now it remains for us to repeat the end of the proof of Theorem 21 quite literally. If $S = S_1 + \cdots + S_r$, then we immediately obtain

$$f_S(x) = f_{S_1}(x) \cdots f_{S_r}(x)$$

from relation (8) by induction. We now set $n(S) = n$ and partition S into n subsets of one element each: $S = S_1 + \cdots + S_n$, where $n(S_i) = 1$. The set S_i with one element has two subsets: the empty set $\varnothing$ with $n(\varnothing) = 1$ and the subset S_i with $n(S_i) = 1$. Therefore, $v(S_i, 0) = 1$, $v(S_i, 1) = 1$, $v(S_i, k) = 0$ for $k > 1$, and $f_{S_i}(x) = 1 + x$. We obtain

$$f_S(x) = (1 + x)^{n(S)}$$

for any finite set S.

We can represent the expression $(1+x)^{n(S)}$ in the form of a polynomial in powers of x using the binomial formula. We saw that for $n = n(S)$,

$$(1 + x)^n = C_n^0 + C_n^1 x + C_n^2 x^2 + \cdots + C_n^n x^n,$$

where $C_n^m = n!/(m!(n - m)!)$ (see formulas (21) and (25) in Chap. 2). Therefore, recalling the definition of the polynomial $f_S(x)$ (see formula (7)), we obtain

$$v(S, m) = C_n^m = \frac{n!}{m!(n - m)!} \quad \text{for } n = n(S). \tag{9}$$

This is the sought formula for $v(S, m)$.   □

Counting the subsets with $0, 1, 2, \ldots, n$ elements, where $n = n(S)$, we count all subsets. Thus, $v(S, 0) + v(S, 1) + \cdots + v(S, n) = v(S)$ or, by formula (9) and Theorem 21, $C_n^0 + C_n^1 + \cdots + C_n^n = 2^n$. This relation between the binomial coefficient is also easily obtained from the binomial formula. We saw this in Sec. 6.

A set of m elements of the set $\{a_1, \ldots, a_n\}$ is sometimes called a *combination* of n elements taken m at a time. The corresponding binomial coefficient C_n^m is called the *number of combinations* of n elements taken m at a time.

The question we considered above about the number of subsets $N \subset S$, where $n(S) = n$ and $n(N) = m$, is connected with certain questions about natural numbers. For example, how many ways can the natural number n be represented as the sum of a given number r of natural numbers? In other words, how many sets of natural numbers $\{x_1, \ldots, x_r\}$ are solutions of the equation $x_1 + \cdots + x_r = n$? Further, solutions that differ in the order of the unknowns will be considered different. For example, for $n = 4$ and $r = 2$, we have $4 = 1+3 = 2+2 = 3+1$ and thus three solutions: $(1, 3)$, $(2, 2)$, and $(3, 1)$.

We consider a line segment AB of length n. We call its points located at integer distances from the origin A integer points. Clearly, each solution of the equation $x_1 + \cdots + x_r = n$ corresponds to a partition of the line segment AB into r segments whose lengths are the integers $x_1, x_2, \ldots, x_r$ (Fig. 11).

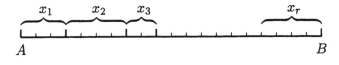

Fig. 11

In turn, each partition defines the right endpoints of the first $r-1$ segments (the end of the last segment is the point B). These endpoints define a subset N of the set S of integer points of the segment different from B. Clearly, $n(S) = r - 1$, and we have thus defined a one-to-one correspondence between solutions of the equation $x_1 + \cdots + x_r = n$ in natural numbers and subsets $N \subset S$, where $n(N) = r - 1$ and $n(S) = n - 1$. This means that the number of such solutions is equal to the number of such subsets. Applying formula (9), we find that the number of these solutions is C_{n-1}^{r-1}. If we do not fix the number of terms into which the number n is partitioned, then the number of partitions is obviously equal to the sum of the partitions into r terms for $r = 1, \ldots, n$. This means that the number of partitions is equal to the sum of all binomial coefficients C_{n-1}^{r-1} for $r = 1, \ldots, n$. We know that this sum is equal to 2^{n-1}. In other words, there are 2^{n-1} ways to partition the natural number n into a sum of natural terms (if any number of terms is allowed and if different orders of terms are considered different).

We now return to the deduction of formula (9). Using the same technique, introducing polynomials $f_S(x)$, is very useful in other questions. We return to this later. But formula (9) itself or, more properly, the connection between the numbers $v(S, m)$ and the binomial coefficients can even be deduced without this technique.

We consider the expression $(1 + x)^n$ as the product of n identical factors:

$$(1 + x)^n = (1 + x)(1 + x) \cdots (1 + x). \tag{10}$$

We analyze the process of completely opening the parentheses in expression (10). We enumerate its factors, that is, we assign them the index numbers $1, 2, \ldots, n$, which form the set $S = \{1, 2, \ldots, n\}$. Opening the parentheses in product (10), we must multiply n factors 1 or x each time choosing one factor from each pair of parentheses. Thus, each term in the expanded expression (10) is determined by indicating from which pairs of pareneteses we take 1 and from which we take x. In a given term, let x be taken from the pairs of parentheses with the m index numbers $i_1, i_2, \ldots, i_m$. Then we take 1 from the remaining $n-m$ pairs of parentheses and obtain the term x^m as a result. We see that each term in the expanded expression (10) is determined by a subset $N = \{i_1, \ldots, i_m\} \subset S$, which indicates the index numbers of the pairs of parentheses from which we take x. From the remaining pairs of parentheses, we take 1. The "remaining" have index numbers composing the complement $\overline{N}$ of the set N. We thus obtain the term x^m as many times as there are subsets $N \subset S$ containing m elements, that is, $v(S, m)$ times. Consequently, expression (10) in the expanded form is the sum of the terms $v(S, m)x^m$:

$$(1 + x)^n = v(S, 0) + v(S, 1)x + \cdots + v(S, n)x^n.$$

Comparing this with the definition of the binomial coefficients (see formula (21) in Chap. 2), we obtain a new proof that $v(S, m) = C_n^m$.

The same argument can even be applied to a more general case. We consider the product of first-degree polynomials $x + a_i$, where we take the coefficient of x to be 1 by bringing a constant factor outside the parentheses. We attempt to represent the product

$$(x + a_1)(x + a_2) \cdots (x + a_n) \tag{11}$$

in the form of a polynomial in x. As previously, we enumerate the n factors. Then each term in the expanded product (11) is obtained as a choice of those factors with the index numbers $i_1, i_2, \ldots, i_m$ from which we take the terms $a_{i_1}, a_{i_2}, \ldots, a_{i_m}$, simultaneously taking x from the remaining $n-m$ factors. The corresponding term has the form $a_{i_1} a_{i_2} \cdots a_{i_m} x^{n-m}$. After combining all such terms of the degree $n - m$, we have $\sigma_m(a_1, \ldots, a_n)x^{n-m}$, where $\sigma_m(a_1, \ldots, a_n)$ is the sum of all terms of the form $a_{i_1} a_{i_2} \cdots a_{i_m}$, where $\{i_1, \ldots, i_m\}$ ranges all combinations of m different indices from the numbers $1, \ldots, n$. The polynomial $\sigma_m(a_1, \ldots, a_n)$ consequently has C_n^m terms. For example,

$$\sigma_1(a_1, \ldots, a_n) = a_1 + \cdots + a_n,$$

and $\sigma_2(a_1, \ldots, a_n) = a_1 a_2 + a_1 a_3 + \cdots + a_2 a_3 + \cdots + a_{n-1} a_n$, which is the sum of all products $a_i a_j$ with $i < j$. The last polynomial σ_n has the form $\sigma_n(a_1, \ldots, a_n) = a_1 \cdots a_n$. We confront a polynomial with an arbitrary number n of unknowns for the first time. The polynomials $\sigma_1, \ldots, \sigma_n$ play a very important role in algebra. In particular, we have proved the formula

$$\begin{aligned}
(x + a_1) \cdots (x + a_n) &= x^n + \sigma_1(a_1, \ldots, a_n)x^{n-1} \\
&\quad + \sigma_2(a_1, \ldots, a_n)x^{n-2} \\
&\quad + \cdots + \sigma_n(a_1, \ldots, a_n).
\end{aligned} \tag{12}$$

It is called the *Viète formula*.

The Viète formula expresses an important property of polynomials. We suppose that the polynomial $f(x)$ of degree n has n roots $\alpha_1, \ldots, \alpha_n$. then, as we already saw more than once, the polynomial is divisible by the product $(x - \alpha_1) \cdots (x - \alpha_n)$, and because this product also has the degree n, we have $f(x) = c(x - \alpha_1) \cdots (x - \alpha_n)$, where c is a number. We suppose that the coefficient of the leading term of the polynomial $f(x)$ is equal to one. Then the number c must be equal to one, and we have the equality

$$f(x) = (x - \alpha_1) \cdots (x - \alpha_n).$$

We can apply Viète formula (12) to this equality, setting $a_i = -\alpha_i$. Because each term of the polynomial σ_k is a product of certain k from the unknowns $a_1, \ldots, a_n$, after substituting $-\alpha_i$ for a_i, we can extract the factor $(-1)^k$:

$$\sigma_k(-\alpha_1, \ldots, -\alpha_n) = (-1)^k \sigma_k(\alpha_1, \ldots, \alpha_n).$$

From formula (12), we then obtain the formula

$$\begin{aligned}
(x - \alpha_1) \cdots (x - \alpha_n) &= x^n - \sigma_1(\alpha_1, \ldots, \alpha_n)x^{n-1} \\
&\quad + \cdots + (-1)^n \sigma_n(\alpha_1, \ldots, \alpha_n).
\end{aligned} \tag{13}$$

It express the coefficients of the polynomial $f(x) = (x - \alpha_1) \cdots (x - \alpha_n)$ in terms of its roots and is also called the Viète formula. We know its particular case for a quadratic equation: then there are only two polynomials σ_1 and σ_2, $\sigma_1 = \alpha_1 + \alpha_2$ and $\sigma_2 = \alpha_1 \alpha_2$.

In conclusion, we again consider formula (9) for the number of subsets (or the number of combinations). We deduced it from the binomial formula, which in turn was proved in Sec. 6 using the properties of derivatives. It was a rather long path. It is desirable to have another proof of our formula based on only combinatorial considerations. We deduce such a proof, incidentally, of an even more general formula. We note that each subset N of the set S defines a partition $S = N + \overline{N}$, where $\overline{N}$ is the complement of the set N. We consider a more general situation: an arbitrary partition $S = S_1 + \cdots + S_r$ into subsets with given numbers of elements $n(S_1) = n_1, \ldots, n(S_r) = n_r$. We call the sequence of numbers $(n_1, \ldots, n_r)$ the *type* of the partition $S = S_1 + \cdots + S_r$. We assume that none of the sets S_i is empty, and that means that all $n_i > 0$.

Because we always consider the same set S with $n(S) = n$, we do not reflect this in our notation. The number of all possible partitions of our set S having a given type $(n_1, \ldots, n_r)$ is denoted by $C(n_1, \ldots, n_r)$. Here, of course, we must have $n_1 + \cdots + n_r = n$. We note that the order of the subsets $S_1, \ldots, S_r$ is taken into account here. For example, for $r = 2$ and given n_1 and n_2, we consider $S = S_1 + S_2$ and $S = S_2 + S_1$ different partitions if $n(S_1) = n_1$ and $n(S_2) = n_2$. If $n_1 \neq n_2$, these partitions even have different types. Because of this, each partition $S = S_1 + S_2$ defines one subset S_1 (the first), and the question is reduced to that considered earlier: $C(n_1, n_2) = v(S, n_1)$. In other words, for any $m < n$, $v(S, m) = C(m, n - m)$.

We deduce an explicit formula for the number $C(n_1, \ldots, n_r)$. We consider an arbitrary partition $S = S_1 + \cdots + S_r$ of type $(n_1, \ldots, n_r)$. We assume that at least one of the numbers $n_1, \ldots, n_r$ is different from 1. To be specific, we assume that $n_1 > 1$, and we choose some element $a \in S_1$. We let S_1' denote the set of all elements of S_1 that are different from a (that is, the complement of $\{a\}$ as a subset of the set S_1). Then we have the partition $S_1 = S_1' + \{a\}$, and our partition $S = S_1 + \cdots + S_r$ corresponds to a new partition $S = S_1' + \{a\} + S_2 + \cdots + S_r$ of the type $(n_1 - 1, 1, n_2, \ldots, n_r)$. From all partitions of the type $(n_1, n_2, \ldots, n_r)$, we thus obtain all partitions of the type $(n_1 - 1, 1, n_2, \ldots, n_r)$. The partition $S = S_1' + \{a\} + S_2 + \cdots + S_r$ is obtained from the partition $S = S_1 + \cdots + S_r$, where $S_1 = S_1' + \{a\}$. One partition of the type $(n_1, n_2, \ldots, n_r)$ yields n_1 different partitions of the type $(n_1 - 1, 1, n_2, \ldots, n_r)$ with the different choices of the element $a \in S_1$ taken into account. Therefore, we obtain the relation

$$n_1 C(n_1, n_2, \ldots, n_r) = C(n_1 - 1, 1, n_2, \ldots, n_r). \tag{14}$$

Applying this relation to partitions of the type $(n_1 - 1, 1, n_2, \ldots, n_r)$, we obtain

$$(n_1 - 1)C(n_1 - 1, 1, n_2, \ldots, n_r) = C(n_1 - 2, 1, 1, n_2, \ldots, n_r),$$

that is,

$$n_1(n_1 - 1)C(n_1, n_2, \ldots, n_r) = C(n_1 - 2, 1, 1, n_2, \ldots, n_r).$$

Continuing this process $n-1$ times, we come to the relation

$$n_1!C(n_1, n_2, \ldots, n_r) = C\left(\underbrace{1, \ldots, 1}_{n_1 \text{ times}}, n_2, \ldots, n_r\right).$$

We now apply the same reasoning to the parameter n_2 in $C(1, \ldots, 1, n_2, \ldots, n_r)$. Absolutely analogously, we obtain the relation

$$n_2!C(1, \ldots, 1, n_2, n_3, \ldots, n_r) = C(1, \ldots, 1, n_3, \ldots, n_r),$$

where 1 occupies n_1+n_2 places in the right-hand side, and this means that

$$n_1!n_2!C(n_1, n_2, \ldots, n_r) = C\left(\underbrace{1, \ldots, 1}_{n_1+n_2 \text{ times}}, n_2, \ldots, n_r\right).$$

Acting thus with all the parameters $n_1, n_2, \ldots, n_r$ in order, we finally obtain the formula

$$n_1!n_2!\cdots n_r!C(n_1, n_2, \ldots, n_r) = C\left(\underbrace{1, \ldots, 1}_{n \text{ times}}\right) \qquad (15)$$

because $n_1 + n_2 + \cdots + n_r = n$. It remains to find the expression for $C(1, \ldots, 1)$. For this, we note that the preceding formula is proved for partitions of all types $(n_1, n_2, \ldots, n_r)$. We apply it to the simplest type (n). There is only one partition of this type: $S = S$. Therefore, $C(n) = 1$. On the other hand, according to formula (15),

$$n!C(n) = C\left(\underbrace{1, \ldots, 1}_{n \text{ times}}\right).$$

Therefore, $C(1, \ldots, 1) = n!$. Substituting this value in formula (15), we obtain the final expression:

$$C(n_1, n_2, \ldots, n_r) = \frac{n!}{n_1!n_2!\cdots n_r!}, \qquad \text{where } n = n_1 + \cdots + n_r. \qquad (16)$$

For $r = 2$, we more often write $(m, n - m)$ instead of (n_1, n_2), where $n_1 + n_2 = n$. Because $C(m, n - m) = v(S, m)$, formula (16) gives our previous relation (9).

Remark 1. We turn our attention to the expression $C(1, \ldots, 1)$, which we met at the end of the preceding argument. What does a partition of the type $(1, \ldots, 1)$ mean? It is a partition of the set S into one-element subsets. But we recall that the order of the sets $S_1, \ldots, S_r$ plays a role in the partition $S = S_1 + \cdots + S_r$. The partition $S = \{a_1\} + \cdots \{a_n\}$ therefore assigns a definite indexing to the elements of the set S. The number $C(1, \ldots, 1)$ indicates how many ways we can enumerate the elements of S. If we most obviously represent the enumeration of the elements $a_1, \ldots, a_n$ by writing them in the order of increasing subscripts, then $C(1, \ldots, 1)$ indicates how many ways we can arrrange the elements of the set S in different orders. As we saw, the number of such arrangements is equal to $n!$. Different arrangements are also called *permutations*. For example, for $S = \{a, b, c\}$, that is, $n = 3$, we have six permutations:

$$(a, b, c), \quad (a, c, b), \quad (b, a, c), \quad (b, c, a), \quad (c, a, b), \quad (c, b, a).$$

Remark 2. In the case where $r = 2$, the expression $C(n_1, n_2)$ coincides with the binomial coefficients—we have even deduced two proofs of this fact. The expression $C(n_1, \ldots, n_r)$ for an arbitrary r has an analogous interpretation. It can be proved that if $x_1, \ldots, x_r$ are unknowns, then complete expansion of the expression $(x_1 + \cdots + x_r)^n$ yields terms of the form $x_1^{n_1} \cdots x_r^{n_r}$ with $n_1 + \cdots + n_r = n$, where the n_i are positive integers. Furthermore, the term in $x_1^{n_1} \cdots x_r^{n_r}$ enters with the coefficient $C(n_1, \ldots, n_r)$. We need only return to our initial definition of a partition and allow the empty set $\varnothing$ to appear among the subsets S_i and allow zero among the numbers n_i. It is easy to see that formula (16) is preserved in this case if we agree to set $0! = 1$. The proof of the generalization of the binomial formula to the case of r terms is completely parallel to the second (combinatorial) proof of the relation $v(S, m) = C_n^m$ (where $n = n(S)$) given above.

For example, from this formula, we find that $(x_1 + x_2 + x_3)^3$ is equal to the sum of the terms $C(n_1, n_2, n_3)x_1^{n_1} x_2^{n_2} x_3^{n_3}$, where (n_1, n_2, n_3) ranges all triples of nonnegative integers for which $n_1 + n_2 + n_3 = 3$ and $C(n_1, n_2, n_3)$ is calculated by formula (16) (with the condition $0! = 1$). Substitution yields

$$(x_1 + x_2 + x_3)^3 = x_1^3 + x_2^3 + x_3^3 + 3x_1^2 x_2 + 3x_1 x_2^2 + 3x_1^2 x_3$$
$$+ 3x_1 x_3^2 + 3x_2^2 x_3 + 3x_2 x_3^2 + 6x_1 x_2 x_3.$$

Problems:

1. Let $I = \{p, q\}$ be a set of two elements and S be a set of n elements that are indexed somehow, $S = \{a_1, \ldots, a_n\}$. With a subset $N \subset S$, associate the element ("text") of I^n in which p is in the ith position if a_i belongs to N and q is in the

*i*th position if a_i does not belong to N. Prove that a one-to-one correspondence is thus established between the sets $U(S)$ and I^n. In the same way, deduce Theorem 21 from Theorem 20.

2. How are the intersection and the union of the subsets N_1 and N_2 of the set S expressed in terms of their corresponding "texts" from I^n (see Problem 1)?

3. Find the number of all partitions $S_1 + \cdots + S_r$ of all types but with a given number r. Convince yourself that for $r = 2$, the answer yields Theorem 21.

4. Find the sum of all numbers $C(n_1, \ldots, n_r)$ for all $n_i \geq 0$ and $n_1 + \cdots + n_r = n$ with a given r and n. Give two solutions, following from Problem 3 and from the assertion formulated in Remark 2.

5. Find the number of decompositions of the natural number n into the product of r factors, $n = a_1 \cdots a_r$, where the factors are pairwise relatively prime.

6. How many solutions of the equation $x_1 + \cdots + x_r = n$ are there for given n and r if integer solutions $x_i \geq 0$ are allowed? Use the following graphic representation of a solution, modifying the one given in Fig. 11. Let AB be a line segment of length $n + r - 1$. With a solution $(x_1, \ldots, x_r)$, associate a partition of that segment consisting of a segment of length x_1 starting from the initial point A, a segment of length x_2 starting from the next integer point after the first segment, and so on (see Fig. 12). In Fig. 12, $x_3 = 0$.

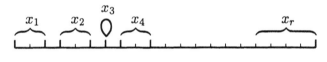

Fig. 12

7. Find the number of partitions $S = S_1 + S_2$ of the type (m, m) if $n(S) = 2m$ and the partitions $S = S_1 + S_2$ and $S = S_2 + S_1$ are not considered different. Solve the analogous problem for partitions $S = S_1 + S_2 + S_3$ of the type (m, m, m) if $n(S) = 3m$ and partitions differing in the order of the set S_1, S_2, and S_3 are not considered different. Finally, solve the problem for partitions of the type (k, k, l, l, l) if $n(S) = 2k + 3l$ and partitions differing in the order of equipotent subsets are not considered different.

8. What terms does the polynomial $(x_1 + \cdots + x_n)^2$ contain? The polynomial $(x_1 + \cdots + x_n)^3$?

9. How many terms are contained in the polynomial $(x_1 + \cdots + x_r)^n$ after like terms are combined?

10. Express the polynomial $a_1^2 + a_2^2 + \cdots + a_n^2$ in terms of the polynomials σ_1 and σ_2. Assume that the polynomial $x^n + ax^{n-1} + bx^{n-2} + \cdots$ has n real roots. Prove that then $a^2 \geq 2b$. When does the equality $a^2 = 2b$ hold? *Hint:* Use the Bézout theorem from Sec. 4 and the fact that the square of a real number cannot be negative.

11. Give a combinatorial proof of the relation $C_n^k = C_{n-1}^k + C_{n-1}^{k-1}$ between binomial coefficients (formula (27) in Chap. 2), interpreting C_n^k as $v(S, k)$, where $n(S) = n$. Find a generalization of this relation for the numbers $C(n_1, \ldots, n_r)$.

12. Find a combinatorial proof of the relation $C_n^m = C_n^{n-m}$ between binomial coefficients.

9. Set Algebra

If the intersection of two subsets $S_1 \subset S$ and $S_2 \subset S$ is empty (that is, $S_1 \cap S_2 = \varnothing$), then the union $S_1 \cup S_2$ comprises elements that belong either to S_1 or to S_2 and each element of the set $S_1 \cup S_2$ can belong to only one of the sets S_1 or S_2. Therefore, $S_1 \cup S_2 = S_1 + S_2$ and consequently

$$n(S_1 \cup S_2) = n(S_1) + n(S_2).$$

The case where $S_1 \cap S_2$ is not empty can be reduced to the previously considered case. Let S_1' denote the complement of $S_1 \cap S_2$ in S_1, that is, the set of elements in S_1 that do not belong to $S_1 \cap S_2$. Then by the previous argument,

$$n(S_1) = n(S_1 \cap S_2) + n(S_1'). \tag{17}$$

Analogously,

$$n(S_2) = n(S_1 \cap S_2) + n(S_2'), \tag{18}$$

where S_2' is the complement of $S_1 \cap S_2$ in S_2. Adding equalities (17) and (18), we obtain

$$n(S_1) + n(S_2) = 2n(S_1 \cap S_2) + n(S_1') + n(S_2'). \tag{19}$$

But the sets $S_1 \cap S_2$, S_1', and S_2' have no elements in common pairwise, and their union is $S_1 \cup S_2$. Therefore, $S_1 \cup S_2 = S_1 \cap S_2 + S_1' + S_2'$, and this means that $n(S_1 \cup S_2) = n(S_1 \cap S_2) + n(S_1') + n(S_2')$. Using this, we can rewrite equality (19) as

$$n(S_1) + n(S_2) = n(S_1 \cap S_2) + n(S_1 \cup S_2)$$

or, equivalently,

$$n(S_1 \cup S_2) = n(S_1) + n(S_2) - n(S_1 \cap S_2). \tag{20}$$

This is the relation we sought. Our further goal is to generalize it and obtain an expression for the number of elements in the union of any number of subsets (and not just two): $n(S_1 \cup \cdots \cup S_r)$. For this, we must formulate some almost obvious properties of intersections and unions of several subsets.

We first note that the union $S_1 \cup S_2 \cup \cdots \cup S_r$ of several subsets $S_1, S_2, \ldots ; S_r$ can be defined based on only the union of two subsets. For example,

$$S_1 \cup S_2 \cup S_3 = (S_1 \cup S_2) \cup S_3.$$

We treat an arbitrary number k of subsets exactly the same,

$$S_1 \cup S_2 \cup \cdots \cup S_k = (S_1 \cup S_2 \cup \cdots \cup S_{k-1}) \cup S_k.$$

If no two of the subsets $S_1, \ldots, S_k$ of the set S contain elements in common, then $S_1 \cup \cdots \cup S_k = S_1 + \cdots + S_k$ and $n(S_1 \cup \cdots \cup S_k) = n(S_1) + \cdots + n(S_k)$. The other formula we need has the form

$$(S_1 \cup S_2 \cup \cdots \cup S_k) \cap N = (S_1 \cap N) \cup (S_2 \cap N) \cup \cdots \cup (S_k \cap N).$$

All these formulas are obvious. It is sufficient to ask yourself what it means for an element $a \in S$ to belong to the left-hand or right-hand side. For example, $a \in (S_1 \cup S_2 \cup \cdots \cup S_k) \cap N$ in the last formula means that $a \in S_1 \cup S_2 \cup \cdots \cup S_k$ and $a \in N$. The second assertion means that $a \in N$, and the first means that $a \in S_i$ for some $i = 1, 2, \ldots, k$. But then $a \in S_i \cap N$ for the same i, and that means that $a \in (S_1 \cap N) \cup (S_2 \cap N) \cup \cdots \cup (S_k \cap N)$. We note that the deduced property is similar to the distributive property for numbers: if we write the numbers $a_1, \ldots, a_k$ and b in place of the sets $S_1, \ldots, S_k$ and N, the sign $+$ in place of $\cup$, and the sign $\times$ in place of $\cap$, then we have the equality $(a_1 + \cdots + a_k)b = a_1 b + \cdots + a_k b$ known as the distributive law for numbers. There are other properties showing that the operations of union and intersection for sets are analogous to addition and multiplication of numbers (see Problem 1). The study of the system of subsets of a given set S with respect to the operations $\cup$ and $\cap$ is called *set algebra*.

We now deduce the formula for $n(S_1 \cup S_2 \cup S_3)$. Because

$$S_1 \cup S_2 \cup S_3 = (S_1 \cup S_2) \cup S_3,$$

we can apply formula (20). We obtain

$$\begin{aligned} n(S_1 \cup S_2 \cup S_3) &= n\big((S_1 \cup S_2) \cup S_3\big) \\ &= n(S_1 \cup S_2) + n(S_3) - n\big((S_1 \cup S_2) \cap S_3\big). \end{aligned}$$

We can directly write the term $n(S_1 \cup S_2)$ using formula (20). To transform the last term, we note that $(S_1 \cup S_2) \cap S_3 = (S_1 \cap S_3) \cup (S_2 \cap S_3)$ as we already saw. We can apply formula (20) again. As a result, we obtain

$$\begin{aligned} n(S_1 \cup S_2 \cup S_3) = &\, n(S_1) + n(S_2) + n(S_3) - n(S_1 \cap S_2) - n(S_1 \cap S_3) \\ &- n(S_2 \cap S_3) + n\big((S_1 \cap S_3) \cap (S_2 \cap S_3)\big). \end{aligned}$$

It is perfectly obvious that

$$(S_1 \cap S_2) \cap (S_2 \cap S_3) = S_1 \cap S_2 \cap S_3,$$

and we can rewrite the last term as

$$n(S_1 \cap S_2 \cap S_3).$$

We obtain the formula

$$n(S_1 \cup S_2 \cup S_3) = n(S_1) + n(S_2) + n(S_3) - n(S_1 \cap S_2)$$
$$- n(S_1 \cap S_3) - n(S_2 \cap S_3) + n(S_1 \cap S_2 \cap S_3).$$

Now you can already guess what the formula for $n(S_1 \cup \cdots \cup S_r)$ should look like. There should be terms in it of the form $n(S_{i_1} \cap \cdots \cap S_{i_k})$ for the intersection $S_{i_1} \cap \cdots \cap S_{i_k}$ for any k sets from the sets $S_1, \ldots, S_r$ for all $k = 1, 2, \ldots, r$. Moreover, terms with a even k have a minus sign, and terms with an odd k have a plus sign. That is, the term $n(S_{i_1} \cap \cdots \cap S_{i_k})$ has the sign $(-1)^{k-1}$.

We can now prove this formula by induction on the number r in exactly the same way as we proved it for $r = 3$. The basis of the induction is formula (20). We write $S_1 \cup S_2 \cup \cdots \cup S_r$ in the form $(S_1 \cup S_2 \cup \cdots \cup S_{r-1}) \cup S_r$ and use formula (20):

$$n(S_1 \cup S_2 \cup \cdots \cup S_r) = n(S_1 \cup S_2 \cup \cdots \cup S_{r-1}) + n(S_r)$$
$$- n\big((S_1 \cup S_2 \cup \cdots \cup S_{r-1}) \cap S_r\big).$$

By the induction assumption, the formula holds for $n(S_1 \cup \cdots \cup S_{r-1})$ and gives just those terms in the formula for $n(S_1 \cup S_2 \cup \cdots \cup S_r)$ that do not contain S_r. We now write

$$(S_1 \cup S_2 \cup \cdots \cup S_{r-1}) \cap S_r = (S_1 \cap S_r) \cup (S_2 \cap S_r) \cup \cdots \cup (S_{r-1} \cap S_r).$$

We can again apply our formula to $n\big((S_1 \cap S_r) \cup \cdots \cup (S_{r-1} \cap S_r)\big)$ in accordance with the induction assumption. The intersection

$$(S_{i_1} \cap S_r) \cap \cdots \cap (S_{i_k} \cap S_r)$$

is obviously simply equal to $S_{i_1} \cap \cdots \cap S_{i_k} \cap S_r$. We thus obtain all terms in the formula that contain S_r. Further, if the corresponding term in the formula for $n\big((S_1 \cap S_r) \cup \cdots \cup (S_{r-1} \cap S_r)\big)$ has the sign $(-1)^{k-1}$, then it has the sign $(-1)^k$ in the formula for $n(S_1 \cup \cdots \cup S_r)$ and depends on $k+1$ sets $S_{i_1} \cap \cdots \cap S_{i_k} \cap S_r$.

It is more convenient to write the deduced formula for $n(S_1 \cup \cdots \cup S_r)$ if we speak of the number of elements in the complement $\overline{S_1 \cup \cdots \cup S_r}$ of the set $S_1 \cup \cdots \cup S_r$, that is, the number of elements in the set S that do not belong to one of the S_i. Because we always have $S = N + \overline{N}$ for any subset $N \subset S$, we have $n(\overline{N}) = n(S) - n(N)$. In our case, $n(\overline{S_1 \cup \cdots \cup S_r})$ is the sum of terms $(-1)^{k-1}n(S_{i_1} \cap \cdots \cap S_{i_k})$, where $S_{i_1}, \ldots, S_{i_k}$ are any k subsets from the total $S_1, \ldots, S_r$ and we assume that the term corresponding to $k = 0$ is $n(S)$. In other words,

$$n(\overline{S_1 \cup \cdots \cup S_r}) = n - n(S_1) - \cdots - n(S_r)$$
$$+ n(S_1 \cap S_2) + \cdots + (-1)^r n(S_1 \cap \cdots \cap S_r), \quad (21)$$

where $n = n(S)$.

In this formula, we have the expressions $n(S_{i_1} \cap \cdots \cap S_{i_k})$, where $i_1, \ldots, i_k$ ranges all collections of indices from the numbers $1, 2, \ldots, r$. We met such expressions in connection with Viète formula (13). It is instructive to compare these two formulas. Formula (21) looks exactly as if we substituted $x = 1$ and $a_i = -x_i$ in formula (12) and then replaced the term $x_{i_1} \cdots x_{i_k}$ everywhere with $n(S_{i_1} \cap \cdots \cap S_{i_k})$. Sometimes, it is even written this way (often called "symbolically"),

$$n(\overline{S_1 \cup \cdots \cup S_r}) = n(1 - S_1) \cdots (1 - S_r), \quad (22)$$

where we assume that the product in the right-hand side is expanded by the Viète formula as if S_i was an unknown and the expression $n \cdot S_{i_1} \cdots S_{i_k}$ (whose sense is unclear) is replaced with the sensible expression $n(S_{i_1} \cap \cdots \cap S_{i_k})$ (and $n \cdot 1$ is replaced with $n = n(S)$).

Such a form can help to more easily recall relation (21). But in algebra, if two relations connected with different questions have completely identical forms, we can **always** invent definitions such that one formula exactly corresponds with the other. We show this with the example of formulas (21), (22), and Viète formula (13).

For this, we must consider functions on the set S. You have undoubtedly met the concept of a function in one or another situation. We understand a function here as any association of each element $a \in S$ of the set S with some number. The process of association itself is denoted by f, and the specific number associated with the element a is denoted by $f(a)$, which is also called the *value* of the function f at the element a. Such a concept is sensible for an arbitrary set, but we are now interested in it in the case where the set S is finite. Then the function can be specified by writing the number $f(a)$ associated with each element a. For example, we show two functions f and g defined on a set with three elements $S = \{a, b, c\}$ in Fig. 13.

Thus, if $S = \{a_1, \ldots, a_n\}$, then a function is simply a collection of numbers $\big(f(a_1), \ldots, f(a_n)\big)$. Functions can be added and multiplied by performing the same operations with their values. In other words, if f

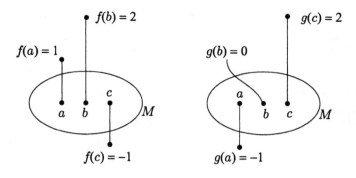

Fig. 13

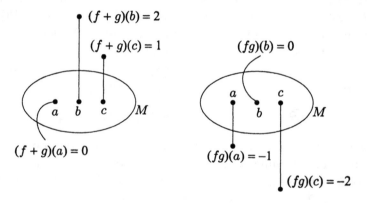

Fig. 14

and g are two functions on the same set, then the functions $f + g$ and fg are defined by $(f + g)(a) = f(a) + g(a)$ and $(fg)(a) = f(a)g(a)$ for any element. For example, the functions $f + g$ and fg for the functions f and g given by Fig. 13 are illustrated in Fig. 14.

Because the operations on functions are in fact operations on their values, these operations have the same properties as the operations on numbers: commutativity, associativity, distributivity, etc. We can apply any identity that is proved for numbers, replacing the quantities in it with arbitrary functions on the set S. The function $f_S(a)$ that associates the number 1 with every element $a \in S$ is denoted by $\mathbf{1}$. It is obvious that $\mathbf{1} \cdot f = f$ for any function f.

We now relate the concept of function and the concept of subset. Any subset $N \subset S$ is corresponds to a definite function that associates the value 1 with elements belonging to the subset and the value 0 with the elements not belonging to the subset. Such a function is called the characteristic function of the subset N and is denoted by f_N. Briefly,

$f_N(a) = 1$ if $a \in N$, and $f_N(a) = 0$ if $a \notin N$. Obviously, the function $f_N(a)$ conversely defines a set N—all $a \in S$ for which $f_N(a) = 1$. (We thus obtain a one-to-one correspondence between subsets $N \subset S$ and functions taking the values 1 and 0. This is the same correspondence that makes it possible to deduce Theorem 21 from Theorem 20; see Problem 1 in Sec. 8, where p and q must be replaced with 0 and 1.)

Certain properties of subsets are simply expressed in terms of their characteristic functions. For example, the characteristic function for the entire set S has all values equal to 1, and this means that $f_S = \mathbf{1}$. If $\overline{N}$ is the complement of the set N, then $f_{\overline{N}} = \mathbf{1} - f_N$: indeed, for $a \in N$, we have $f_N(a) = 1$, that is, $1 - f_{\overline{N}}(a) = 0$ as it should be for $f_{\overline{N}}$. The situation is exactly analogous for $a \in \overline{N}$. If N_1 and N_2 are any subsets, then $f_{N_1 \cap N_2} = f_{N_1} f_{N_2}$ because if $a \in N_1$ and $a \in N_2$, then $f_{N_1} f_{N_2}(a) = 1 \cdot 1 = 1$ as it should for $f_{N_1 \cap N_2}$, and if a does not belong to one of the sets N_1 or N_2, then one of the factors $f_{N_1}(a)$ or $f_{N_2}(a)$ is equal to 0, which means that $f_{N_1} f_{N_2}(a) = 0$ just as $f_{N_1 \cap N_2}(a) = 0$.

Clearly, this also holds for a larger number of subsets,

$$\text{if } N' = N_1 \cap \cdots \cap N_r, \quad \text{then } f_{N'} = f_{N_1} \cdots f_{N_r}. \tag{23}$$

We can now rewrite formula (21) in the language of characteristic functions. We first note that the set $\overline{S_1 \cup \cdots \cup S_r}$ we are interested in is equal to $\overline{S_1} \cap \cdots \cap \overline{S_r}$. This is obvious: the element a does not belong to the set $S_1 \cup \cdots \cup S_r$ if it does not belong to a single one of the sets S_i, that is, it belongs to all $\overline{S_i}$. We now use formula (23) to write the characteristic function of the desired set $\overline{S_1 \cup \cdots \cup S_r}$ in the form

$$f_{\overline{S_1 \cup \cdots \cup S_r}} = f_{\overline{S_1} \cap \cdots \cap \overline{S_r}} = f_{\overline{S_1}} \cdots f_{\overline{S_r}}.$$

Moreover, we know that $f_{\overline{S_i}} = \mathbf{1} - f_{S_i}$. As a result, we obtain

$$f_{\overline{S_1 \cup \cdots \cup S_r}} = (\mathbf{1} - f_{S_1})(\mathbf{1} - f_{S_2}) \cdots (\mathbf{1} - f_{S_r}).$$

We can now apply Viète formula (13) substituting $x = 1$ and $\alpha_i = f_{S_i}$ in it. We explained above why this formula can be applied to functions. As a result, we obtain

$$f_{\overline{S_1 \cup \cdots \cup S_r}} = \mathbf{1} - \sigma_1(f_{S_1}, \ldots, f_{S_r})$$
$$+ \sigma_2(f_{S_1}, \ldots, f_{S_r}) - \cdots + (-1)^r \sigma_r(f_{S_1}, \ldots, f_{S_r}).$$

Here, $\sigma_k(f_{S_1}, \ldots, f_{S_r})$ is the sum of all products $f_{S_{i_1}} \cdots f_{S_{i_k}}$ for all collections of different indices from the numbers $1, \ldots, r$. We know that $f_{S_{i_1}} \cdots f_{S_{i_k}} = f_{S_{i_1} \cap \cdots \cap S_{i_k}}$, and we obtain

$$f_{\overline{S_1 \cup \cdots \cup S_r}} = 1 - f_{S_1} - \cdots - f_{S_r} + \cdots + (-1)^r f_{S_1 \cap \cdots \cap S_r}, \qquad (24)$$

that is, the sum of all functions $f_{S_{i_1} \cap \cdots \cap S_{i_k}}$ taken with the plus sign for even k and with the minus sign for odd k.

We note that we have obtained properly something more than formula (21); we have found an expression not for the number of elements $n(\overline{S_1 \cup \cdots \cup S_r})$ of the set $\overline{S_1 \cup \cdots \cup S_r}$ but for its characteristic function, which determines not only the number of elements of the subset but also the entire subset itself. Furthermore, formula (21) has meaning only when the containing set S is finite, and our formula (24) holds for a finite number of subsets of any set S.

To deduce formula (21) from it, we must return from functions to numbers. Here it is already consequential that the set S is finite. For any function f, we define the number Sf as the sum of all values $f(a)$ of the function f for all elements $a \in S$: if $S = \{a_1, \ldots, a_n\}$, then $Sf = f(a_1) + \cdots + f(a_n)$. For example, for the functions f and g in Fig. 12, $Sf = 2$ and $Sg = 1$. Obviously, $S(f + g) = Sf + Sg$ for any two functions f and g. Indeed, the value of the function $f + g$ at the element a_i is equal to $f(a_i) + g(a_i)$. Therefore, $S(f+g) = (f(a_1)+g(a_1)) + \cdots + (f(a_n)+g(a_n)) = (f(a_1)+\cdots+f(a_n)) + (g(a_1)+\cdots+g(a_n)) = Sf+Sg$. If f_N is the characteristic function of the subset N, then $f_N(a) = 1$ at exactly the elements $a \in N$ and is zero for all other elements a. Therefore, $Sf_N = n(N)$.

It remains to consider the number Sf for functions in the left-hand and right-hand sides of equality (24). By virtue of the listed properties, we obtain exactly relation (21).

We now consider two applications of formula (21). The first is a question already considered by Euler. It concerns permutations of a set. We said at the end of the previous section (see Remark 1) that a distribution of the elements of a set S in a definite order is called a permutation. The number of permutations is equal to $n!$ if $n(S) = n$. For example, all six permutations of the set $S = \{a, b, c\}$ of three elements are written at the end of Sec. 8. In the general case, we similarly write all $n!$ permutations, the first being denoted by $(a_1, \ldots, a_n)$. We pose the question: how many permutations are there in which not a single element occupies the same place as in the first permutation? This is Euler's question. Answer this question for the case $n = 3$ and the six permutations written at the end of Sec. 8. Convince yourself that only two permutations satisfy the required condition: (c, a, b) and (b, c, a).

In the general case, we have an obvious exercise with formula (21). Let P denote the set of all permutations of elements of the set S. Then $n(P) = n!$. We consider the permutations in which the element a_i occupies the same position as in the first permutation, that is, the ith position. Let P_i denote the set of such permutations. Then the sense of our question is: find $n(\overline{P_1 \cup \cdots \cup P_n})$. (Convince yourself that this is so: there is nothing to prove here.) We thus have the situation considered above with the enveloping set P and the subsets $P_1, \ldots, P_n$ (in formula (21), the set is denoted by S and the subsets by S_i). To apply the formula, we must find the numbers $n(P_{i_1} \cap \cdots \cap P_{i_k})$. But the set $P_{i_1} \cap \cdots \cap P_{i_k}$ exactly contains those permutations in which $a_{i_1}, \ldots, a_{i_k}$ occupy the same positions as in the first permutation, that is, the corresponding positions $i_1, \ldots, i_k$. Such a permutation is distinguished from the first only by the distribution of the elements in the remaining positions. That is, there are as many such permutations as there are permutations of the set $\overline{\{a_{i_1}, \ldots, a_{i_k}\}}$. Because $n(\overline{\{a_{i_1}, \ldots, a_{i_k}\}}) = n - k$, $n(P_{i_1} \cap \cdots \cap P_{i_k}) = (n - k)!$ in the general formula. All sets $P_{i_1} \cap \cdots \cap P_{i_k}$ with the same k consequently yield an identical term in formula (21), and the number of such terms is equal to the number of subsets $\{a_{i_1}, \ldots, a_{i_k}\} \subset \{a_1, \ldots, a_n\}$ with a given value of k, that is, C_n^k according to Theorem 22. This means that the contribution of terms with a given value of k is equal to $C_n^k(n - k)!$. Substituting the value of the binomial coefficient, we obtain

$$\frac{n!}{k!(n - k)!}(n - k)! = \frac{n!}{k!},$$

and formula (21) in our case yields

$$n(\overline{P_1 \cup \cdots \cup P_n}) = n! - \frac{n!}{1!} + \frac{n!}{2!} - \cdots + (-1)^n \frac{n!}{n!}$$
$$= n!\left(1 - \frac{1}{1!} + \frac{1}{2!} - \cdots + \frac{(-1)^n}{n!}\right).$$

This is the formula found by Euler. He was interested in the "proportion" of such permutations among all permutations, that is, the ratio of this number to the number $n!$. It is equal to $1 - 1/1! + 1/2! - \cdots + (-1)^n/n!$, which can be shown to approach a definite number as n increases, namely, $1/e$, where e is the base of the natural logarithms (for those who know what that is). The irrational number $1/e$ is approximately equal to $0.36787 \ldots$.

The second application of formula (21) relates to properties of natural numbers. Let n be a given natural number, and let $p_1, \ldots, p_r$ be some of its prime factors, which differ from each other. How many natural numbers are there that are not greater than n and are not divisible by a single one of the numbers p_i? This is again an exercise with formula (21).

Let S denote the set of natural numbers $1, 2, \ldots, n$, and let S_i denote the set of these numbers that are divisible by p_i. Clearly, our task is equivalent to calculating $n(\overline{S_1 \cup \cdots \cup S_r})$. We find the value of the term $n(S_{i_1} \cap \cdots \cap S_{i_k})$ in formula (21). The set $S_{i_1} \cap \cdots \cap S_{i_k}$ comprises those natural numbers $t \leq n$ that are divisible by the prime numbers $p_{i_1}, p_{i_2}, \ldots, p_{i_k}$. This is equivalent to t being divisible by their product $p_{i_1} \cdots p_{i_k}$. In general, let m be some divisor of n. How many natural numbers $t \leq n$ are there that are divisible by m? Such numbers have the form $t = mu$, where u is a natural number, and the condition $t \leq n$ is equivalent to $u \leq n/m$. The possible values for u are therefore $1, 2, \ldots, n/m$, and the number of such numbers is equal to n/m. If $m = p_{i_1} \cdots p_{i_k}$, then we have $n(S_{i_1} \cap \cdots \cap S_{i_k}) = n/(p_{i_1} \cdots p_{i_k})$, and formula (21) takes the form

$$n(\overline{S_1 \cup \cdots \cup S_r}) = n - \frac{n}{p_1} - \cdots - \frac{n}{p_r} + \frac{n}{p_1 p_2} + \cdots + (-1)^r \frac{n}{p_1 \cdots p_r}.$$

The right-hand side can be written in the form

$$n\left(1 - \frac{1}{p_1} - \frac{1}{p_2} - \cdots + \frac{1}{p_1 p_2} + \cdots + (-1)^r \frac{1}{p_1 \cdots p_r}\right).$$

The expression in the parentheses can be transformed according to the Viète formula (now applied simply to numbers) in which we set $x = 1$ and $\alpha_i = 1/p_i$. This expression is equal to (see formula (13))

$$\left(1 - \frac{1}{p_1}\right)\left(1 - \frac{1}{p_2}\right) \cdots \left(1 - \frac{1}{p_r}\right).$$

For the total number of natural numbers not greater than n and not divisible by $p_1, p_2, \ldots, p_r$, we obtain the expression

$$n\left(1 - \frac{1}{p_1}\right)\left(1 - \frac{1}{p_2}\right) \cdots \left(1 - \frac{1}{p_r}\right). \tag{25}$$

Especially often, we meet the case where $p_1, p_2, \ldots, p_r$ are all the prime divisors of the number n. Then t is not divisible by a single one of the p_i if and only if it is relatively prime to n: if t and n had a common divisors d, then that divisor would have a prime divisor p_i that would divide t and would be a divisor of n. Therefore, formula (25) gives the number of all natural numbers not greater than n and relatively prime to n if we take all prime divisors of n for the $p_1, p_2, \ldots, p_r$. Expression (25) in this case was found by Euler; it is denoted by $\varphi(n)$ and is called the *Euler function*. For example, for $n = 675 = 3^3 \cdot 5^2$, we have $n(1 - 1/3)(1 - 1/5) =$

$3^2 \cdot 5(3 - 1)(5 - 1) = 360$ numbers not greater than 675 and relatively prime to it.

We now assume that $p_1, p_2, \ldots, p_n$ do not necessarily divide n. How many natural numbers $t \leq n$ are there that are not divisible by $p_1, p_2, \ldots, p_r$? We can repeat the previous argument with a change in one place. We must find the number of natural numbers $t \leq n$ that are divisible by $p_{i_1} \cdots p_{i_k}$. Let m be an arbitrary natural number. How many natural numbers $t \leq n$ are there that are divisible by m? We again set $t = mu$ and write the condition $mu \leq n$. Therefore, we must take all numbers $u = 1, 2, \ldots$ such that $mu \leq n$. Let u be the last of such numbers. Then $r = n - mu < m$. Otherwise, there would be another number $mu + m = m(u + 1)$. But then $n = mu + r$, where $0 \leq r < m$. This is the formula for division of n by m with a remainder (see Theorem 5). This means the sought number u is equal to the quotient from dividing n by m. This quotient is denoted by $[n/m]$. Therefore, the number of natural numbers not greater than n and divisible by m is equal to $[n/m]$. Now we can repeat the previous argument literally and apply formula (21). For the number of natural numbers not greater than n and divisible by $p_1, p_1, \ldots, p_r$, we obtain the expression

$$n - \left[\frac{n}{p_1}\right] - \left[\frac{n}{p_2}\right] - \cdots + \left[\frac{n}{p_1 p_2}\right] + \cdots + (-1)^r \left[\frac{n}{p_1 \cdots p_r}\right]. \qquad (26)$$

It is not as elegant as expression (25), but we can write it in the same form approximately. For this, we recall the formula for division with a remainder: $n = mu + r$, where $0 \leq r < m$ and $u = [n/m]$ is the partial quotient. Dividing this equality by m, we obtain $n/m = u + r/m$, and because $0 \leq r < m$, we have $n/m - 1 < [n/m] \leq n/m$. That is, the partial quotient can be replaced with the full quotient with an error less than 1. We make such a replacement in all terms in expression (26). What error results? Each term in expression (26) corresponds to a subset $\{i_1, \ldots, i_k\}$ of the set $\{1, \ldots, r\}$. According to Theorem 21, the number of such subsets is equal to 2^r. This means that this is the number of terms in expression (26). It each term yields an error less than 1 when the partial quotient is replaced by the full quotient, then the total error is less than 2^r. That is, expression (26) differs from the expression

$$n - \frac{n}{p_1} - \frac{n}{p_2} - \cdots + \frac{n}{p_1 p_2} + \cdots + (-1)^r \frac{n}{p_1 \cdots p_r} \qquad (27)$$

by less than 2^r. We already met the last expression, and we saw that it is equal to

$$n\left(1 - \frac{1}{p_1}\right) \cdots \left(1 - \frac{1}{p_r}\right).$$

We thus find that the number N of natural numbers not greater than n and not divisible by the given primes $p_1, p_2, \ldots, p_r$ satisfies the inequality

$$\left| N - n\left(1 - \frac{1}{p_1}\right) \cdots \left(1 - \frac{1}{p_r}\right) \right| \leq 2^r. \tag{28}$$

For example, if we have three primes p, q, and r, then N is equal to $n(1 - 1/p)(1 - 1/q)(1 - 1/r)$ with an error less than 8.

Problems:

1. Convince yourself of the correctness of the relations $S_1 \cap \cdots \cap S_k = (S_1 \cap \cdots \cap S_{k-1}) \cap S_k$ and $(S_1 \cap \cdots \cap S_k) \cup N = (S_1 \cup N) \cap \cdots \cap (S_k \cup N)$. The second of them is again an analogue of the distributive law for numbers $(a_1 + \cdots + a_k)b = a_1 b + \cdots + a_k b$, but the operation $\cup$ now plays the role of multiplication and the operation $\cap$ plays the role of addition.

2. Convince yourself that, in general, to any relation between subsets in which there are the operations $\cup$ and $\cap$, there corresponds another relation with the operations interchanged. For this, prove that $\overline{S_1 \cup S_2} = \overline{S_1} \cap \overline{S_2}$ and $\overline{S_1 \cap S_2} = \overline{S_1} \cup \overline{S_2}$.

3. How many times does the function $\sin ax$ take the value zero on the segment from 0 to $2\pi b$, where $0 < a < b$ and a and b are natural numbers?

4. For natural numbers $a_1, \ldots, a_m$, $\max(a_1, \ldots, a_m)$ denotes the largest of them, and $\min(a_1, \ldots, a_m)$ denotes the least. Let $N = \max(a_1, \ldots, a_m)$. For the set $S = \{1, \ldots, N\}$, define S_i as the subset of those j for which $a_j < a_i$. Applying formula (21), find a relation between $\max(a_1, \ldots, a_m)$ and $\min(a_{i_1}, \ldots, a_{i_k})$, where the $\{a_{i_1}, \ldots, a_{i_k}\}$ are all subsets of the set $\{a_1, \ldots, a_m\}$.

5. Apply formula (21) setting $S_i = \overline{\{a_i\}}$. Directly calculating all terms in it, obtain the relation

$$n - C_n^1(n - 1) + C_n^2(n - 2) - \cdots + (-1)^r C_n^{n-1} \cdot 1 = 0.$$

6. Let S be a finite set and h be an arbitrary function on S. For the subset $N \subset S$, we introduce the number $S_h(N)$ as the sum of all the values $h(a)$ for all $a \in N$. Prove a formula analogous to formula (21), where $n(N)$ is replaced with $S_h(N)$ everywhere.

7. Find the sum of all natural numbers not greater than n and relatively prime to n. *Hint:* Apply the result of Problem 7 setting $h(k) = k$.

8. Find the similar sum for the squares of those numbers.

9. Prove that 2^r can be replaced with 2^{r-1} in the right-hand side of inequality (28).

10.   The Language of Probability

The theory of probability like all other areas of mathematics has its own fundamental concepts that are not subject to definition (such as point or number). The first such concept is *event*. In this section, we consider situations in which the number of events is finite. Usually each event is the result of the occurrence of several simplest events, which are said to be *elementary*. For example, six elementary events are possible when a die is cast: the number 1, the number 2, the number 3, the number 4, the number 5, the number 6. The event that we obtain an even number comprises three elementary events: either the number 2 or the number 4 or the number 6. The set of elementary events is a simple set (finite in this section) whose elements are given a new name (elementary events). Events are subsets of the set of elementary events.

The second fundamental concept is the concept of *probability*: it is a real number assigned to each elementary event. Thus, if $S = \{a_1, \ldots, a_n\}$ is a set of elementary events, then the assignment of probability is assigning each elementary event $a_i \in S$ a real number p_i, which is called the probability of the event a_i. It is required the probabilities satisfy two conditions: they must be nonnegative, and the sum of the probabilities of all elementary events must be equal to 1:

$$p_i \geq 0, \qquad p_1 + \cdots + p_n = 1. \tag{29}$$

In other words, probability is a function $p(a)$ on a set of elementary events S with real values and satisfying the conditions that $p(a) \geq 0$ for $a \in S$ and the sum of all numbers $p(a)$ for $a \in S$ is equal to 1. These conditions play the role of the axioms of probability.

If N is an arbitrary event (we recall that an event is a subset of S), then the sum of the numbers $p(a)$ for all $a \in N$ is called the *probability* of the event N. This probability is denoted by $p(N)$. In the particular case where $N = S$, the corresponding event is said to be *certain*. Condition (29) indicates that the probability of a certain event is equal to 1. The condition $p(S) = 1$ is not essential; it is important only that $p(S) > 0$. An arbitrary case can be reduced to the case where $p(S) = 1$ by dividing all probabilities by $p(S)$. We simply choose the probability of a certain event as the unit of measure for the probabilities of other events. We emphasize that the object studied by the theory of probability is a set (finite for us) of elementary events with given probabilities. This set and probabilities are taken from the specific circumstances of one or another problem. After they are given, it is possible to calculate the probabilities of other events. Specialists in the theory of probability therefore say that their task is to find the probability of some events in terms of the probabilities of other events.

If two events are given (and we recall that this means simply two subsets N_1 and N_2 of the set S), their union $N_1 \cup N_2$ and intersection $N_1 \cap N_2$ are also events. It follows from the definition that $p(N_1 \cup N_2) \leq p(N_1) + p(N_2)$. Here, the equality may not hold, because we can have the term $p(a)$ twice in $p(N_1) + p(N_2)$, where $a \in N_1 \cap N_2$. We can say more precisely that

$$p(N_1 \cup N_2) = p(n_1) + p(N_2) - p(N_1 \cap N_2).$$

We met such a relation earlier (see Problem 6 in Sec. 9). In particular, if $N_1 \cap N_2 = \varnothing$, that is, N_1 and N_2 do not intersect, then the events N_1 and N_2 are said to be *inconsistent*. Then $p(N_1 \cup N_2) = p(N_1) + p(n_2)$. An even more particular case is where $N_1 = N$ is an arbitrary subset and $N_2 = \overline{N}$ is its complement. We obtain $p(N_1) + p(N_2) = 1$ or $p(\overline{N}) = 1 - p(N)$. The event $\overline{N}$ is called the *opposite* of the event N.

The basic object, the set S and the given function p on it satisfying probability axioms (29), is called a *probability scheme*. It is denoted by $(S; p)$.

One important case of specifying a probability scheme is where all elements of the set S are equally probable (for example, in view of the symmetry of the given problem). It follows from relation (29) that all $p_i = 1/n$. If $N \subset S$ is any event, then $p(N) = n(N)/n$. This is the situation with casting a die, for example, if the die is symmetric (cheaters move its center of gravity). Therefore, all six elementary events corresponding to one or another number have the equal probability $1/6$, and the event of an even number has the probability $3 \cdot 1/6 = 1/2$.

If the die is not homogeneous, then we do not have a basis for considering all elementary events equally probable. Then their probabilities are determined experimentally, casting the die many times and recording the results. After a large number n of throws, if the number i comes up k_i times, then the probability of the elementary event that i comes up is assumed to be equal to k_i/n. Obviously, conditions (29) will be satisfied. The number n is chosen depending on the degree of accuracy that we want. A different probability scheme $(S; p)$ is thus obtained.

The situation with a favorite problem in the theory of probability is analogous to casting a symmetric cube (die): taking balls from an urn. Let there be n identical balls in an urn. Without looking, we take one ball from the urn. Taking one or another ball is an elementary event. We can consider that we identify the elementary events with the balls. The phrase "identical balls" is formulated mathematically as the condition that the probability of these events are equal. This means that they are equal to $1/n$. Now let the urn contain balls of different colors: a black and b white, $a + b = n$. Then the event "take a white ball from the urn" is the subset $N \subset S$ of white balls. Because $n(N) = b$, $p(N) = b/n$ is the probability of taking a white ball.

The situation with casting a die is somewhat more complicated if we throw two times. An elementary event is now given by two numbers (a, b), $1 \le a \le 6$, $1 \le b \le 6$, which indicate that a came up on the first throw and b came up on the second throw. The total number of elementary events is 36. They can be represented in the form of a table (Fig. 15).

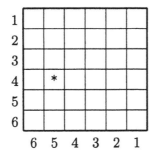

Fig. 15

Here the possible results of the first throw are written horizontally below, and the possible results of the second throw are written vertically on the left. For example, the elementary event that 5 comes up on the first throw and 4 comes up on the second throw corresponds to the cell marked with the asterisk. The event that 5 comes up on the first throw again has the probability 1/6. But now it is not elementary: it comprises six elementary events corresponding to the vertical column over the number 5. They correspond to one or another number i, $1 \leq i \leq 6$, coming up on the second throw if 5 comes up on the first throw. Because the result of the first throw does not influence the second throw and the die is symmetric as before, we conclude that all six elementary events have the same probability, and because the probability of the event comprising them is equal to 1/6, the probability of each elementary event is 1/36. We are thus convinced that the probability of any elementary event is equal to 1/36.

a

	6	5	4	3	2	1	
1	7	6	5	4	3	2	
2	8	7	6	5	4	3	
3	9	8	7	6	5	4	
4	10	9	8	7	6	5	
5	11	10	9	8	7	6	
6	12	11	10	9	8	7	
	6	5	4	3	2	1	b

Fig. 16

We consider the event N_k: "the sum of the numbers coming up on the first and second throws is equal to k" ("the total points are equal to k"). We go through all the pairs (a, b) writing the sum $a + b$ in each cell (Fig. 16).

We see that 12 occurs in only one cell. Therefore, $n(N_{12}) = 1$, and in exactly the same way, $n(N_{11}) = 2$, $n(N_{10}) = 3$, $n(N_9) = 4$, $n(N_8) = 5$, $n(N_7) = 6$, $n(N_6) = 5$, $n(N_5) = 4$, $n(N_4) = 3$, $n(N_3) = 2$, and $n(N_2) = 1$. The largest value is for $n(N_7)$, and because $p(N_k) = n(N_k)/36$, $p(N_7)$ has the largest value among all $p(N_k)$. In other words, the most probable total points for two throws is equal to 7.

What is the answer for n throws? Here the elementary event is given by a sequence of n numbers $(a_1, \ldots, a_n)$, each of which can take a value $1, 2, \cdots, 6$. The same reasoning shows that the probability of each is equal to $1/6^n$. The event N_k "the total points for all throws is equal to k" comprises all sequences such that $a_1 + \cdots + a_n = k$. Therefore, we must determine which number k has the most representations of the form

$$k = a_1 + \cdots + a_n, \quad 1 \le a_i \le 6. \tag{30}$$

For this, we consider the polynomial $F(x) = (x + x^2 + \cdots + x^6)^n$. Opening parentheses, we take the term x^{β_i} from the ith pair of parentheses. As a result, we obtain the term $x^{\beta_1 + \cdots + \beta_n}$. There will be several such terms, and we combine them. Therefore, the number of different representations (30) is equal to the coefficient of x^k in the polynomial $F(x)$, and our problem is thus reduced to determining which term has the largest coefficient. Because $F(x) = x^n G(x)$, where $G(x) = (1 + x + \cdots + x^5)^n$, the coefficient of x^k in $F(x)$ is equal to the coefficient of x^{k-n} in $G(x)$. It is sufficient to determine the term with the largest coefficient in $G(x)$.

The polynomial $G(x)$ has two properties from which the answer follows by itself.

An arbitrary polynomial $f(x) = c_0 + c_1 x + \cdots + c_n x^n$ is said to be *reciprocal* if its terms at the same distance from the ends have the same coefficient: $c_k = c_{n-k}$. If the coefficients c_i are represented by points with the coordinates (i, c_i) on the plane, then this property means that they are distributed symmetrically with respect to the middle, the line $x = n/2$. The case where n is even is shown in Fig. 17a, and the case where n is odd is shown in Fig. 17b.

The polynomial $x^n f(1/x)$ has the same coefficients as $f(x)$ but in the reverse order. Indeed, if $f(x) = a_0 + a_1 x + \cdots + a_n x^n$, then $f(1/x) = a_0 + a_1/x + \cdots + a_n/x^n$ and $x^n f(1/x) = a_0 x^n + a_1 x^{n-1} + \cdots + a_n$. It follows that if $f(x)$ is reciprocal, then $x^n f(1/x) = f(x)$. Hence, the product of two reciprocal polynomials is reciprocal. Indeed, if $f(x)$

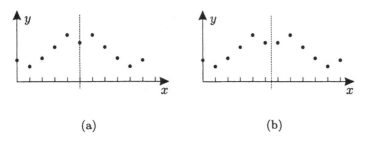

(a)   (b)

Fig. 17

and $g(x)$ are reciprocal polynomials of the respective degrees n and m, then $x^n f(1/x) = f(x)$ and $x^m g(1/x) = g(x)$. Multiplying, we obtain $x^n f(1/x) x^m g(1/x) = f(x)g(x)$, that is, $x^{n+m} f(1/x)g(1/x) = f(x)g(x)$. And this means that the polynomial $f(x)g(x)$ is reciprocal. We conclude by induction that the product of any number of reciprocal polynomials is a reciprocal polynomial. Finally, because the polynomial $1 + x + \cdots + x^5$ is reciprocal, the same is true for $G(x) = (1 + x + \cdots + x^5)^n$.

The polynomial $f(x) = c_0 + c_1 x + \cdots + c_n x^n$ is said to be *unimodal* if for some $m \leq n$, the inequality $c_0 \leq c_1 \leq \cdots \leq c_m \geq c_{m+1} \geq \cdots \geq c_n$ holds. That is, the coefficients c_i initially do not decrease, and then from some position, they do not increase. If we again represent them with the points (i, c_i), then they are distributed with "one hump" (Fig. 18).

For example, the polynomial $(1 + x)^n$ is reciprocal, which is expressed in the property $C_n^m = C_n^{n-m}$ of the binomial coefficients (see Sec. 6). It is unimodal, which is expressed in a property of the binomial coefficients that was proved in Sec. 6.

Fig. 18

It can be proved that if the polynomials $f(x)$ and $g(x)$ have non-negative coefficients and are reciprocal and unimodal, then $f(x)g(x)$ is unimodal. The proof is completely elementary but somewhat long. It follows from this theorem that $G(x)$ is unimodal. However, you can establish this property for the specific polynomial $G(x)$ yourself (Problem 3). But for a reciprocal unimodal polynomial, it is easy to determine the term with the largest coefficient. Namely, if the term $c_k x^k$ has the largest coefficient, then we have $c_{n-k} = c_k$ by the reciprocity property. As a result, we have the symmetric term $c_k x^{n-k}$. We can consider that $k \leq n/2$ and $n - k \geq n/2$. By the unimodal property, not one term

$c_i x^i$ with $k \leq i \leq n - k$ can have a smaller coefficient; otherwise, there would be two "humps" in the graph. Therefore, the largest must be the middle coefficient $c_{n/2}$ for even n or the two middle coefficients $c_{(n-1)/2}$ and $c_{(n+1)/2}$ for odd n (however, there may be other coefficients equal to them). In particular, we see that if n is even, then the term in $G(x)$ with $n^{5n/2}$ has the largest coefficient, and if n is odd, then the terms in $G(x)$ with $x^{(5n-1)/2}$ and $x^{(5n+1)/2}$ have identical coefficients, which are the largest.

In the polynomial $F(x)$, this term is multiplied by x^n and therefore has the index $5n/2 + n = 7n/2$ for even n. For odd n, there are two terms with equal coefficients and the indices $(5n-1)/2 + n = (7n-1)/2$ and $(5n+1)/2 + n = (7n+1)/2$. Thus, for n throws of the die, the most probable total points is equal to $7n/2$ for even n; for odd n, there are two equally probable total points, $(7n-1)/2$ and $(7n+1)/2$, with the highest probability.

We consider one more similar problem. Some number m of physical particles are captured by n instruments such that each particle is captured by some instrument. The capture of a specific particle by one or another instrument is considered equally probable. What is the probability that each instrument registers the capture of at least one particle? Here, the elementary events are the result of one experiment that indicates which instrument captured which particles. Let the instruments be denoted by the elements a of the set S. By assumption, $n(S) = n$. We number the particles $1, 2, \ldots, m$. Then an elementary event is a sequence $(a_1, \ldots, a_m)$, where $a_i \in S$ and the sequence indicates that the ith particle was captured by the instrument a_i. In other words, the set of elementary events is S^m in the sense defined in Sec. 7. The conditions of the problem state in another way that the elementary events have equal probabilities. Because $n(S^m) = n^m$ by Theorem 20, the probability of each elementary event is equal to $1/n^m$. The event we are interested in is the subset $N \subset S$ comprising those sequences $(a_1, \ldots, a_m)$ that contain each element of S at least once. For example, if $S = \{a, b, c\}$ and $m = 4$, then $(a, b, c, a) \in N$, but the sequence (a, b, a, b) is not contained in N, because it does not contain c. Our task is to calculate the value $n(N)$.

Let N_a denote the subset of the set S^m comprising all sequences $(a_1, \ldots, a_m)$ in which not one a_i is equal to a. It is then obvious that $N = \overline{\cup N_a}$, the complement of the union of all subsets N_a for all $a \in S$. Therefore, $n(N) = n(\overline{\cup N_a})$, and the value of the number $n(\overline{\cup N_a})$ is given by formula (21). We find the number $n(N_{a_1} \cap N_{a_2} \cap \cdots \cap N_{a_r})$, where $a_1, \ldots, a_r$ are some different elements of the set S. We are speaking here, consequently, of sequences $(c_1, \ldots, c_m)$ in which none of the elements c_i

is equal to any of the $a_1, \ldots, a_r$. In other words, each c_i is an arbitrary element of the set $\overline{\{a_1, \ldots, a_r\}}$, the complement of the subset $\{a_1, \ldots, a_r\}$ in the set S. Such sequences compose the set $\left(\overline{\{a_1, \ldots, a_r\}}\right)^m$, and the number of them is equal to $\left(n\left(\overline{\{a_1, \ldots, a_r\}}\right)\right)^m$ by Theorem 20. Because

$$n(\{a_1, \ldots, a_r\}) = r, \qquad n(S) = n,$$

we have

$$n\left(\overline{\{a_1, \ldots, a_r\}}\right) = n - r$$

and

$$n(N_{a_1} \cap N_{a_2} \cap \cdots \cap N_{a_r}) = (n - r)^m.$$

Consequently, each term $n(S_{i_1} \cap S_{i_2} \cap \cdots \cap S_{i_r})$ in formula (21) is equal to $(n - r)^m$ in our case. The number of terms with a given r, as we saw, is equal to C_n^r. Therefore, formula (21) yields

$$n(N) = n^m - C_n^1(n - 1)^m + \cdots + (-1)^{n-1}C_n^{n-1} \cdot 1^m.$$

The sought probability is equal to

$$\frac{n(N)}{n^m} = 1 - C_n^1 \left(\frac{n-1}{n}\right)^m + \cdots + (-1)^{n-1}C_n^{n-1}\left(\frac{1}{n}\right)^m. \qquad (31)$$

In all the preceding examples, elementary events had identical probabilities, consequently equal to $1/n$, where n is the number of elementary events. As a result, calculating the probabilities of other events was reduced to determining the number of elements of a subset, that is, a problem in combinatorics. We now analyze examples that are more typical for the theory of probability.

Let $(M; p)$ and $(N; q)$ be two probability schemes. We assume that they are determined by multiple repetitions of some experiment, one for each scheme. We call the experiment defining probability scheme $(M; p)$ experiment A and similarly for $(N; q)$ the experiment B. We now consider the experiment consisting of the consecutive performance of first experiment A and then experiment B, and we try to define a certain new probability scheme in accordance with it. We consider such a situation earlier in connection with the consecutive throws of a die (see Fig. 15). Let $n(M) = m$, $M = \{a_1, \ldots, a_m\}$, $p(a_i) = p_i$, $n(N) = n$, $N = \{b_1, \ldots, b_n\}$, and $q(b_i) = q_1$. Then the new experiment defines certain elementary events: the event $a \in M$ occurred in the first experiment, and the event $b \in N$ occurred in the second. The new elementary events thus comprise pairs (a, b), where $a \in M$ and $b \in N$, or elements of the set $X = M \times N$. What probabilities can we assign to these events? We can do this uniquely in a reasonable manner if we allow one more assumption. We consider that the experiments A and B that are used to determine the probability schemes $(M;p)$ and $(N;q)$ are *independent*. This means that the result of the second experiment (that is, B) does not depend on the result of the first experiment (that is, A). With this condition, we can define the probabilities $p(a,b)$ of the elementary events (a,b). Our argument closely follows that used for two throws of a die (see Fig. 15).

The same as there (and as we did in Sec. 7), we represent the elements of the set in the form of a rectangular table (Fig. 19).

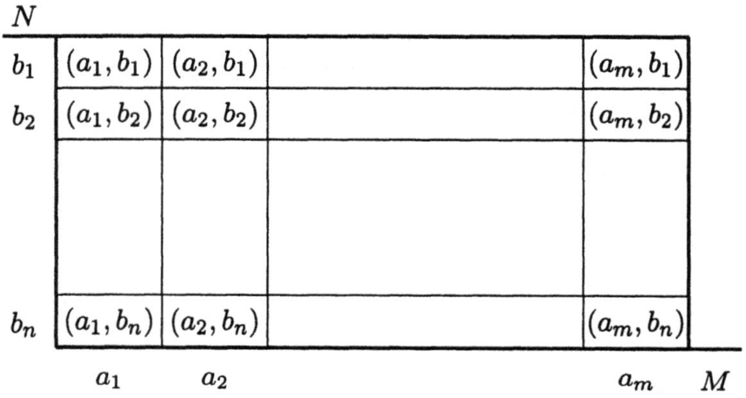

N

b_1	(a_1,b_1)	(a_2,b_1)			(a_m,b_1)
b_2	(a_1,b_2)	(a_2,b_2)			(a_m,b_2)
b_n	(a_1,b_n)	(a_2,b_n)			(a_m,b_n)
	a_1	a_2		a_m	M

Fig. 19

The events that include the occurrence of the event a_i in the first experiment have the given probability p_i. Now it is not an elementary event but comprises the elementary events $(a_i,b_1),(a_i,b_2),\ldots,(a_i,b_n)$ distributed in the ith column in Fig. 19. By our convention, the probability of these events must be the same as if experiment A were not performed, that is, the same as the probability of the events $b_1,\ldots,b_n$ in the probability scheme $(N;q)$. But a contradiction arises here: the sum of the probabilities of the events $(a_i,b_1),(a_i,b_2),\ldots,(a_i,b_n)$ is equal to p_i, and the sum of the probabilities $b_1,\ldots,b_n$ is equal to 1. In other words, the column with the index i itself composes a probability scheme that should be "the same as" the scheme $(N;q)$. But condition (29) in the definition of a probability scheme is not satisfied by this "scheme." We must "renormalize" it, dividing the probabilities of all elementary events by the probability of the entire event p_i. We then obtain a genuine probability scheme with the probabilities $p((a_i,b_j))/p_i$. It is reasonable to demand that namely this scheme coincide with probability scheme $(N; q)$. This leads to the equality $p((a_i, b_j))/p_i = q_i$, that is, $p((a_i, b_j)) = p_i q_j$. In view of this, we assume that

$$p((a_i, b_j)) = p_i q_j \qquad (32)$$

by definition.

We thus indeed obtain a new probability scheme: the sum of the probabilities of elementary events in the ith column in Fig. 19 is equal to $p_i q_1 + \cdots + p_i q_n = p_i(q_1 + \cdots + q_n) = p_i$, and the sum of all elementary events is equal to $p_1 + \cdots + p_m = 1$. Condition (29) is therefore satisfied.

The newly obtained probability scheme $(X; p)$ is called the *product* of the probability schemes $(M; p)$ and $(N; q)$. We can write this different-ly: if the initial probability schemes have the forms (M, p) and $(N; q)$, then $X = M \times N$ and $p((a, b)) = p(a)q(b)$. The product of probabili-ty schemes corresponds to our intuitive representation of a probability scheme determined by two consecutive experiments that are *indepen-dent* of each other. The preceding argument was needed only to *explain* the reasonability of this definition. The formal *definition* is contained in the one line of equality (32).

We now define the product of several probability schemes $(M_1; p_1), \ldots, (M_r; p_r)$ by induction:

$$M_1 \times \cdots \times M_r = (M_1 \times \cdots \times M_{r-1}) \times M_r, \qquad (33)$$

where $M_1 \times \cdots \times M_{r-1}$ is considered already defined by induction and the product of the two schemes $(M_1 \times \cdots \times M_{r-1})$ and M_r we have just defined previously. We find an explicit form of this probability scheme. The set $M_1 \times \cdots \times M_r$ is the product of the sets $M_1, \ldots, M_r$ defined in Sec. 7. It consequently comprises arbitrary sequences $(a_1, \ldots, a_r)$, where a_i can be any element of the set M_i. The probability of the elementary event $(a_1, \ldots, a_r)$ is equal to

$$p((a_1, \ldots, a_r)) = p_1(a_1)p_2(a_2) \cdots p_r(a_r). \qquad (34)$$

We can also verify this immediately by induction on r. Indeed, ac-cording to definition (33) and definition (32),

$$p((a_1, \ldots, a_r)) = p(((a_1, \ldots, a_{r-1}), a_r)) = p((a_1, \ldots, a_{r-1}))p_r(a_r),$$

and by the induction assumption,

$$p((a_1, \ldots, a_{r-1})) = p_1(a_1)p_2(a_2) \cdots p_{r-1}(a_{r-1}),$$

whence we obtain equality (34) by substitution. This equality can be ex-pressed as follows: we must replace each element in $(a_1, \ldots, a_r)$ with its

probability and multiply the numbers obtained. This is the probability
of the sequence.

We now apply this general construction to the particular case of the
probability scheme I^n, where $I = \{a, b\}$ is a probability scheme consisting of two elementary events with the probabilities $p(a) = p$ and
$p(b) = q$ and the conditions $p \geq 0$, $q \geq 0$, and $p + q = 1$ are of course
satisfied. We already discussed I^n as a set in Sec. 7. It comprises all
possible "texts" of length n of the type $aabbbab$ in an "alphabet" of the
two letters a and b. Consequently, these will be the elementary events.
Their probabilities, in accordance with what we just established, are
defined as follows: if the letter a occurs k times in the "text" and the
letter b occurs $n-k$ times, then the probability of the "text" is equal
to $p^k q^{n-k}$. Such a probability scheme is called a *Bernoulli scheme*. As
we saw, it reflects the probability of the occurrence of the events a or b
for n repetitions of an experiment in which each time a occurs with the
probability p and b occurs with the probability q. It is further assumed
that the result of each experiment does not influence the results of the
following experiments.

For example, for $n = 3$, we have eight elementary events: (a, a, a),
(a, a, b), (a, b, a), (b, a, a), (a, b, b), (b, a, b), (b, b, a), and (b, b, b) with the
respective probabilities p^3, $p^2 q$, $p^2 q$, $p^2 q$, pq^2, pq^2, pq^2, and q^3. We note
that the letter p here denotes not a probability but a specific number
$0 < p < 1$. The probability of an elementary event with k signs a and
$n-k$ signs b is equal to $p^k q^{n-k}$. Both notations are too rooted to change
them, but it should be clear what p denotes each time.

We calculate the probability of the event A_k, which is the event that a
occurred k times in a series of n experiments. This event comprises the
elementary events that yield those "texts" $babbbaa\ldots$ in which a occupies exactly k positions. The remaining $n-k$ positions are consequently
occupied by the event b. By the general theory, such an elementary event
has the probability $p^k q^{n-k}$. How many elementary events does the event
A_k comprise? This number is equal to the ways to choose k indices from
n indices $1, \ldots, n$, that is, the number of subsets of k elements in a set
of n elements. According to Theorem 22, the sought number is equal to
the binomial coefficient C_n^k. Therefore, we have found the probability of
the event A_k:

$$p(A_k) = C_n^k p^k q^{n-k} = \frac{n!}{k!(n-k)!} p^k q^{n-k}. \tag{35}$$

We can use this to find the most probable number of occurrences of
the event a. It is that value of k for which the expression in formula (35)
takes the largest value. We write these numbers in order:

$$1q^n, \quad npq^{n-1}, \quad \frac{n(n-1)}{2}p^2q^{n-2}, \quad \ldots, \quad 1p^n.$$

We consider the ratio of two neighboring numbers:

$$\frac{p(A_{k+1})}{p(A_k)} = \frac{\frac{n!}{(k+1)!(n-k-1)!}p^{k+1}q^{n-k-1}}{\frac{n!}{k!(n-k)!}p^kq^{n-k}} = \frac{(n-k)p}{(k+1)q}$$

(after all simplifications, which you can easily do yourself).

If this number is greater than 1, then the $(k+1)$th number is greater than the kth; if it is equal to 1, then those two numbers are equal; and if it is less than 1, then the $(k+1)$th number is less than the kth. The number is greater than 1 if $(n-k)p/((k+1)q) > 1$, that is, $(n-k)p > (k+1)q$. Recalling that $p+q=1$, we can write this inequality in the form $np > k+1-p$ or $(n+1)p-1 > k$. If $k > (n+1)p-1$, then the ratio of the consecutive $p(A_{k+1})/p(A_k)$ is less than 1. Finally, if $k = (n+1)p-1$, then $p(A_{k+1}) = p(A_k)$. Thus, while k takes values less than $(n+1)p-1$, we obtain always larger numbers moving from the kth to the $(k+1)$th. Further, we can meet two cases.

Case 1: The number $(n+1)p-1$ is not integer. Then for the largest integer m not greater than $(n+1)p-1$, we obtain the largest number $p(A_m)$. Moreover, $m \neq (n+1)p-1$, and for larger values of k, each number $p(A_k)$ is less than the preceding one, and thus to the last, the nth number. This means that in this case, there is one most probable number of occurrences of the event a, and it is the largest integer m not greater than $(n+1)p-1$.

Case 2: The number $(n+1)p-1$ is integer. Then the numbers $p(A_k)$ increase for $k < m = (n+1)p-1$. Further, $p(A_{m+1}) = p(A_m)$, and for $k > m+1$, the numbers $p(A_k)$ decrease.

Thus, the numbers $p(A_k)$ increase until they reach a maximum, then one or two identical maximums occur, and then they decrease. In other words, they are distributed with "one hump" as in Fig. 18. This means that the polynomial formed using them,

$$q^n + np^{n-1}qt + \frac{n(n-1)}{2}p^{n-2}q^2t^2 + \cdots + p^nt^n,$$

is unimodal. We can use the binomial formula to write it in the form $(q+pt)^n$. How can we see its unimodality using this simple expression? I do not know any way to do this.

In the simplest case where $p = q = 1/2$, we find that if $(n+1)/2 - 1$ is not integer, that is, if n is even, then $(n+1)/2 - 1 = n/2 - 1/2$ and $m = n/2$. Hence, there is one most probable number of occurrence of the event a, and that is $m = n/2$. That is, it is most likely that a occurs

$n/2$ times and b occurs $n/2$ times. There is nothing surprising in this; such an answer is suggested by symmetry considerations. If n is odd, then $m = (n + 1)/2 - 1 = (n - 1)/2$ is integer, and there are two most probable numbers of the occurrence of the event a: $(n - 1)/2$ (then the event b occurs $(n + 1)/2$ times) and $(n + 1)/2$ (then the event b occurs $(n - 1)/2$ times). This is also completely natural. But for other values of p, we obtain answers that would be difficult to foresee. Here is a problem from a textbook in the theory of probability.

As a result of many years of observation at a certain location, it was established that the probability that rain falls on the Fourth of July is equal to 4/17. Find the most probable number of rainy Fourths of July in the next 50 years.

Here, $n = 50$, $p = 4/17$, and $m = (n + 1)p - 1 = 11$. This means that the most probable numbers of rainy days are the equally probable numbers 11 and 12.

The values of the probabilities $C_n^k p^k (1 - p)^{n-k}$, $k = 0, 1, \ldots, n$, that we found possess remarkable properties. In Fig. 20, taken from a course in the theory of probability, they are represented for the case where $p = 1/3$ and $n = 4, 9, 16, 36$, and 100.

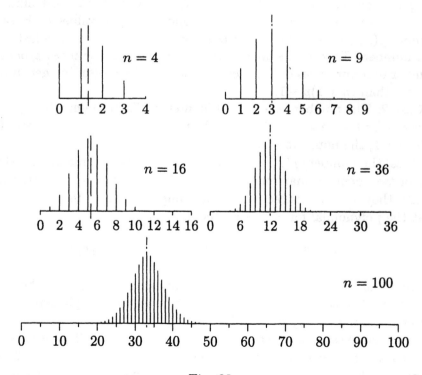

Fig. 20

You can see that as n increases, they are distributed not every which way but approach some smooth curve. To see this, we must shift the picture each time so that the largest number is located on the y axis, moreover reduce the distance between the points on the x axis over which is marked the number (such a change of scale for the x axis has been performed in Fig. 20), and finally proportionally reduce the marked numbers such that the largest numbers stay at more or less the same height. After this, it seems that our points more and more closely approach a single curve as x increases, the graph of the function

$$y = \frac{1}{\sqrt{2\pi}} c^{x^2},$$

where π is the usual ratio of the circumference of a circle to its diameter and c is equal to $1/\sqrt{e}$ (for those who know that e is the base of the natural logarithms).

This assertion, called the *Laplace theorem*, is strictly speaking a complicated expression of a property of binomial coefficients. But to prove it requires explicating the words "more and more closely approaches," that is, the concept of passing to the limit, and we will not do this.

Problems:

1. In an arbitrary probability scheme $(M; p)$, there are k events given: $M_1 \subset M, \ldots, M_k \subset M$. Express the probability

$$p(M_1 \cup M_2 \cup \cdots \cup M_k)$$

   of the event $M_1 \cup M_2 \cup \cdots \cup M_k$ in terms of the probabilities $p(M_{i_1} \cap \cdots \cap M_{i_r})$ of the events $M_{i_1} \cap \cdots \cap M_{i_r}$.
2. Prove that if the polynomial $f(x)$ is reciprocal and unimodal, then the polynomial $f(x)(1 + x)$ has the same properties.
3. Prove that if the polynomial $f(x)$ is reciprocal and unimodal, then the polynomial $f(x)(1 + x + x^2 + x^3 + x^4 + x^5)$ has the same properties. Deduce from this that the polynomial $(1 + x + x^2 + x^3 + x^4 + x^5)^n$ is unimodal.
4. Convince yourself that in the problem about m particles and n instruments with $n = m$, the answer is $n!/n^n$. What relation between binomial coefficients follows from comparison with formula (31)?
5. There are n identical balls in an urn of which m are white and $n - m$ are black. From the urn, r balls are taken at random. What is the probability that k white and $r - k$ black balls are taken? *Hint:* The phrase "at random" means that the probabilities of taking different selections of r balls are considered equal.
6. Prove that if the probability p in the Bernoulli scheme is an irrational number, then there exists exactly one most probable number of the occurrences of the event a.
7. The ratio of the most probable number of occurrences of the event a to the number n in the Bernoulli scheme is called the most probable proportion. Prove that the most probable proportion becomes ever closer to the probability p of the event a as the number n increases without limit.

Supplement: The Chebyshev Inequality

We discuss one question concerning the Bernoulli scheme that was examined at the end of Sec. 10. As was said there, the Bernoulli scheme arises in practice in situations where there are multiple repetitions of an experiment that has only two outcomes. For example, we have an asymmetric coin. The question is does it come up heads or tails when tossed. For this, a large series of experiments is conducted, for example, 1000. If heads came up k times, then the probability of heads is considered to be $p = k/1000$. After this, we can apply our definition of the Bernoulli scheme $(I^n; p)$ and use it to calculate various other probabilities, for example, given by formula (35). But is our abstraction satisfying? Does it sufficiently accurately reflect the reality that we dealt with in the first place: a long series of independent experiments?

In our abstraction, the Bernoulli scheme, we cannot ask how many times the event a actually occurred in the scheme I^n. We have only the language of probability at our disposal, and it therefore only makes sense to ask about some probabilities. But the concept of probability is connected with reality based on confidence that events with a very small probability practically do not occur. In other words, if the probability of a specific event is sufficiently small, then we can act practically as if we knew it would not happen. Certainly, the sense of the words "sufficiently small" should be made precise in each concrete situation. From this standpoint, we can specify a certain number $\varepsilon > 0$ and consider the event A_ε consisting of the occurrence of the event a in our Bernoulli scheme $(I^n; p)$ a number of times k such that $|k/n - p| > \varepsilon$. In other words, the occurrence of the event A_ε means that the "frequency" of the occurrence of the event a differs from its supposed probability p by more than ε. It is naturally hoped that for any fixed ε, the probability $p(A_\varepsilon)$ of the event A_ε will become arbitrarily small as n increases without limit. This would mean that the deviation of the "frequency" k/n from the probability p can be neglected for large n.

Jakob Bernoulli came to this problem at the beginning of the 18th century and understood that investigating the probability $p(A_\varepsilon)$ is a purely mathematical question connected with the properties of binomial coefficients. He proved that the probability $p(A_\varepsilon)$ indeed becomes arbitrarily small as n increases. This assertion is also called the *law of large numbers*. In the 19th century, Chebyshev not only proved the very simple qualitative assertion of Bernoulli but also found a simple explicit inequality that is satisfied by the probability $p(A_\varepsilon)$ we are interested in. We set forth his theorem here. (In this place in our book, we first meet a contribution from a Russian mathematician. P. L. Chebyshev lived from 1821 to 1894 and was the founder of the St. Petersburg school of mathematics.)

We write the expression that we want to investigate in the form of an algebraic formula. In Sec. 10, we examined the Bernoulli scheme $(I^n; p)$ and the event A_k that the event a for which $p(a) = p$ occurs k times in a series of experiments, and we calculated its probability (see formula (35)):

$$p(A_k) = C_n^k p^k q^{n-k}. \tag{36}$$

We are now given a number ε, and we are interested in the event A_ε, which is that one of the events A_k for which the index k satisfies the inequality $|k/n - p| > \varepsilon$ occurs. We must find the probability $p(A_\varepsilon)$ of the event A_ε. We recall that an event (in particular, A_k or A_ε) is a subset of the set I^n. It is obvious that the subsets A_k with different indices k do not intersect, and A_ε is the union of the subsets A_k for all k such that $|k/n - p| > \varepsilon$. Therefore, the probability $p(A_\varepsilon)$ is equal to the sum of the probabilities $p(A_k)$ with such indices k. Because the quantity $p(A_k)$ is given by formula (36), we obtain an explicit, albeit somewhat complicated, expression for the probability $p(A_\varepsilon)$ of the event A_ε. It is more convenient to write it if we rewrite the condition defining the indices k we are interested in, $|k/n - p| > \varepsilon$, in the equivalent form

$$|k - np| > \varepsilon n. \tag{37}$$

We thus come to the sum S_ε of the expressions $C_n^k p^k q^{n-k}$ for all k,

$$1 \leq k \leq n, \tag{38}$$

satisfying condition (37). We see that the probability $p(A_\varepsilon)$ of the event A_ε is equal to S_ε.

We can now formulate the Chebyshev theorem.

Theorem 23 (Chebyshev Theorem). *The probability $p(A_\varepsilon)$ of the event A_ε, which is that the number k of occurrences of the event a in the Bernoulli scheme $(I^n; p)$ satisfies $|k/n - p| > \varepsilon$, satisfies the inequality*

$$p(A_\varepsilon) \leq \frac{pq}{\varepsilon^2 n}. \tag{39}$$

Inequality (39) is sometimes written in the form

$$p\left(\left|\frac{k}{n} - p\right| > \varepsilon\right) < \frac{pq}{\varepsilon^2 n}.$$

Here, the p in the left-hand side denotes the probability of an event, and the p in the right-hand side is the number in the definition of the Bernoulli scheme.

Obviously, for a given p (then $q = 1 - p$) and ε, the right-hand side of inequality (39) decreases as n increases, which we wanted to establish. This qualitative result is called the *Bernoulli theorem*.

As we saw, the probability $p(A_\varepsilon)$ is equal to the sum S_ε determined by condition (38). This means that inequality (39) is equivalent to

$$S_\varepsilon \leq \frac{pq}{\varepsilon^2 n}.$$

The proof of the Chebyshev theorem is based on the explicit calculation of certain sums, which we set apart in the form of a lemma.

Lemma 7. *The probabilities $p(A_k)$ defined by relation (36) satisfy the equalities*

$$p(A_0) + p(A_1) + p(A_2) + \cdots + p(A_n) = 1, \tag{40}$$
$$p(A_1) + 2p(A_2) + 3p(A_3) + \cdots + np(A_n) = np, \tag{41}$$
$$p(A_1) + 2^2 p(A_2) + 3^2 p(A_3) + \cdots + n^2 p(A_n) = n^2 p^2 - npq. \tag{42}$$

Proof. In general, for any $r \geq 0$, we let σ_r denote the sum of the terms $k^r p(A_k)$ for $k = 0, 1, 2, \ldots, n$ and let $f_r(t)$ denote the polynomial equal to the sum of the terms $k^r C_n^k t^k$ for $k = 0, 1, 2, \ldots, n$. We recall that $p(A_k) = C_n^k p^k q^{n-k}$ according to formula (36). Bringing the factor q^n outside the parentheses in the sum σ_r and setting $\alpha = p/q$, we obtain

$$\sigma_r = q^n f_r(p/q) = q^n f_r(\alpha). \tag{43}$$

It is immediately evident from the binomial formula that

$$f_0(t) = (t + 1)^n. \tag{44}$$

(We note that the polynomial $f_0(t)$ has a free term equal to 1 and for $r > 0$, the free term of the polynomial $f_r(t)$ is equal to $0^r C_n^0$, that is, is equal to 0.) There is a simple way to consecutively calculate the polynomials $f_r(t)$ using the properties of derivatives. We now apply it for $r = 0$, 1, and 2.

We consider the derivative $f_r'(t)$ of the polynomial $f_r(t)$. By definition

$$f_0(t) = 1 + C_n^1 t + C_n^2 t^2 + \cdots + t^n,$$
$$f_r(t) = 1^r C_n^1 t + 2^r C_n^2 t^2 + \cdots + n^r C_n^n t^n, \quad r > 0,$$

and according to formula (16) in Chap. 2, we have

$$f'_r(t) = 1^{r+1}C_n^1 + 2^{r+1}C_n^2 t + \cdots + n^{r+1}C_n^n t^{n-1}.$$

Multiplying the right-hand side of this equality by t, we obtain exactly the polynomial $f_{r+1}(t)$. Therefore,

$$f_{r+1}(t) = f'_r(t)t. \tag{45}$$

This relation allows calculating the infinite sequence of polynomials $f_r(t)$ for $r = 1, 2, \ldots$ ($f_0(t)$ is given by formula (44)). We apply it for $r = 0$ and 2. Applying differentiation rule (20) in Chap. 2 to formula (44), we obtain

$$f'_0(t) = n(t+1)^{n-1} \tag{46}$$

because $(t+1)' = 1$ in accordance with formula (16) in Chap. 2. Applying formula (45), we obtain

$$f_1(t) = n(t+1)^{n-1}t. \tag{47}$$

We consider the polynomial $f_1(t)$ as the product of the two factors $(t+1)^{n-1}$ and nt and calculate the derivative $f'_1(t)$ using differentiation rule (18) in Chap. 2. We obtain

$$f'_1(t) = \left((t+1)^{n-1}\right)'nt + (t+1)^{n-1}(nt)'. \tag{48}$$

But $\left((t+1)^{n-1}\right)' = (n-1)(t+1)^{n-2}$, and $(nt)' = n$. Because $f_2(t) = f'_1(t)t$ (by formula (45)), we obtain

$$f_2(t) = n(n-1)(t+1)^{n-2}t^2 + n(t+1)^{n-1}t. \tag{49}$$

To obtain expressions for σ_0, σ_1, and σ_2, we must substitute $t = \alpha = p/q$ in accordance with formula (43). We note that $\alpha + 1 = p/q + 1 = (p+q)/q = 1/q$ because $p + q = 1$. We obtain

$$\sigma_0 = q^n(\alpha+1)^n = q^n(1/q)^n = 1,$$
$$\sigma_1 = q^n f_1(\alpha) = q^n n(\alpha+1)^{n-1}\alpha = q^n n(1/q)^{n-1} \cdot p/q = np,$$
$$\sigma_2 = q^n f_2(\alpha) = q^n\left(n(n-1)(\alpha+1)^{n-2}\alpha^2 + n(\alpha+1)^{n-1}\alpha\right)$$
$$= n(n-1)p^2 + np.$$

The last formula can be rewritten in the form $\sigma_2 = n^2p^2 - np^2 + np = n^2p^2 + np(1-p) = n^2p^2 + npq$ because $1 - p = q$. We have obtained formulas (40), (41), and (42). $\qquad\square$

We now turn to the proof of the Chebyshev theorem. Chebyshev's trick consists in writing inequality (37), which characterizes the indices we need, as

$$\left|\frac{k - np}{\varepsilon n}\right| > 1,$$

that is,

$$\left(\frac{k - np}{\varepsilon n}\right)^2 > 1,$$

and then multiplying each term $p(A_k)$ in the sum S_ε by the number $((k - np)/(\varepsilon n))^2$, which being greater than unity only increases the sum. After that, he examines the *complete sum* $\overline{S_\varepsilon}$ of *all* terms $((k - np)/(\varepsilon n))^2 p(A_k)$ for $k = 0, 1, \ldots, n$ and does not limit himself to only the indices k that satisfy condition (37). Clearly, the sum $\overline{S_\varepsilon}$ differs from the previously obtained sum by additional positive terms and can therefore only be greater. As a result, we see that $S_\varepsilon \le \overline{S_\varepsilon}$. After this, a very elementary transformation (based on the lemma) shows that the sum $\overline{S_\varepsilon}$ can be calculated exactly, which then proves the needed inequality for the sum S_ε.

Therefore, we must find the sum $\overline{S_\varepsilon}$ of all terms $((k - np)/(\varepsilon n))^2 p(A_k)$ for $k = 0, 1, \ldots, n$. The common divisor $(\varepsilon n)^2$ can be taken outside the summation sign, and the expression $(k - np)^2$ in each term can be expanded: $(k - np)^2 = k^2 - 2npk + n^2 p^2$. Each term in the sum $\overline{S_\varepsilon}$ (after taking out the common divisor $(\varepsilon n)^2$) yields three terms. The sum of all first terms is the sum σ_2 in the left-hand side of (42); the sum of the second terms, after factoring out $-2np$, coincides with the sum σ_1, defined by (41); and the sum of the third terms, after factoring out $n^2 p^2$, finally yields σ_0, defined by (40). Combining all the obtained equalities together, we find the expression for the sum $\overline{S_\varepsilon}$:

$$\overline{S_\varepsilon} = \frac{1}{\varepsilon^2 n^2}(\sigma_2 - 2np\sigma_1 + n^2 p^2 \sigma_0).$$

Substituting the expressions for σ_2, σ_1, and σ_0 obtained in the lemma, we have

$$\overline{S_\varepsilon} = \frac{1}{\varepsilon^2 n^2}(n^2 p^2 + npq - 2n^2 p^2 + n^2 p^2) = \frac{pq}{\varepsilon^2 n}. \tag{50}$$

As we saw, $S_\varepsilon \le \overline{S_\varepsilon}$, and therefore $S_\varepsilon \le (pq)/(\varepsilon^2 n)$. The Chebyshev theorem is thus proved. □

We direct our attention to the trick at the heart of the proof. The terms of the sum S_ε, whose value we must estimate, have a sufficiently simple form. The difficulty in the estimate consists in the fact that the terms to be included in the sum are chosen by a rather whimsical principle (their indices must satisfy condition (37)). It first comes to mind to ignore this condition and take the sum of *all* such terms. Such a sum can be easily calculated: according to the lemma, it is equal to 1. It is too large, and does not yield the inequality we want. Chebyshev's find consists in introducing the additional multiplier $\left((k - np)/(\varepsilon n)\right)^2$ and only *after* that examining the sum of all terms, rejecting the "life-spoiling" limitation (37). With this, the terms that enter S_ε even increase, but simultaneously the terms not entering S_ε decrease so much that the total sum $\overline{S_\varepsilon}$ is sufficiently small (namely, $\left((k - np)/(\varepsilon n)\right)^2 < 1$ for the terms not entering the sum S_ε).

We here encounter a manifestation that is often met in mathematics. Namely, an important, interesting inequality usually appears as a consequence of some identity from which it is obtained by an obvious estimate. Such an obvious estimate in our case is the inequality $S_\varepsilon \leq \overline{S_\varepsilon}$, and the identity is relation (50), which gives the sum $\overline{S_\varepsilon}$ in explicit form. The greatest portion of inequalities playing a principal role in mathematics are proved in just this way. But sometimes they are proved differently— then this may be a hint that some as yet unknown identity relation is hiding behind them.

We return once more to the formulation of the Chebyshev theorem. As we just now established, there the event is investigated in which the event a in the Bernoulli scheme I^n occurs k times, where either $k > np + n\varepsilon$ or $k < np - n\varepsilon$. In other words, the event in which the event a in the Bernoulli scheme I^n now occurs k times with $np - n\varepsilon \leq k \leq np + n\varepsilon$. The theorem asserts that the first event has a small probability (for large n) that does not exceed $pq/(\varepsilon^2 n)$, and this means the second has a large probability not less than $1 - pq/(\varepsilon^2 n)$. As an example, we consider a series of a large number of repetitions of the same experiment under constant conditions. We assume that one experiment can have only one of two outcomes a or b with the probability of the outcome a being p. This situation (if the number of experiments is equal to n) is described by the Bernoulli scheme $(I^n; p)$, as we saw. The experiments might consist of the examination of a large collection of objects (animals, technical parts, etc.) for the presence of some characteristic, and it is known that a pth portion of the collection has this characteristic. The set I^n describes the possible results of our examination of a set of n objects for the characteristic we are interested in. According to the Chebyshev theorem, the number of occurrences of the event a in a series of n experiments falls between $np - n\varepsilon$ and $np + n\varepsilon$ with a probability greater than $1 - p(1 - p)/(\varepsilon^2 n)$. Here, ε can be any number, which we can chose arbitrarily. For example, let $p = 3/4$. Choosing $\varepsilon = 1/100$, we see that the number k of occurrences of the event a in a series of n experiments satisfies the inequality $3n/4 - n/100 \leq k \leq 3n/4 + n/100$ with a probability not less than

$$1 - \frac{\frac{3}{4} \cdot \frac{1}{4}}{\left(\frac{1}{100}\right)^2 n}.$$

Because $3/4^2 < 2/10$, this probability is not less than

$$1 - \frac{\frac{2}{10}}{\left(\frac{1}{100}\right)^2 n} = 1 - \frac{2000}{n}.$$

Setting $n = 200\,000$, we make this probability not less than 0.99. With such a large probability, the number of outcomes a for $200\,000$ experiments will be included between $148\,000$ and $152\,000$ (because $3n/4 = 150\,000$, $n/100 = 2\,000$, $np - n\varepsilon = 148\,000$, and $np + n\varepsilon = 152\,000$).

Conversely, we can use the Chebyshev theorem to estimate the number of experiments needed to determine the probability p with sufficient accuracy. We suppose that we want to determine it to the accuracy of $1/10$ and that it should equal the found value with a probability not less than 0.99. According to the Chebyshev theorem, we must set $\varepsilon = 1/10$ and satisfy the inequality

$$\frac{pq}{\left(\frac{1}{10}\right)^2 n} < 0.01.$$

We note that $q = 1 - p$, and for any p such that $0 \leq p \leq 1$, we have $pq = p(1 - p) \leq 1/4$. This follows because the geometric mean of the values p and q does not exceed their arithmetic mean, which is equal to $1/2$. Therefore, it is sufficient for us that n satisfies the inequality

$$\frac{\frac{1}{4}}{\left(\frac{1}{10}\right)^2 n} < 0.01,$$

whence $n > 2500$.

Problems:

1. In a collection of certain objects, an average of 5% have definite characteristic. Prove that among $200\,000$ objects, the number that have this characteristic is included between $9\,000$ and $11\,000$ with a probability not less than 0.99.
2. In the situation given in Problem 1, the portion of objects with the given characteristic is unknown. What is the probability that by examining 100 objects we determine this portion to an accuracy of 0.1?
3. For any natural number $r \leq n$, find the sum of all terms $k(k-1)\cdots(k-r+1)p(A_k)$ for $k = 1, \ldots, n$, where $p(A_k)$ is calculated by formula (36).
4. For $r \leq 4$, calculate the sums σ_r consisting of the terms $k^r p(A_k)$ for all $k = 1, 2, 3, 4$. Perform the calculation two ways and verify that the results coincide: (1) repeating the reasoning in the proof of the lemma and (2) expressing the sum σ_r in terms of the sums calculated in Problem 3 for $r = 1, 2, 3, 4$.
5. Attempt to perfect inequality (39) in the Chebyshev theorem, applying the multiplier $\left((k-np)/(n\varepsilon)\right)^4$ instead of $\left((k-np)/(n\varepsilon)\right)^2$. The perfection should appear in the denominator in the right-hand side of the inequality containing n^2 instead of n.

4

Prime Numbers

Topic: Numbers

11.   The Number of Prime Numbers is Infinite

In this chapter, we return to a question examined in Chap. 1. It was shown there that a natural number has a unique decomposition into prime factors. From the standpoint of the operation of multiplying, therefore, prime numbers are the simplest elements from which we can obtain all natural numbers, similar to how we obtain them all from the number 1 using the operation of adding. From this standpoint, the interest in the collection of prime numbers is understandable. Four prime numbers are found in the first decade of natural numbers: 2, 3, 5, 7. Further, we can find prime numbers, in turn dividing each number by all previously found smaller primes to determine if it is prime. We thus find 25 prime numbers in the first century:

$$2, 3, 5, 7, \quad 11, 13, 17, 19, \quad 23, 29, \quad 31, 37, \quad 41, 43, 47,$$
$$53, 59, \quad 61, 67, \quad 71, 73, 79, \quad 83, 89, \quad 97.$$

How far does this sequence continue?

This question already arose in antiquity. We find the answer to this question in Euclid. It is formulated in Theorem 24.

Theorem 24. *The number of prime numbers is infinite.*

We present several proofs of this theorem. The *first proof* is the one contained in Euclid's *Elements*. Suppose we have found n primes in all: $p_1, p_2, \ldots, p_n$. We consider the number $N = p_1 p_2 \cdots p_n + 1$. As we saw in Sec. 2, each number has at least one prime divisor. In particular, N

has a prime divisor. But it cannot be one of the numbers $p_1, \ldots, p_n$.
Indeed, suppose it were p_i. Then $N - p_1 \cdots p_n$ must be divisible by p_i,
and because $N - p_1 \cdots p_n = 1$, this is impossible. Therefore, the prime
divisor of N is different from p_i, $i = 1, \ldots, n$. This means that for each
n prime numbers, there follows one more prime number. This proves
the theorem.   □

Second proof. It was proved in Sec. 9 (see formula (25)) that the
number of numbers less than a given number N and relatively prime to
it is given by the formula

$$N \left(1 - \frac{1}{p_1}\right) \left(1 - \frac{1}{p_2}\right) \cdots \left(1 - \frac{1}{p_n}\right), \qquad (1)$$

where $p_1, \ldots, p_n$ are all the prime divisors of the number N.

Again suppose we have found n prime numbers $p_1, \ldots, p_n$. We set
$N = p_1 \cdots p_n$. Substituting this expression in formula (1), we obtain the
simple factor $p_i - 1$ from each factor $p_i(1 - 1/p_i)$, and we thus obtain
the expression $(p_1 - 1)(p_2 - 1) \cdots (p_n - 1)$ for the whole of formula (1).
Because we know that there exists a prime number greater than 2 (for
example, 3), this expression must be a number *greater than* 1. Therefore,
there exists a number a less than N and relatively prime to it that is
different from 1. But a has at least one prime divisor that cannot be
contained among the numbers $p_1, \ldots, p_n$, because a is relatively prime
to N. We have obtained one more prime number, and this proves the
theorem.   □

The endless sequence of prime numbers is rather sparsely distributed
among the natural numbers. For example, there is a "gap" in it however
large you want, that is, we can find any given number of consecutive
numbers (sufficiently far out) that are not prime. For example, the n
numbers

$$(n + 1)! + 2, \quad (n + 1)! + 3, \quad \ldots, \quad (n + 1)! + n + 1$$

are obviously not prime: the first is divisible by 2, the second is divisible
by 3, and the last is divisible by $n + 1$. For some time, people tried to
find a formula expressing prime numbers. Euler found the remarkable
polynomial $x^2 + x + 41$, which has a prime value for 40 values of x from
0 to 39. It is obvious, however, that for $x = 40$, it takes the nonprime
value 41^2. It is easy to verify that there cannot exist a polynomial $f(x)$
that would yield prime values for all integer values $x = 0, 1, 2, \ldots$ (not
to speak of it yielding *all* prime numbers). We demonstrate this with
the example of a second-degree polynomial $ax^2 + bx + c$ with the integer
coefficients a, b, c.

We suppose that the value c, which the polynomial yields for $x = 0$, is prime. Then for an arbitrary positive integer k, we take $x = kc$ and find that the polynomial yields the value $ak^2c^2 + bkc + c$, which is obviously divisible by c. This value is either not prime or is exactly c. You can easily verify that for given a and b, there is at least one positive integer k for which $ak^2c^2 + bkc + c$ is equal to c. Therefore, all such values except possibly two are not prime.

Furthermore, there does not exist a polynomial $f(x)$ of arbitrary degree with integer coefficients such that all its values for integer x are prime numbers, *beginning from some boundary*. Indeed, suppose that the values of the polynomial $f(x) = a_0 + a_1x + \cdots + a_nx^n$ are prime for all integers $x \geq m$, where m is some natural number. We set $x = y + m$, $f(y + m) = g(y)$. The polynomial $g(y) = a_0 + a_1(y + m) + \cdots + a_n(y + m)^n = b_o + b_1y + \cdots + b_ny^n$ is obtained by opening parentheses and combining like terms. Therefore, its coefficients b_i are again integers, but it already yields prime values for all $y \geq 0$. In particular, $g(0) = b_0 = p$ is a prime number. Then for any integer k, the value $g(kp) = p + b_1kp + \cdots + b_n(kp)^n$ is divisible by p. They can coincide with p only if $p + b_1kp + \cdots + b_n(kp)^n = p$, that is,

$$b_1 + b_2kp + \cdots + b_n(kp)^{n-1} = 0.$$

This is a polynomial of degree $n - 1$ in k. According to Theorem 14, it has at most $n - 1$ roots. For all other values of k, the number $g(kp)$ is divisible by p and is different from p, that is, it is not prime.

It can be proved that for any number k of unknowns, there cannot exist a polynomial in k unknowns with integer coefficients such that all its values for all natural values of the unknowns are prime numbers. Nevertheless, it turns out that there exists a 25th-degree polynomial in 26 unknowns that has the following property: if we select the values it yields for nonnegative integer values of the unknowns such that the values themselves are positive, then their set coincides with the set of prime numbers. Because 26 is equal to the number of letters in the English alphabet, the unknowns can be denoted by those letters: $a, b, c, \ldots, x, y, z$. Then the polynomial has the form

$$F(a,b,c,d,e,f,g,h,i,j,k,l,m,n,o,p,q,r,s,t,u,v,w,x,y,z) =$$
$$= (k+2)\Big\{1 - \big[wz + h + j - q\big]^2 - \big[(gk + 2g + k + 1)(h + j) + h\big]^2$$
$$- \big[2n + p + q + z - e\big]^2 - \big[16(k+1)^2(k+2)(n+1)^2 + 1 - f^2\big]^2$$
$$- \big[e^3(t+2)(a+1)^2 + 1 - o^2\big]^2 - \big[(a^2 - 1)y^2 + 1 - x^2\big]^2$$
$$- \big[16r^2y^4(a^2 - 1) + 1 - u^2\big]^2$$
$$- \big[(a + u^2(u^2 - a^2) - 1)(n + 4dy)^2 + 1 - (x - cu)^2\big]^2$$
$$- \big[n + l + v - y\big]^2 - \big[(a-1)l^2 + 1 - m^2\big]^2 - \big[ai + k + 1 - l - i\big]^2$$
$$- \big[p + l(a - n - 1) + b(2an + 2a - n^2 - 2n - 2) - m^2\big]$$
$$- \big[q + y(a - p - 1) + s(2ap + 2a - p^2 - 2p - 2) - x\big]^2$$
$$- \big[z + pl(a - p) + t(2ap - p^2 - 1) - pm\big]^2\Big\}.$$

This polynomial is written here only to make the reader's eyes pop. The number of variables in it is very large. It can be proved that it also yields negative values $-m$, where m is not prime. Therefore, it does not give us a representation of the sequence of prime numbers.

Long efforts inclined the majority of mathematicians to the conviction that more or less simple formulas describing the sequence of prime numbers do not exist. "Explicit formulas" describing prime numbers exist, but they use objects about which we know less than about prime numbers. The mathematicians' attention therefore focused on characteristics of prime numbers "collectively" and not "individually." We clarify this posing of the question in the next section.

Problems:

1. Prove that number of prime numbers of the form $3s + 2$ is infinite.
2. Prove that number of prime numbers of the form $4s + 3$ is infinite.
3. Prove that any two numbers $2^{2^n} + 1$ and $2^{2^m} + 1$, where $n \neq m$, are relatively prime. From this, once more deduce the infiniteness of the number of prime numbers. *Hint:* Suppose that p is a common divisor of two such numbers and find the remainder from dividing 2^{2^n} and 2^{2^m} by p.
4. Let $f(x)$ be a polynomial with integer coefficients. Prove that among the prime divisors of its values $f(1), f(2), \ldots$, there exist an infinite number of different ones. (If the problem is not solved quickly, solve it first for first-degree polynomials $f(x)$, then second-degree.)
5. Let p_n denote the nth prime in ascending order. Prove that $p_{n+1} < p_n^n + 1$.
6. In the notation in Problem 5, prove that $p_n < 2^{2^n}$. Deduce the close inequality $p_{n+1} \leq 2^{2^n} + 1$ from the result of Problem 3.
7. In the notation in Problem 5, prove that $p_{n+1} < p_1 p_2 \cdots p_n$.

12. Euler's Proof That the Number of Prime Numbers is Infinite

We give yet another proof, belonging to Euler, of the infiniteness of the number of prime numbers, which elucidates certain general properties of this sequence.

We begin with the "prehistory," that is, with certain simple facts that were known before Euler began to study the question of prime numbers. The matter concerns the magnitude of the sums

$$ 1, \quad 1+\frac{1}{2}, \quad 1+\frac{1}{2}+\frac{1}{3}, \quad \ldots, \quad 1+\frac{1}{2}+\cdots+\frac{1}{n}, \quad \ldots. $$

In the notation in Sec. 6, these are the sums $(Sa)_n$, where a is the sequence of inverse natural numbers $1, 1/2, 1/3, \ldots$. Because the sums of the mth power of the natural numbers from 1 to n is denoted by $S_m(n)$ in our notation (see formula (29) in Chap. 2), our sums here are naturally denoted by $S_{-1}(n)$.

We come upon a concept here that we meet often in what follows. We therefore discuss it in more detail. It generally relates to properties of an *infinite* sequence of positive numbers $s_1, s_2, \ldots, s_n, \ldots$ (for us, it arose as the sequence of sums of another sequence, but this is not important now). One type of sequence is called an *bounded* sequence. This means that there exists a single number C for the whole sequence such that $s_n < C$ for all $n = 1, 2, 3, \ldots$. If the sequence does not have this property, then it is said to be *unbounded*. This means that no number C has that property, that is, for any number C, an index n can be found such that $s_n \geq C$. Finally, it can happen that for any number C, an index n can be found such that *all* $s_m \geq C$ for all $m = n, n+1, \ldots$. In other words, the number s_n becomes however large we want for sufficiently large n. In this case, the sequence is said to *increase without limit*. For example, the sequence $1, 1, 1, 2, 1, 3, \ldots$, in which the odd positions contain 1 and the even positions contain the sequence of natural numbers, is unbounded but does not increase without limit, because we can still find the number 1 no matter how far out we go.

If a sequence $a = a_1, a_2, \ldots, a_n, \ldots$ of positive numbers is given and $s = Sa$, then $s_{n+1} > s_n$ (because $s_{n+1} = s_n + a_{n+1}$, $a_{n+1} > 0$), and, more generally, $s_m > s_n$ for all $m > n$. Therefore, such a sequence increases without limit if it is not bounded. For example, if all $a_i = 1$, then $s_n = n$, and the sequence $s_1, s_2, \ldots$ is unbounded. But it might be bounded in other cases. An example is illustrated in Fig. 21, where we first divide the segment from 0 to 1 in half and set $a_1 = 1/2$, then divide

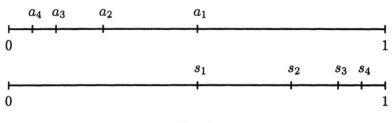

Fig. 21

the segment from 0 to $1/2$ in half and set $a_2 = 1/4$, and so on. Thus, $a_n = 1/2^n$. The result of adding these numbers is shown in Fig. 21: we can see that their sums s_n always stay inside our segment because s_2 is the middle of the segment from s_1 to 1, s_3 is the middle of the segment from s_2 to 1, and so on. That is, $s_n < 1$. This is easily verified by calculating. If $a_n = 1/2^n$, then

$$(Sa)_n = \frac{1}{2} + \frac{1}{4} + \cdots + \frac{1}{2^n} = \frac{1}{2}\left(1 + \frac{1}{2} + \cdots + \frac{1}{2^n}\right),$$

and by formula (12) in Chap. 1,

$$(Sa)_n = \frac{1}{2}\frac{1/2^n - 1}{1/2 - 1} = 1 - \frac{1}{2^n}.$$

It follows that $(Sa)_n < 1$ for any n.

We show that the *first* case holds for the sequence $1, 1/2, 1/3, \ldots$. Although the terms of the sequence decrease, they do not decrease sufficiently rapidly, and their sum (i.e., $S_{-1}(n)$) increases without limit.

Lemma 8. *For sufficiently large n, the sum $S_{-1}(n)$ is greater than any fixed number given in advance.*

Let the number k be given. We show that for some n (and this means for all indices following it also), $S_{-1}(n) > k$. We take n such that $n = 2^m$ for some m. We subdivide the sum

$$S_{-1}(n) = 1 + \left(\frac{1}{2}\right) + \left(\frac{1}{3} + \frac{1}{4}\right) + \left(\frac{1}{5} + \frac{1}{6} + \frac{1}{7} + \frac{1}{8}\right)$$
$$+ \cdots + \left(\frac{1}{2^{m-1} + 1} + \cdots + \frac{1}{2^m}\right)$$

into subtotals enclosed in parentheses as shown in the formula. Each set of parentheses encloses a sum of the general form

$$\frac{1}{2^{k-1} + 1} + \frac{1}{2^{k-1} + 2} + \cdots + \frac{1}{2^k},$$

and there are m sets of parentheses. Within each set of parentheses, we replace each term with the least term, that is, the last. Because the number of terms in each set of parentheses is equal to $2^k - 2^{k-1} = 2^{k-1}$, we find that the sum in the kth set of parentheses is greater than $2^{k-1}/2^k = 1/2$. As a result, we obtain $S_{-1}(n) > 1+m/2$. This inequality holds for any n if $n = 2^m$. It remains for us to set $1 + m/2 = k$, that is, $m = 2k - 1$. Then we take $n = 2^{2k-1}$; it follows that $S_{-1}(n) > k$. $\square$

We now turn to Euler's proof. His idea is connected with the method for calculating the sums of powers of divisors of a natural number described in Sec. 3 (see formula (13) in Chap. 1). The sum of the kth powers of all divisors (including 1 and n) of the natural number n is denoted by $\sigma_k(n)$. According to formula (13) in Chap. 1,

$$\sigma_k(n) = \frac{p_1^{k(\alpha_1+1)-1}}{p_1^k - 1} \frac{p_2^{k(\alpha_2+1)-1}}{p_2^k - 1} \cdots \frac{p_r^{k(\alpha_r+1)-1}}{p_r^k - 1} \qquad (2)$$

for the number n with the canonical decomposition $n = p_1^{\alpha_1} \cdots p_r^{\alpha_r}$. Formula (2) was already known from the time of antiquity, but it was tacitly assumed that k is a positive number in it. It finally fell into Euler's circle of interests, and he posed the question "what if k is an integer, but negative." The answer, of course, is that there is no difference; the deduction of formula (2) is perfectly formal and works equally for negative just as for positive numbers k. In particular, it holds for $k = -1$. Retaining the previous notation, we write the sum of the (-1)th powers (i.e., the inverse values) of the divisors of a given number n as $\sigma_{-1}(n)$. Formula (2) then yields

$$\sigma_{-1}(n) = \frac{1 - 1/p_1^{\alpha_1+1}}{1 - 1/p_1} \cdots \frac{1 - 1/p_r^{\alpha_r+1}}{1 - 1/p_r}$$

(we change the order of the terms in the numerator and denominator of each fraction). Hence (because all expressions in the numerators are less than 1),

$$\sigma_{-1}(n) < \frac{1}{(1 - 1/p_1)(1 - 1/p_2) \cdots (1 - 1/p_r)}. \qquad (3)$$

We now replace n with $n!$ in this formula ($p_1, \ldots, p_r$ are now the prime divisors of $n!$, that is, simply speaking, all the prime numbers not exceeding n). Among the divisors of $n!$, we must certainly have $1, 2, \ldots, n$. We must therefore find the terms $1, 1/2, 1/3, \ldots, 1/n$ included in the sum $\sigma_{-1}(n!)$, and the sum of these terms is equal to $S_{-1}(n)$. According to Lemma 8, for sufficiently large n, the sum $S_{-1}(n)$ is already greater than any fixed number k given in advance. Because the other terms in $\sigma_{-1}(n!)$ are also positive, this assertion is still more applicable to the full sum. If the number of prime numbers were finite and if $p_1,\ldots,p_r$ were the complete list of them, then we would have

$$\frac{1}{(1-1/p_1)(1-1/p_2)\cdots(1-1/p_r)} > k,$$

where k is any number. This, of course, is a contradiction.   □

In giving this proof, it is valuable that the proposition that the number of prime numbers is finite not only leads to a contradiction but also yields a certain quantitative characteristic of the sequence of prime numbers. Namely, rephrasing the result obtained, we can now state that if $p_1,p_2,\ldots,p_n,\ldots$ is the infinite sequence of all prime numbers, then the expression

$$\frac{1}{(1-1/p_1)(1-1/p_2)\cdots(1-1/p_r)}$$

for a sufficiently large n becomes greater than any number given in advance. Finally, this is equivalent to the fact that the denominator of this fraction for sufficiently large n becomes *less than* any number given in advance. We have proved Theorem 25.

Theorem 25. *If $p_1,p_2,\ldots,p_n,\ldots$ is the sequence of all prime numbers, then the product $(1-1/p_1)(1-1/p_2)\cdots(1-1/p_n)$ for sufficiently large n becomes less than any positive number given in advance.*

This is a first approximation to our goal. We now try to give the result obtained a more customary form.

Theorem 26. *If $p_1,p_2,\ldots,p_n,\ldots$ is the sequence of all prime numbers, then the sequence of sums $1/p_1+1/p_2+\cdots+1/p_n$ increases without limit.*

The deduction of Theorem 26 from Theorem 25 is purely formal. It does not depend on $p_1,p_2,\ldots,p_n,\ldots$ being the sequence of *prime* numbers; it could be any sequence of natural numbers for which the conclusion in Theorem 25 holds.

Lemma 9. *The inequality*

$$1-\frac{1}{n} \geq \frac{1}{4^{1/n}} \tag{4}$$

holds for any natural number $n > 1$.

Because both sides of inequality (4) are positive, we raise them to the power n and obtain the *equivalent* inequality

$$\left(1 - \frac{1}{n}\right)^n \geq \frac{1}{4},$$ (5)

which we prove. Expanding the left-hand side of inequality (5) in accordance with the binomial formula, we obtain

$$\left(1 - \frac{1}{n}\right)^n = 1 - n\frac{1}{n} + \frac{n(n-1)}{2!}\frac{1}{n^2} -$$
$$- \frac{n(n-1)(n-2)}{3!}\frac{1}{n^3} + \cdots + (-1)^n\frac{1}{n^n}.$$ (6)

The absolute value of the terms in the right-hand side of equality (6) form the sequence C_n^k/n^k. We examined such a sequence of numbers in connection with the Bernoulli scheme in Sec. 10 (formula (7) in Chap. 3). More precisely, if we set $p = 1/(n+1)$ and $q = 1 - 1/(n+1) = n/(n+1)$ in those formulas, then we obtain $p + q = 1$ and $p^k q^{n-k} = (n+1)^{-n}n^{n-k}$, that is, the numbers we obtain differ from those in formula (6) only in the factor $(n/(n+1))^n$ that is common to all. In our case, the expression $(n+1)p - 1$ is equal to zero. We proved in Sec. 10 that if $k > (n+1)p - 1$ (if $k > 0$ in our case), then the $(k+1)$th term is less than the kth. This means that all numbers in the sequence C_n^k/n^k for $k = 1, 2, \ldots, n$ decrease monotonically. (We here refer to Chap. 3 to show how the questions we consider are connected with each other. It would be easy to write the ratio of the $(k+1)$th term to the kth term directly and verify that it is less than one.)

We see that the first two terms in the right-hand side of formula (6) cancel. The second two terms (after a reduction that you can easily perform) yield $1/3 - 1/(3n^2)$. This number is not less than $1/4$ for $n \geq 2$ (verify this!). And the remaining terms group into pairs in which the first term is positive and the second term is negative. But as we saw, the absolute value of the second term in each pair is less than the first. Therefore, each pair yields a positive contribution to sum (6). If n is odd, then the number of terms in the right-hand side of formula (6) is even (it is equal to $n + 1$), and the sum exactly subdivides into $(n + 1)/2$ pairs. And if n is even, then one positive term $1/n^n$ remains after pairing the terms. In either case, the right-hand side thus consists of a term that is not less than $1/4$ plus some additional positive terms. This proves inequality (5) and consequently proves the lemma.   □

Theorem 26 is now almost obvious. For any p_i, we have

$$1 - \frac{1}{p_i} \geq \frac{1}{4^{1/p_i}}$$

according to the lemma. Multiplying these inequalities for $i = 1, \ldots, n$, we obtain

$$\left(1 - \frac{1}{p_1}\right)\left(1 - \frac{1}{p_2}\right) \cdots \left(1 - \frac{1}{p_n}\right) \geq \frac{1}{4^{1/p_1 + 1/p_2 + \cdots + 1/p_n}}.$$

If the sum $1/p_1 + \cdots + 1/p^n$ did not exceed a certain value k for all n, then it would follow that

$$\left(1 - \frac{1}{p_1}\right)\left(1 - \frac{1}{p_2}\right) \cdots \left(1 - \frac{1}{p_n}\right) \geq \frac{1}{4^k}.$$

This contradicts Theorem 25.   □

According to Theorem 26, the sum $1/p_1 + 1/p_2 + \cdots + 1/p_n$ will be greater than any previously specified number C if we take all prime numbers less than a certain number N whose choice depends on the number C. However, calculations show that the sequence of sums $1/p_1 + 1/p_2 + \cdots + 1/p_n$ grows exceptionally slowly, that is, N must be chosen very large for the sum to be greater than even a fairly small number C. For example, the first term yields the value $1/2$. The sum of three terms, corresponding to the prime numbers 2, 3, and 5, is already equal to $31/30$, that is, already greater than 1. But the sum first becomes greater than 2 only when we add the values $1/p$ for all prime numbers not exceeding 277. However, for $N = 10\,000$, that is, when we include $1/p$ for all prime numbers $p < 10\,000$, we obtain a value less than 3. For $N = 10^7$ (that is, ten million), this sum is still less than 3, and for the enormous value $N = 10^{18}$ (a million trillion), it is less than 4. Nevertheless, Theorem 26 confirms that the sum becomes greater than any, even extremely large, given number C, but then the number N must be chosen to be simply humongous! This is a curious example showing that numerical experiments can suggest a totally wrong answer—and the situation is obviously the same with a physical experiment.

In connection with Theorem 26, we come upon a new type of question. If N is a subset of a finite set S, then we can say how much less than S N is by comparing the numbers of their elements, for example, calculating the ratio $n(N)/n(S)$. But now we have two infinite subsets: the set of natural numbers and the set of prime numbers contained within it. How to compare them? Theorem 26 offers one possibility of comparison, not very simple at first glance. It can be applied to any sequence of natural numbers $a = a_1, a_2, \ldots, a_n, \ldots$. According to Lemma 8, the sums of the inverse values for the sequence of all natural numbers (that is, the sums

$S_{-1}(n)$) increase without limit. We can consider a sequence a "densely" distributed among the natural numbers if the same property is preserved for it, that is, the sums

$$\frac{1}{a_1}, \quad \frac{1}{a_1} + \frac{1}{a_2}, \quad \ldots, \quad \frac{1}{a_1} + \frac{1}{a_2} + \cdots + \frac{1}{a_n}, \quad \ldots$$

increase without limit. This means that sufficiently many natural numbers are retained in the sequence a for the sums of the inverses of its terms to be not much less than the sums $S_{-1}(n)$ of the inverses of all natural numbers. But if the sums of the inverse values of a sequence a remain bounded, then we can consider it "sparsely" distributed in the ranks of the natural numbers. Theorem 26 confirms that the sequence of prime numbers is "dense." The most extreme "sparse" case is where the sequence a consists of only a finite number of terms.

But there do exist intermediate cases. For example, the sequence of squares: $1, 4, 9, \ldots, n^2, \ldots$. The corresponding sums $1 + 1/4 + 1/9 + \cdots + 1/n^2$ are naturally denoted by $S_{-2}(n)$. We prove that it is bounded, irrespective of n. For this, we use the same approach used to prove Lemma 8. Let m be such that $2^m \geq n$. Then $S_{-2}(n) \leq S_{-2}(2^m)$. We subdivide the sum $S_{-2}(2^m) = 1 + 1/2^2 + 1/3^2 + \cdots + 1/2^{2m}$ into parts:

$$(1) + \left(\frac{1}{2^2}\right) + \left(\frac{1}{3^2} + \frac{1}{4^2}\right) + \cdots + \left(\frac{1}{(2^{m-1}+1)^2} + \cdots + \frac{1}{2^{2m}}\right).$$

Each part

$$\frac{1}{(2^{k-1}+1)^2} + \cdots + \frac{1}{2^{2k}}$$

again contains 2^{k-1} terms, and the first term here is the largest. Therefore, each such part does not exceed $2^{k-1}/(2^{k-1}+1)^2 < 2^{k-1}/(2^{k-1})^2 = 1/2^{k-1}$. Hence,

$$S_{-2}(2^m) \leq 1 + 1 + \frac{1}{2} + \frac{1}{2^2} + \cdots + \frac{1}{2^{m-1}} = 1 + \frac{1 - 1/2^m}{1 - 1/2} \leq 1 + \frac{1}{1 - 1/2} = 3,$$

that is, $S_{-2}(n)$ does not exceed 3.

Theorem 26 thus shows that the prime numbers, for example, are more densely distributed among the natural numbers than the squares.

Problems:

1. Prove that for any $k > 1$ and all natural numbers n, the sums $S_{-k}(n) = 1/1^k + 1/2^k + \cdots + 1/n^k$ are bounded.
2. Let the sequence a be an arithmetic progression: $a_0 = p$, $a_1 = p + q$, $a_2 = p + 2q, \ldots, a_n = p + nq, \ldots$ for some natural numbers p and q. Prove that the sums $1/a_1$, $1/a_1 + 1/a_2$, $\ldots$, $1/a_1 + 1/a_2 + \cdots + 1/a_n$, $\ldots$ increase without limit.

3. Let the sequence a be a geometric progression: $a_0 = c$, $a_1 = cq$, $a_2 = cq^2$, ..., $a_n = cq^n$, ..., where c and q are some natural numbers. Is it "dense" or "sparse" in the sequence of natural numbers?
4. Let $p_1, \ldots, p_n, \ldots$ be the sequence of all prime numbers. Prove that the expression

$$\frac{1}{\left(1 - \frac{1}{p_1^2}\right)\left(1 - \frac{1}{p_2^2}\right) \cdots \left(1 - \frac{1}{p_n^2}\right)}$$

is bounded for all n.

13. Distribution of Prime Numbers

In this section, we again attempt to estimate how much the sequence of prime numbers differs from the entire sequence of natural numbers. For this, we replace the more elaborate method of comparing "dense" and "sparse" sequences, which arose by itself from Euler's proof in the preceding section, with a more naive method that first comes to mind. Namely, we try to answer the naive question "what portion of the natural numbers consists of prime numbers" by determining how many prime numbers there are that are less than 10, how many less than 100, how many less than 1000, and so on. For any natural number n, the number of prime numbers not exceeding n is denoted by $\pi(n)$: $\pi(1) = 0$, $\pi(2) = 1$, $\pi(4) = 2, \ldots$. What can we say about the ratio $\pi(n)/n$ when n increases without limit?

We first consider what a table can tell us. Any assertion or question about natural numbers can be checked for all natural numbers not exceeding a certain limit N. Such a situation plays a role in number theory (the study of the properties of natural numbers) that is played by the possibility of an actual experiment in physics. In particular, we can calculate the value of $\pi(n)$ for $n = 10^k$, $k = 1, 2, \ldots, 10$. We obtain the table on the next page.

We see that the ratio $n/\pi(n)$ constantly increases, and this means that $\pi(n)/n$ constantly decreases. That is, the portion of natural numbers that are prime numbers comes closer and closer to zero as n increases. According to the table, we can say that "prime numbers comprise a zero portion of all natural numbers." Euler thus formulated it, although his considerations did not include a complete proof. We formulate this assertion precisely and then prove it.

Theorem 27. *For sufficiently large n, the ratio $\pi(n)/n$ is less than any positive number given in advance.*

n	$\pi(n)$	$\dfrac{n}{\pi(n)}$
10	4	2.5
100	25	4.0
1 000	168	6.0
10 000	1 229	8.1
100 000	9 592	10.4
1 000 000	78 498	12.7
10 000 000	664 579	15.0
100 000 000	5 761 455	17.4
1 000 000 000	50 847 534	19.7
10 000 000 000	455 059 512	22.0

To prove the theorem, we must somehow estimate the value of the expression $\pi(n)$. For actual calculation of its value, we begin with the prime number 2 and cross out all other numbers divisible by 2 and not exceeding n. We then take the first remaining number—in this case, 3—and cross out all other numbers divisible by 3 and not exceeding n. We repeat this process until all numbers not exceeding n have been crossed out or used. The numbers not crossed out (2, 3, and so on) are all the prime numbers not exceeding n. This approach was already used in antiquity and is called the *sieve of Eratosthenes*.

We apply this approach to our problem. Suppose we have already found r prime numbers: $p_1, p_2, \ldots, p_r$. Then the next prime numbers not exceeding n are contained among the numbers not exceeding n that are "not crossed out," that is, among those numbers $m \le n$ that are not divisible by one of the numbers $p_1, p_2, \ldots, p_r$. But we investigated the number of numbers not exceeding n and not divisible by one of the prime numbers $p_1, p_2, \ldots, p_r$ in Chap. 3—it is given by formula (25) in Sec. 9. The expression in that formula can be replaced with the simpler expression $n(1-1/p_1) \cdots (1-1/p_n)$, as was proved there, and the resulting error does not exceed 2^r (formula (28) in Chap. 3). Therefore, the number s of numbers $m \le n$ not divisible by one of the prime numbers $p_1, p_2, \ldots, p_r$ satisfies the inequality

$$s \le n \left(1 - \frac{1}{p_1}\right) \cdots \left(1 - \frac{1}{p_r}\right) + 2^r. \tag{7}$$

All the $\pi(n)$ prime numbers not exceeding n are included either among the r prime numbers $p_1, p_2, \ldots, p_r$ or among the s numbers covered by inequality (7). Hence, $\pi(n) \le s + r$, and this means that

$$\pi(n) \le n \left(1 - \frac{1}{p_1}\right) \cdots \left(1 - \frac{1}{p_r}\right) + 2^r + r. \tag{8}$$

That inequality (8) contains the product $(1 - 1/p_1) \cdots (1 - 1/p_r)$ is remarkable, and Theorem 25 already gives us information about its magnitude.

We can now turn directly to the proof of Theorem 27. Let an arbitrarily small positive number ε be given. We must find a number N such that $\pi(n)/n < \varepsilon$ for all $n > N$. In inequality (8), we replace r with the larger value 2^r (see Problem 6 in Sec. 2) to obtain the simpler inequality

$$\pi(n) \le n \left(1 - \frac{1}{p_1}\right) \cdots \left(1 - \frac{1}{p_r}\right) + 2^{r+1}. \tag{9}$$

There are two terms in the right-hand side of inequality (9), and we choose N such that each term does not exceed $\varepsilon n/2$ for $n \ge N$. It then follows from inequality (9) that $\pi(n) < \varepsilon n$ and hence $\pi(n)/n < \varepsilon$. But we recall that the number r has so far been arbitrary in our considerations. We first choose r such that the first term does not exceed $\varepsilon n/2$ and then choose N such that the second term does not exceed $\varepsilon n/2$. The first choice is possible by virtue of Theorem 25. It states that for sufficiently large r, the product $(1 - 1/p_1) \cdots (1 - 1/p_r)$ is less than any positive number given in advance. We can take $\varepsilon/2$ for such a positive number. Then the first term in inequality (9) does not exceed $\varepsilon n/2$. The matter is even simpler for the second term. Now, we have already chosen r. We choose N such that $2^{r+1} < \varepsilon N/2$. For this, we must choose $N > 2^{r+2}/\varepsilon$. Then $2^{r+1} < \varepsilon N/2 \le \varepsilon n/2$ for any $n \ge N$. Theorem 27 is proved.   □

We note that if we take an arithmetic progression $am + b$, even with a very large difference a, that is, appearing very rarely, then the number of terms in this progression not exceeding n coincides with the number of integers m for which $am \le n - b$, that is, $[(n-b)/a]$. In Sec. 9, we saw that $[(n - b)/a]$ differs from $(n - b)/a$ by not more than 1. Therefore, the number of terms in the progression not exceeding n is not less than $(n - b)/a - 1$. Its ratio to n is not less than

$$\frac{1}{n} \left(\frac{n - b}{a} - 1\right) = \frac{1}{a} - \frac{1}{n}\frac{b}{a} - \frac{1}{n}.$$

As n increases, this number approaches $1/a$ and does not become arbitrarily small. Therefore, Theorem 27 would not be true if we took any arithmetic progression for the sequence. This shows that the prime numbers are distributed more sparsely than any arithmetic progression.

Problems:

1. Let p_n denote the nth prime number. Prove that for any arbitrarily large positive number C, the inequality $p_n > Cn$ holds for sufficiently large n. *Hint*: Use the fact that $\pi(p_n) = n$.

2. Consider the natural numbers with the property that their representation in the decimal system does not contain a specified digit (for instance, 0). Let $q_1, q_2, \ldots, q_n, \ldots$ be these numbers written in ascending order, and let $\pi_1(n)$ denote the number of such numbers not exceeding n. Prove that for sufficiently large n, the ratio $\pi_1(n)/n$ is less than any positive number given in advance. Prove that the sums

$$\frac{1}{q_1}, \quad \frac{1}{q_1} + \frac{1}{q_2}, \quad \ldots, \quad \frac{1}{q_1} + \cdots + \frac{1}{q_n}, \quad \ldots$$

are bounded. *Hint*: Do not try to copy the proof of Theorem 27. Subdivide the sum into parts with the denominator ranging from 10^k to 10^{k+1}. Find the number of numbers q_i in such an interval. The answer depends on the digit chosen for exclusion: $r = 0$ or $r \neq 0$.

Supplement: The Chebyshev Inequality for $\pi(n)$

We place this material in a supplement first for a formal reason: we must use logarithms here, and the rest of the text does not assume familiarity with them. We recall that the *logarithm of a number x to the base a* is a number y such that

$$a^y = x.$$

This is written as

$$y = \log_a x.$$

Always in what follows, we assume that $a > 1$, and we consider positive numbers x. The basic properties of logarithms follow directly from the definition:

$$\log_a(xy) = \log_a x + \log_a y, \quad \log_a c^n = n \log_a c, \quad \log_a a = 1.$$

We have $\log_a x > 0$ if and only if $x > 1$. The logarithm function is monotonic, that is, $\log_a x \leq \log_a y$ if and only if $x \leq y$.

If the base of the logarithm is not shown here, then we assume that it is 2: $\log x$ means $\log_2 x$.

The second reason for segregating the following considerations in a supplement consists in the following. In the other parts of the book, the logic of the arguments is clear, why we go along namely that path (so I hope, at least). We here encounter a case, not rare in mathematical research, where some new thought seems to fall out the blue sky, as it were, and even the author is often unable to explain where it came from. About such situations, Euler said, "It sometimes seems to me that my pencil is smarter than I." Understandably, it is the result of uncounted trials, much cogitation, and the working of the subconscious mind.

We continue to study the question of the ratio $\pi(n)/n$ as n increases without limit. We once more examine the table on page 129, which shows the values of $\pi(n)$ for $n = 10^k$, $k = 1, 2, \ldots, 10$. We focus on the last column of the table, which gives the ratios $n/\pi(n)$ for certain values of n. We notice that when passing from $n = 10^k$ to $n = 10^{k+1}$, that is, when dropping down one line in the table, the values $n/\pi(n)$ change by almost the same amount. Namely, the first number is equal to 2.5; the second differs from it by 1.5; and the differences are equal to 2, 2.1, 2.3, 2.3, 2.3, 2.4, 2.3, and 2.3. We see that all these numbers are very close to one value: 2.3. Not trying to solve the riddle of why this value for the time being, we propose that even further beyond the bounds of our table, the number $n/\pi(n)$ when passing from $n = 10^k$ to $n = 10^{k+1}$ will increase by an amount even closer to a certain fixed constant α. This would mean that $n/\pi(n)$ for $n = 10^k$ would be very close to αk. But if $n = 10^k$, then $k = \log_{10} n$ by definition. Then it is natural to propose that for other values of n, the value of $n/\pi(n)$ is very close to $\alpha \log_{10} n$. This means that $\pi(n)$ is very close to $cn/\log_{10} n$, where $c = \alpha^{-1}$.

Many mathematicians were fascinated by the secret of the distribution of prime numbers and tried to discover it based on tables. In particular, Gauss was interested in this question almost in childhood. His interest in mathematics evidently began with a childhood interest in numbers and constructing tables. In general, great mathematicians were virtuosos of calculation and were able to perform enormous calculations, sometimes mentally. (Euler even struggled with insomnia in this way!) When Gauss was 14 years old, he constructed a table of prime numbers (true, less comprehensive than our table on page 129) and came to the same proposition we just formulated. It was later considered by many mathematicians. But the first result was proved more than half a century later, by Chebyshev in 1850.

Theorem 28. *There exist constants c and C such that for all $n > 1$*

$$c \frac{n}{\log n} \leq \pi(n) \leq C \frac{n}{\log n}. \tag{10}$$

We present a proof that is a result of simplications of Chebyshev's original proof subsequently given by many mathematicians. The principal idea of the proof is unchanged. Before turning to the proof, we make a few remarks concerning the formulation of the theorem. What is the base of the logarithm considered here? Answer: any base. It follows immediately from the definition of a logarithm that $\log_b x = \log_b a \log_a x$ (we need only replace a with $b^{\log_b a}$ in the relation $a^{\log_a x} = x$, and we obtain $b^{\log_b a \log_a x} = x$, which shows that $\log_b x = \log_b a \log_a x$). Therefore, if inequality (10) is proved for $\log_a n$, then it also holds for $\log_b n$ with c replaced with $c / \log_b a$ and C replaced with $C / \log_b a$.

Inequality (10) indeed expresses the thought suggested by the table that $\pi(n)$ is "close" to $cn / \log n$ for some constant c. Why are there two constants in the theorem (c and C) when there was only one constant c in our hypothetical considerations? Is it impossible to replace the two constants in the theorem with one in some sense? We consider these questions after proving the theorem.

The secret key to the proof of the Chebyshev theorem is properties of the binomial coefficients C_n^k: primarily the fact that they are integers and some properties of their divisibility by prime numbers. We list the properties that we need in the proof.

First is the assertion (proved in Sec. 6) that the sum of all binomial coefficients C_n^k for $k = 0, 1, \ldots, n$ is equal to 2^n. Because the sum of positive terms is not less than each term, we obtain

$$C_n^k \leq 2^n. \tag{11}$$

Large binomial coefficients will be especially useful for us. We saw in Chap. 2 that for even $n = 2m$, the coeeficient C_{2m}^m is larger than the others. For odd $n = 2m + 1$, there are two equal coefficients C_{2m+1}^m and C_{2m+1}^{m+1} that are larger than the others. We pay special attention to them. In particular,

$$C_{2n}^n = \frac{2n(2n - 1) \cdots (n + 1)}{1 \cdot 2 \cdots n}. \tag{12}$$

If we group the factors of the numerator with the factors of the denominator in reverse order, then we obtain

$$C_{2n}^n = \frac{2n}{n} \frac{2n - 1}{n - 1} \cdots \frac{n + 1}{1}.$$

Obviously, each factor in this formula is not less than 2; therefore,

$$C_{2n}^n \geq 2^n. \tag{13}$$

We now consider properties of the divisibility of binomial coefficients by prime numbers. In expression (12), the factors in the numerator are obviously divisible by all prime numbers not exceeding $2n$ and greater than n. Such prime numbers cannot divide the factors of the denominator. Therefore, they do not cancel and are divisors of C_{2n}^n. The number of prime numbers distributed between $2n$ and n is equal to $\pi(2n) - \pi(n)$, and all of them are greater than n. Therefore,

$$C_{2n}^n \geq n^{\pi(2n) - \pi(n)}. \tag{14}$$

An analogous assertion holds, of course, for the "middle" coefficients $C_{2n+1}^n = C_{2n+1}^{n+1}$ with an odd lower index. Writing them in the form

$$C_{2n+1}^n = \frac{(2n+1)\cdots(n+2)}{1 \cdot 2 \cdots n},$$

we see that $\pi(2n+1) - \pi(n+1)$ prime numbers not exceeding $2n+1$ and greater than $n+1$ divide the numerator and cannot be canceled with the denominator. Because they are greater than $n+1$, we have

$$C_{2n+1}^n > (n+1)^{\pi(2n+1) - \pi(n+1)}. \tag{15}$$

A remarkable connection between binomial coefficients and prime numbers is already revealed in inequalities (14) and (15).

Finally, we introduce the last property of binomial coefficients needed for the proof. Although it is entirely simple, in contrast to the previous properties, it is not entirely obvious.

Lemma 10. *For any binomial coefficient C_n^k, the power of a prime number dividing it does not exceed n.*

We stress that we are speaking not of the *degree* but of the *power itself*. That is, we assert that if p^r divides C_n^k, where p is a prime number, then $p^r \leq n$. For example, $C_9^2 = 9 \cdot 4$ is divisible by 9 and by 4, and both numbers do not exceed 9.

We write the binomial coefficient in the form

$$C_n^k = \frac{n(n-1)\cdots(n-k+1)}{1 \cdot 2 \cdots k}. \tag{16}$$

The prime number p we are considering must divide the numerator of this fraction. We let m denote the factor containing the maximum power of p (or one of them if there are several) and let p^r denote that maximum power. It is obvious that $n \geq m \geq n - k + 1$ for $k \geq 1$. We set $n - m = a$ and $m - (n - k + 1) = b$. Then $a + b = k - 1$, and C_n^k can be written as

$$C_n^k = \frac{(m+a)(m+a-1)\cdots(m+1)m(m-1)\cdots(m-b)}{k!}. \tag{17}$$

The factor m is now fundamental for us, and we write the product in the numerator with a factors to the left of it and b factors to the right. We transform the denominator analogously:

$$k! = (1 \cdot 2 \cdots a)(a+1) \cdots (a+b)(a+b+1).$$

Because $(a+1)(a+2) \cdots (a+b)$ as a product of b consecutive natural numbers is divisible by $b!$, this product can be written as $a!b!l$, where l is an integer.

We can now write C_n^k in the convenient form

$$C_n^k = \frac{m+a}{a} \frac{m+a-1}{a-1} \cdots \frac{m+1}{1} \frac{m-1}{1} \cdots \frac{m-b}{b} \frac{m}{l}, \qquad (18)$$

where we move the factor m/l to the end.

We note that in each of the factors $(m+i)/i$ or $(m-j)/j$, where $i = 1, \ldots, a$ and $j = 1, \ldots, b$, the power of p in the numerator completely cancels with the denominator; therefore, after canceling the common factor in the numerator and denominator, only the denominator can be divisible by p (although it can also be relatively prime to p). Indeed, we consider the fraction $(m+i)/i$ as an example (the fraction $(m-j)/j$ is treated in exactly the same way). Let i by exactly divided by p^s, that is, $i = p^s u$, where u is relatively prime to p. If $s < r$, then $m+i$ is also exactly divisible by p^s: setting $m = p^r v$ (we recall that m is divisible by p^r), we obtain $m + i = p^s(u + p^{r-s}v)$. And if $s \geq r$, then in exactly the same way, $m + i$ is divisible by p^r. Recalling the choice of m (it is divisible by the largest power of p among all numbers from n to $n - k + 1$, and this power is p^r), we conclude that a larger power of p than the rth power cannot divide $m + i$. Therefore, p^r cancels in the numerator and denominator, and a number remains in the numerator that is not divisible by p. As a result, we see that of all the factors in expression (18), p can be retained only in the numerator of the last one, that is, in m. But the power of p dividing m is p^r, and this means that product (18) cannot be divided by a larger power of p than p^r. Because p^r divides m and $m \leq n$, we have $p^r \leq n$. The lemma is proved. $\qquad \square$

We consider what this tells us about the canonical decomposition $C_n^k = p_1^{\alpha_1} \cdots p_m^{\alpha_m}$. First, the prime numbers $p_1, \ldots, p_m$ can appear only from the numerator of expression (16), which means that all $p_i \leq n$ and the number of them m is therefore less than $\pi(n)$. According to the lemma, $p_i^{\alpha_i} \leq n$ for $i = 1, \ldots, m$. As a result, we obtain

$$C_n^k \leq n^{\pi(n)}. \qquad (19)$$

We can now begin the actual proof of the Chebyshev theorem, that is, inequality (10). We note that it is sufficient for us to prove the satisfaction of the inequality just for all n beginning from some fixed boundary n_0. For all $n < n_0$, satisfaction of the inequality can be achieved by decreasing the constant c and increasing the constant C. If you want to obtain the explicit value of these constants most economically, then you can verify that inequality (10) is satisfied for $n \leq n_0$ by constructing a table of prime numbers (in our considerations here, n_0 turns out to be not very large).

We begin by combining inequalities (13) and (20) for the binomial coefficient C_{2n}^n. We obtain $2^n \leq C_{2n}^n \leq (2n)^{\pi(2n)}$ and consequently

$$2^n \leq (2n)^{\pi(2n)}. \tag{20}$$

Taking the logarithm to the base 2 of both sides (we recall that we write $\log_2 x = \log x$) and using the monotonicity of logarithms, we obtain $n \leq \pi(2n) \log 2n$, which means

$$\pi(2n) \geq \frac{n}{\log 2n} = \frac{1}{2} \frac{2n}{\log 2n},$$

that is, the left inequality in (10) with the constant $c = 1/2$. But so far, it is proved only for even values n. For odd values of the form $2n + 1$, we use the monotonicity of logarithms and the function $\pi(n)$. It follows that

$$\pi(2n + 1) \log(2n + 1) \geq \pi(2n) \log 2n.$$

Substituting the inequality obtained for $\pi(2n)$ in this expression, we see that

$$\pi(2n + 1) \geq \frac{n}{\log 2n} \frac{\log 2n}{\log(2n + 1)} = \frac{n}{\log(2n + 1)}.$$

Because always $n \geq (2n + 1)/3$, it follows that

$$\pi(2n + 1) \geq \frac{1}{3} \frac{2n + 1}{\log(2n + 1)}.$$

The left inequality in (10) is thus proved for odd n and $c = 1/3$. This means that the left inequality in (10) holds for all n and $c = 1/3$.

We turn to the proof of the right inequality in (10). We prove it by induction on n. First let n be even. Instead of n, we write $2n$. We combine inequality (11) for the coefficient C_{2n}^n (that is, we replace n with $2n$ and k with n) with inequality (14). As a result, we obtain

$$n^{\pi(2n) - \pi(n)} \leq 2^{2n}.$$

Passing to logarithms, we have

$$\pi(2n) - \pi(n) \leq \frac{2n}{\log n},$$

$$\pi(2n) \leq \pi(n) + \frac{2n}{\log n}. \tag{21}$$

In accordance with the induction assumption, we can consider the inequality we need already proved: $\pi(n) \leq Cn/\log n$ with a constant C, whose value we later determine more precisely. Substituting in formula (21), we obtain

$$\pi(2n) \leq C\frac{n}{\log n} + \frac{2n}{\log n} = \frac{(C+2)n}{\log n}.$$

But we wanted to prove the inequality $\pi(2n) \leq C \cdot 2n/\log 2n$. For this, it remains to select a constant C such that the inequality

$$\frac{(C+2)n}{\log n} \leq \frac{2Cn}{\log 2n} \qquad (22)$$

is satisfied for all n beginning from some point.

This is already a simple school exercise, not connected with the properties of prime numbers. We cancel n on both sides of the inequality, and noting that $\log 2n = \log 2 + \log n = 1 + \log n$, we let x denote $\log n$. Then inequality (22) becomes

$$\frac{C+2}{x} \leq \frac{2C}{1+x}.$$

Multiplying both sides by $x(1+x)$ (because $x > 0$) and combining like terms, we write it in the form

$$(C-2)x \geq C+2.$$

Obviously, we must choose C such that $C - 2 > 0$. Taking $C = 3$, for example, we find that the inequality is satisfied for $C = 3$ and all $x \geq 5$. Because x denotes $\log n$, this means that the needed inequality is satisfied for $n \geq 2^5 = 32$, $2n \geq 64$.

It only remains to consider the case with an odd value having the form $2n + 1$. For this, we combine inequality (11) (replacing n with $2n + 1$ and k with n) with inequality (15). We obtain the inequality

$$2^{2n+1} \geq (n+1)^{\pi(2n+1)-\pi(n+1)}.$$

Taking the logarithms, we obtain the inequality

$$2n + 1 \geq \big(\pi(2n+1) - \pi(n+1)\big)\log(n+1).$$

From this, we use the induction assumption about $\pi(n+1)$ as previously to obtain

$$\pi(2n+1) \leq C\frac{n+1}{\log(n+1)} + \frac{2n+1}{\log(n+1)}.$$

The needed inequality $\pi(2n+1) \leq C(2n+1)/\log(2n+1)$ will be proved if we verify that

$$C\frac{n+1}{\log(n+1)} + \frac{2n+1}{\log(n+1)} \leq C\frac{2n+1}{\log(2n+1)} \tag{23}$$

for an appropriate choice of the constant C and for all n beginning from some point. This is again a school exercise, although slightly more complicated than the previous one. To make it easier to compare the two sides, we replace $2n+1$ in the left-hand side with the larger value $2(n+1)$:

$$C\frac{n+1}{\log(n+1)} + \frac{2n+1}{\log(n+1)} \leq \frac{(C+2)(n+1)}{\log(n+1)}. \tag{24}$$

To transform the right-hand side, we note that $2n+1 \geq (3/2)(n+1)$ for $n \geq 1$ and that $\log(2n+1) \leq \log(2n+2) = 1+\log(n+1)$. Therefore,

$$\frac{2n+1}{\log(2n+1)} \geq \frac{(3/2)(n+1)}{1+\log(n+1)}. \tag{25}$$

Combining inequalities (24) and (25), we see that inequality (23) will be proved if we prove that

$$\frac{(C+2)(n+1)}{\log(n+1)} \leq \frac{(3/2)C(n+1)}{1+\log(n+1)}.$$

We cancel $n+1$ in both sides and set $\log(n+1) = x$. We obtain the inequality

$$\frac{C+2}{x} \leq \frac{(3/2)C}{1+x},$$

which is solved in exactly the same way as the previously analyzed case. We must multiply both sides by $x(1+x)$ and combine like terms. We obtain the inequality $(C+2)x + C + 2 \leq (3/2)Cx$ or

$$\left(\frac{1}{2}C - 2\right)x \geq C+2.$$

Setting $C = 6$, we see that the inequality holds for $x \geq 8$, that is, $n+1 \geq 2^8$, $2n+1 \geq 511$. The right inequality in (10) is thus proved for the constant $C = 6$ and all values of n beginning with 511. The theorem is proved. $\square$

We note that Theorem 27 is a very simple consequence of the theorem just proved. Indeed, if $\pi(n) \leq Cn/\log n$, then $\pi(n)/n \leq C/\log n$.

And because logarithms change monotonically and increase without limit ($\log 2^k = k$), $\pi(n)/n$ becomes less than any positive number. On the other hand, the proof of the Chebyshev theorem is based on completely different ideas than those used to prove Theorem 27.

In conclusion, we return once more to the propositions that can be made from examining the table on page 129. From it, we guessed that $n/\pi(n)$ is close to $C \log_{10} n$ with some definite constant C: the first two digits in the decimal representation of C^{-1} have the form 2.3. Hence, we can conclude that $\pi(n)$ is close to $C^{-1} n / \log_{10} n$. This expression can be given the simpler form $n/\log_e n$ if a new logarithm base e is chosen such that $C \log_{10} n = \log_e n$. But as was mentioned previously, always $\log_b x = \log_b a \log_a x$, and our relation is therefore satisfied if $C = \log_e 10$. Substituting the value $x = b$ in the relation $\log_b x = \log_b a \log_a x$, we obtain $\log_b a \log_a b = 1$, and the relation $C = \log_e 10$ that interests us can be rewritten as $C^{-1} = \log_{10} e$.

Fourteen-year-old Gauss certainly paid attention to these relations and guessed what the number e is for which $\log_{10} e$ is close to $(2.3)^{-1}$. Such a number was well known by that time specifically because logarithms to such a base have many useful properties (e is its conventionally accepted symbol). Logarithms to the base e are called *natural* logarithms and are denoted by ln: $\log_e x = \ln x$. Here, we are compelled to assume that the reader is familiar with natural logarithms.

The natural proposition following from studying the table is thus that $\pi(n)$ becomes ever closer to $n/\ln n$. The proved Chebyshev theorem (if natural logarithms are used) confirms the existence of two constants c and C such that $cn/\ln n \leq \pi(n) \leq Cn/\ln n$ beginning from some n. That hypothetical sharpening, which can be obtained from the table, asserts that the inequality $cn/\ln n \leq \pi(n) \leq Cn/\ln n$ is satisfied beginning with some n *whatever* constants $c < 1$ and $C > 1$ we might choose. This assertion is called the asymptotic law of the distribution of prime numbers. It was stated by Gauss and other mathematicians at the end of the 18th and beginning of the 19th century. After the proof of the Chebyshev theorem in 1850, the matter seemed to be only determining the constants c and C more precisely and bringing them closer together. However, the asymptotic law of the distribution of prime numbers was proved only half a century later, at the very end of the 19th century, on the basis of completely new ideas proposed by Riemann.

Problems:

1. Prove that $p_n > an \log n$ for some constant $a > 0$. *Hint:* Use the fact that $\pi(p_n) = n$.
2. Prove that $\log n < \sqrt{n}$ beginning from some point (determine it). *Hint:* Reduce the problem to proving that the inequality $2^x > x^2$ holds for real x beginning

from some point. Let $n \leq x \leq n+1$, where n is an integer. Reduce it to proving the inequality $2^n \geq (n+1)^2$ and use induction.

3. Prove that $p_n < Cn^2$ for some constant C. *Hint*: Apply the inequality in the preceding problem, and use the fact that $n = \pi(p_n)$.

4. Prove that $p_n < An \log n$ for some constant A.

5. Prove that the degree a of the highest power p^a that divides $n!$ is equal to

$$\left[\frac{n}{p}\right] + \left[\frac{n}{p^2}\right] + \cdots + \left[\frac{n}{p^k}\right].$$

Here, $[r/s]$ denotes the integer quotient of r divided by s, the sum includes all k for which $p^k \leq n$, p denotes an arbitrary prime number, and n denotes an arbitrary natural number.

6. Using the result of Problem 5, give a different proof of Lemma 10 in the supplement.

7. Prove that if $p_1, \ldots, p_r$ are prime numbers included between m and $2m+1$, then their product does not exceed 2^{2m}.

8. Determine the constants c and C for which inequality (10) is satisfied for all n.

9. Try to find the largest possible c and the smallest possible C for which inequality (10) is satisfied for all n beginning from some point. (Chebyshev himself used a very ingenious sharpening of his arguments to prove that it is possible to set $c = 0.694$ and $C = 1.594$.)

5

Real Numbers and Polynomials

Topic: Numbers and Polynomials

14.   Axioms of the Real Numbers

In this chapter, we try to sharpen our conception of real numbers. We do not strive toward any special rigor in our considerations and only try to give our ideas and arguments sufficient precision to be able to *prove* statements about real numbers.

Choosing a starting point on a straight line and a unit of measure, we can represent real numbers as points on the line. Therefore, clarifying our understanding of real numbers, we simultaneously offer a more precise description of a straight line and the points lying on it. In what follows, we often use the one-to-one correspondence between real numbers and points on a straight line for illustrative purposes.

We take geometry as an exemplar and try to reach the level of exactness of definition and demonstration that a school course in geometry has. There, at the base of the whole structure lie certain axioms, which are used to prove all the remaining assertions. The axioms are not proved: we accept them based on experience or intuition.

For definiteness, we take the construction of plane geometry on the basis of axioms as a model. In this construction, we can distinguish three types of logical concepts. The first is *fundamental notions* of geometry: point, straight line. The second is *fundamental relations* between them: a point lies on a line, a point on a line lies between two given points on that line. Neither the first nor the second is defined. We present the matter to ourselves as if somewhere there is a "list" of all points and all lines and it is known which points lie on which lines and, for example, that for the three points A, B, and C on the line l, point B lies between A and C. The third type of concept is *axioms*, that is, assertions about the fundamental notions and the relations between them. For example, "every two distinct points belong to one and only one straight line," or "among three distinct points on a line, one and only one of them lies between two others."

The situation with real numbers is absolutely analogous. The *fundamental notions* are the real numbers themselves. At this moment, we offer nothing more than that real numbers compose some set. The *fundamental relations* between real numbers are of two types: operations on them and inequalities. We describe them in more detail.

1. Operations on Real Numbers. For every two real numbers a and b, there is determined a third number c called their *sum*. It is written as $a + b = c$.

For every two real numbers a and b, there is determined a third number d called their *product*. It is written as $ab = c$.

2. Inequalities Between Real Numbers. For some pairs of real numbers a and b, it is determined that a is less than b. This is written as $a < b$.

The same property is also written as $b > a$. If we wish to express that either $a < b$ or $a = b$, then we write $a \leq b$ (or $b \geq a$).

Before formulating the axioms that connect the fundamental notions and the fundamental relations between them, we once more emphasize the analogy with what is known from geometry. We combine analogous concepts in a table:

Algebra	Geometry
Fundamental Notions	
Real number	Point, Line (straight line)
Fundamental Relations	
Sum: $ab = c$ Product: $ab = c$ Inequality: $a < b$	A point lies on a line Point C lies between points A and B $\ldots\ldots$
Axioms	
$\ldots\ldots$	$\ldots\ldots$

There is no need for us to recall the axioms of geometry, and we now list the axioms of real numbers. They are formulated in terms of the fundamental notions and the relations between them that are listed in the table. We attempt to group the axioms in accordance with the fundamental relations to which they relate.

I (axioms of addition)

I_1. Commutative law: $a + b = b + a$ for any real numbers a and b.

I_2. Associative law: $a + (b + c) = (a + b) + c$ for any real numbers a, b, and c.

I_3. There exists a number, denoted by 0 and called zero, such that $a + 0 = a$ for any real number a.

(*Remark*: There is only one such number. If $0'$ were another number with the same property, then we would have $0' + 0 = 0'$ by the definition of zero, $0' + 0 = 0 + 0'$ by the commutative law, and $0 + 0' = 0$ by the definition of $0'$. We finally obtain $0' = 0' + 0 = 0 + 0' = 0$, that is, $0' = 0$.)

I_4. For any real number a, there exists a number, denoted by $-a$ and called negative a, such that $a + (-a) = 0$.

(*Remark*: For a given number a, there exists only one such number. If a' were another number with the same property $a + a' = 0$, then we would have $(a + (-a)) + a' = 0 + a' = a'$. Moreover, $(a + (-a)) + a' = ((-a)+a)+a'$, and by the associative law, $((-a)+a)+a' = (-a)+(a+a')$. By the property of the number a', $a + a' = 0$ and $(-a) + 0 = -a$. Combining these equalities, we find that $a' = -a$.)

II (axioms of multiplication)

II_1. Commutative law: $ab = ba$ for any real numbers a and b.

II_2. Associative law: $a(bc) = (ab)c$ for any real numbers a and b.

II_3. There exists a number, denoted by 1 and called unity, such that $a \cdot 1 = a$ for any real number a.

(*Remark*: There is only one such number. This is proved the same way as in the remark to axiom I_3—we need only replace addition with multiplication and 0 with 1.)

II_4. For any real number a different from 0, there exists a number, denoted by a^{-1} and called the inverse, such that $a \cdot a^{-1} = 1$.

(*Remark*: For each nonzero real number a, there is only one such number. The proof is exactly the same as in the remark to axiom I_4.)

III (axiom of addition and multiplication)

III_1. Distributive law: $(a + b)c = ac + bc$ for any real numbers a, b, and c.

IV (axioms of order)

IV_1. For any two real numbers a and b, one and only one of the following relations holds: either $a = b$ or $a < b$ or $b < a$.

IV_2. If $a < b$ and $b < c$ hold for three real numbers a, b, and c, then $a < c$.

IV_3. If $a < b$, then $a + c < b + c$ holds for any three real numbers a, b, and c.

IV_4. If $a < b$ and $c > 0$, then $ac < bc$ for any three real numbers a, b, and c.

V (real and rational numbers)

Rational numbers are contained among the real numbers, and operations and inequalities defined for real numbers when applied to rational numbers give the usual operations and inequalities.

VI (axiom of Archimedes)

For any real number a, there exists a natural number n such that $a < n$.

VII (axiom of included intervals)

Let $a_0, a_1, a_2, \ldots$ and $b_0, b_1, b_2, \ldots$ be two sequences of real numbers with the properties that $a_0 \le a_1 \le a_2 \le \cdots$ and $b_0 \ge b_1 \ge b_2 \ge \cdots$ and $b_n \ge a_n$ for all n. Then there exists a real number c such that $b_n \ge c$ and $c \ge a_n$ for all n.

If we use the representation of real numbers as points on a straight line, then the numbers x satisfying the conditions $a \le x$ and $x \le b$ (in brief, $a \le x \le b$) are represented by a set, which is called an *interval* and is denoted by $[a, b]$. Therefore, the premise of this axiom asserts that the intervals $I_n = [a_n, b_n]$ are successively included in one another: $I_0 \supset I_1 \supset I_2 \supset \ldots$. The axiom states that there exists a point (or number) that is common to all these intervals successively included in one another (and hence the name of the axiom).

It is easy to deduce the usual properties of real numbers from the given axioms. It would be boring to devote several pages to these absolutely obvious arguments. We therefore only formulate a few statements that are necessary in what follows and limit ourselves to isolated remarks regarding proofs (see also Problems 2, 3, and 4).

It follows from Axiom II that for any nonzero number a and any number b, the number $c = a^{-1}b$ is the unique solution of the equation $ax = b$. It is called the *ratio* of b to a and is written as b/a. All the customary rules for opening parentheses and operating with fractions are deduced from the axioms.

Because the equality $n = 1 + \cdots + 1$ (with n terms) holds for any natural number n, it follows from Axiom III that for any number a, the number na (the product of n and a) is equal to the sum $a + \cdots + a$ (with n terms).

It follows from Axiom IV_3 that if $a < b$ and $c < d$, then $a + c < a + d < b + d$. If $a < 0$, then $-a > 0$ (because $0 < 0$ would follow from $-a < 0$). As a result, we see that any real number is either positive ($a > 0$) or has the form $-b$ with $b > 0$ (we say it is negative) or is equal to zero. Multiplication obeys the usual "sign" rule. As usual, $|a|$ denotes the number a itself if $a \geq 0$ or $-a$ if $a < 0$.

The axiom of included intervals (Axiom VII) is especially useful when the length of the intervals I_n (that is, the difference $b_n - a_n$) becomes arbitrarily small as n increases. In other words, for any real number $\varepsilon > 0$, there exists an index N such that $b_n - a_n < \varepsilon$ for all $n \geq N$. In such a case, we can assert more than the axiom.

Lemma 11. *If the difference $b_n - a_n$ becomes arbitrarily small as the index n increases, then the number c whose existence is guaranteed by Axiom VII is unique.*

Proof. We suppose that there exist two such numbers c and c' and, for example, $c < c'$. Then $a_n \leq c < c' \leq b_n$, and $c' - c = b_n - a_n - (c - a_n) - (b_n - c') \leq b_n - a_n$. For sufficiently large n, we obtain $c' - c < \varepsilon$ for any given number $\varepsilon > 0$. For example, such a relation must hold for $\varepsilon = (c' - c)/2$, whence $(c' - c)/2 < 0$. This contradicts our supposition that $c' - c > 0$ because $1/2 > 0$. $\qquad\qquad\square$

We meet just such a case when we use rational numbers to measure a real number approximately with a deficit and a surplus. Then a_n and b_n are rational numbers. An example is $\sqrt{2}$, discussed in Sec. 1. Axiom VII thus formulates our intuitive idea when we speak of "ever more precise measure." Together with the preceding lemma, it gives us the possibility of constructing real numbers with the needed properties. We use this possibility often in what follows.

With regard to Axioms V and VI, we note that we presuppose that natural and, more generally, rational numbers are known. We do not analyze this concept in more detail.

In conclusion, we note that the axioms introduced above are not independent. This means that some of them could be deduced as a theorem based on the remaining axioms (see Problem 6 for an example). We collected those properties of the real numbers that are already customary and intuitively convincing for you. At the cost of increasing the number of axioms, we gain the right to skip uninteresting proofs of a series of intuitively obvious facts.

Problems:

1. Of the Axioms I–VII, which are also true for the collection of rational numbers, and which are specific to the real numbers?

2. On the basis of Axioms I–III, prove that $0a = 0$ for any real number a.

3. Prove that the equation $a + x = b$ has a unique solution for any real numbers a and b.

4. For real numbers a, b, c, and d, prove that if $a < b$ and $0 < c < d$, then $ac < bd$, and if $0 < a < b$, then $a^n < b^n$ for any natural number n.

5. Consider the set of rational numbers as a part of the set of real numbers on the basis of Axiom V. Prove that the rational number 0 coincides with the real number 0 whose existence is guaranteed by Axiom I_3. Prove that the rational number 1 coincides with the real number 1 whose existence is guaranteed by Axiom II_3.

6. Without using Axiom V, prove that the numbers $0, 1, 1+1, \ldots, 1+1+\cdots+1$ (with n terms) are different for all natural numbers n. Here 1 denotes the real number whose existence is guaranteed by Axiom II_3. Use this to prove that the natural numbers are contained among the real numbers and that the operations and inequalities defined for real numbers give the customary operations and inequalities when applied to natural numbers. Then prove the assertion in Axiom V. Introducing this axiom would thus be unnecessary: it could be proved on the basis of the other axioms.

7. In addition to the operation of multiplication defined for real numbers, we define a new operation, denoted by $\otimes$ and given by the formula $a \otimes b = a + b + ab$. Does it satisfy Axiom II?

15. Limits and Infinite Sums

To illustrate the role of the axiom of included intervals as a method for constructing new real numbers, we introduce a few concepts that are useful later.

In Chap. 4, we met bounded sequences and sequences that increase without limit. We now consider sequences not from the standpoint of increase but from the standpoint of decrease. For simplicity, we first consider sequences of positive numbers and take the notion of decrease without limit to mean an unlimited approach to zero. The exact definition is constructed in exact analogy with the definition of increase without limit given in Sec. 12.

A sequence a_n of nonnegative real numbers is said to *approach zero without limit* if for any arbitrarily small positive number ε, there exists a natural number N such that $a_n < \varepsilon$ for all $n > N$. In this case, we also say that the sequence a_n *goes to* zero, and we write $a_n \to 0$ as $n \to \infty$ (read: as n goes to infinity). For example, in the formulation of Lemma 11, we could write $b_n - a_n \to 0$ as $n \to \infty$.

A typical example of a sequence that approaches zero without limit is given by the sequence $a_n = 1/n$.

We consider a somewhat less obvious example.

Lemma 12. *If a is any positive number less than 1, then the sequence $a_n = a^n$ approaches zero without limit, that is, $a^n \to 0$ as $n \to \infty$.*

Indeed, we set $a = 1/A$. Then $A > 1$, and it can be written in the form $A = 1 + x$ with $x > 0$. According to the binomial formula, $A^n = (1 + x)^n = 1 + nx + y$, where y is the sum of positive terms, that is, $y > 0$. This means that $A^n > 1 + nx$. Consequently, for any $\varepsilon > 0$, there exists N such that $A^n > 1/\varepsilon$ for all $n \geq N$ (you can calculate this N exactly). Therefore, $a^n < \varepsilon$, which means that $a^n \to 0$ as $n \to \infty$. $\square$

We generalize the definition given above to sequences $a_1, a_2, \ldots, a_n, \ldots$ whose numbers can be negative. Then the numbers $|a_1|, |a_2|, \ldots, |a_n|, \ldots$ are nonnegative, and the previous definition can be applied to them. A sequence of numbers a_n is said to approach zero without limit if the sequence of numbers $|a_n|$ approaches zero without limit. In this case, we write $a_n \to 0$ as $n \to \infty$.

We now come to our main definition. For a sequence $a = (a_1, a_2, \ldots, a_n, \ldots)$, if there exists a real number α such that $a_n - \alpha \to 0$ as $n \to \infty$, then α is called the *limit* of the sequence a. We also say that the sequence a *goes to* α and write $a_n \to \alpha$ as $n \to \infty$.

Not every sequence necessarily has a limit. For example, if a sequence has a limit, then it is bounded. Indeed, let $a_n \to \alpha$ as $n \to \infty$. Then there exists N such that $|a_n - \alpha| < 1$ for $n > N$. Because $a_n = \alpha + (a_n - \alpha)$, it follows that $|a_n| \leq |\alpha| + 1$ for $n > N$. And this means that $|a_n| \leq C$ for all n, where C is the largest of the numbers $|a_1|, \ldots, |a_N|, |\alpha| + 1$. But even if a sequence is bounded, it might not have a limit. The sequence $0, 1, 0, 1, \ldots$ in which 0 and 1 alternate is such a sequence. If it has a limit α, then by the definition of limit, we could take $\varepsilon = 1/2$ and would have $|a_n - \alpha| < 1/2$ for all $n > N$. But among the a_n with $n > N$, we have both 0 and 1. Therefore, we would have $|\alpha| < 1/2$ and $|1 - \alpha| < 1/2$. Clearly, there is no such number α.

But if a sequence has a limit, then it has only one limit. Indeed, we suppose that a sequence $a_1, a_2, \ldots, a_n, \ldots$ has two limits α and β and

$\alpha \neq \beta$. Then for every $\varepsilon > 0$, there exist N and N' such that $|a_n - \alpha| < \varepsilon$ for $n > N$ and $|a_n - \beta| < \varepsilon$ for $n > N'$. Let $n > N$ and $n > N'$. Then $|a_n - \alpha| < \varepsilon$ and $|a_n - \beta| < \varepsilon$. Hence, $|\alpha - \beta| < 2\varepsilon$. Because ε is an arbitrary positive number, we can choose $\varepsilon < |\alpha - \beta|/2$ and thus obtain a contradiction.

Because not every bounded sequence has a limit, considering such sequences still does not lead to constructing new real numbers. Our fundamental result indicates one simple type of sequence that always has a limit and thus gives a method for constructing new numbers.

A sequence $a_1, a_2, \ldots, a_n, \ldots$ is said to be increasing if $a_n \leq a_{n+1}$ for all n, that is, $a_1 \leq a_2 \leq a_3 \leq a_4 \leq \ldots$.

Theorem 29. *An increasing bounded sequence has a limit.*

The proof follows the logic of an anecdote that was popular at the time when I was a student (that is, before World War II). It was a series of witticisms about who proposes which method for capturing a lion in the desert. There were the Frenchman's method, the NKVD investigator's method, and the mathematician's method. The mathematician's method consisted of the following. He divides the desert into two equal parts. The lion is located in one of them. That part is then divided into two equal parts—and the process continues until the lion is located in a part of the desert whose size does not exceed the size of the cage. It only remains to surround the lion with bars. This was a parody on a certain type of proof of an existence theorem, of which we now give one example.

Let $a = (a_1, a_2, \ldots, a_n, \ldots)$ be an increasing sequence. By condition, it is bounded, that is, there exists a number C such that all $|a_n| < C$. All numbers a_i are therefore contained in the interval $I_0 = [a_1, C]$. Let D denote the length of this interval, that is, $C - a_1$.

We use the number $C_1 = (a_1 + C)/2$ to divide the interval $I_0 = [a_1, C]$ into two equal parts. Then one of two possibilities holds: either there exist m such that $a_m \geq C_1$ (and then all a_n with $n \geq m$ are contained in the interval $[C_1, C]$ because the sequence is increasing) or $a_n \leq C_1$ for all n (and then all terms of the sequence are contained in the interval $[a_1, C_1]$). We let I_1 denote that interval $[a_1, C_1]$ or $[C_1, C]$ containing all terms of the sequence beginning from some point. We then divide the interval I_1 into two equal parts and repeat our argument.

Obviously, we can continue thus indefinitely and obtain a sequence of included intervals $I_0 \supset I_1 \supset I_2 \supset \cdots \supset I_n \supset \cdots$ in which the interval I_k has the length $D/2^k$ and the property that it contains all terms of the sequence a beginning from some point. According to the axiom of included intervals (Axiom VII), there exists a real number α that belongs to all the intervals. It is the limit of the sequence a. Indeed, as we saw, all terms of the sequence a beginning from some point are contained in the interval I_k. This means that for any natural number k, there exists an N such that $a_n \in I_k$ for all $n > N$. But also $\alpha \in I_k$. Because the length of the interval I_k is equal to $D/2^k$, it follows that $|a_n - \alpha| < D/2^k$ for $n > N$. And this gives us the property in the definition of limit if we choose k such that $D/2^k < \varepsilon$. We emphasize that such a choice is always possible: clearly, the sequence $D, D/2, D/4, D/8, \dots$ approaches zero without limit.   □

Theorem 29 is particularly useful when a sequence $a = (a_1, a_2, \dots, a_n, \dots)$ is the sequence of sums of a sequence of nonnegative numbers $c = (c_1, c_2, \dots, c_n, \dots)$ $(c_n \geq 0)$, that is, when $a_1 = c_1$, $a_2 = c_1 + c_2$, $\dots$, $a_n = c_1 + c_2 + \cdots + a_n, \dots$. Then the sequence a is obviously increasing. But it must be verified that it is bounded (this is not always simple). For example, if all $c_n = 1$ in a sequence c, then $a_n = n$, and the sequence a is not bounded. We considered a less obvious example in Sec. 12: the sequence c where all $c_n = 1/n$. We saw that the sequence a is not bounded in that case. But if it is verified that the sequence of sums a is bounded, then according to the theorem, it has a unique limit α. This limit is called the *sum of the sequence* $c_1, c_2, \dots, c_n, \dots$ and is written in the form

$$c_1 + c_2 + \cdots + c_n + \cdots = \alpha.$$

An infinite sequence c is sometimes also called a *series*, and its sum is called the *sum of the series*.

If the sequence of sums a_n is bounded, then, as we saw, the sum of the series $c_1 + c_2 + \cdots + c_n + \cdots$ exists. If it is not bounded, then we say that the sum of the series does not exist. Thus, Lemma 8 confirms that the sum of the series $1 + 1/2 + 1/3 + \cdots$ does not exist.

We consider an example. Let a nonnegative number a less than 1 be given. We consider the sequence $c = (1, a, a^2, \dots, a^n, \dots)$. Then $a_n = 1 + a + a^2 + \cdots + a^{n-1}$ (the nth position in the sequence c contains a^{n-1}). The sum $1 + a + a^2 + \cdots + a^{n-1}$ can be calculated according to the formula for the sum of a geometric progression (formula (12) in Chap. 1):

$$a_n = 1 + a + a^2 + \cdots + a^{n-1} = \frac{1 - a^n}{1 - a} = \frac{1}{1-a} - \frac{a^n}{1-a}. \qquad (1)$$

We saw that $a^n \to 0$ as $n \to \infty$. It follows directly from this that $a^n/(1-a) \to 0$ as $n \to \infty$. It therefore follows from formula (1) that $a_n \to 1/(1-a)$. We can write this differently:

$$1 + a + a^2 + \cdots + a^{n-1} + \cdots = \frac{1}{1-a} \quad \text{for } a < 1. \tag{2}$$

The series in the left-hand side of equality (2) is called an *infinite geometric progression*, and formula (2) is the formula for the sum of an infinite geometric progression.

But there are examples of series for which it is easy to establish the existence their sums, but much harder to calculate the sums explicitly. For example, we proved in Sec. 12 that the sums $1/1^2 + 1/2^2 + \cdots + 1/n^2$ are bounded. This means that the sum of the series $1 + 1/2^2 + 1/3^2 + \cdots + 1/n^2 + \cdots$ exists. But what is it equal to? J. Bernoulli posed this problem in the 17th century. It greatly fascinated mathematicians in the middle of the 18th century. Euler solved this problem, establishing the amazing equality

$$1 + \frac{1}{2^2} + \frac{1}{3^2} + \cdots + \frac{1}{n^2} + \cdots = \frac{\pi^2}{6}. \tag{3}$$

This was one of the most sensational discoveries of Euler. News of it quickly spread among mathematicians, and the majority of Euler's correspondents asked him to explain how he proved this equality. But Euler had gone further, having calculated the sum of the series $1 + 1/2^k + 1/3^k + \cdots + 1/n^k + \cdots$ for any *even* k. It turned out that this sum is connected with the Bernoulli numbers, which we discussed in the supplement to Chap. 2. Namely, the formula

$$1 + \frac{1}{2^k} + \frac{1}{3^k} + \cdots + \frac{1}{n^k} + \cdots = \pi^k (-1)^{k/2-1} \frac{2^{k-1} B_k}{k!} \tag{4}$$

holds for any even k. To this time, almost nothing is known about the analogous sums for odd k. It was proved only relatively recently (in 1978) that the sum $1 + 1/2^3 + 1/3^3 + \cdots + 1/n^3 + \cdots$ is an irrational number. It has been proved that the number of odd k for which the corresponding sum is irrational is infinite, but no such particular k except 3 is known. To this time, this is the only known fact about such sums for odd values of k.

We note that some useful consequences can be deduced from the existence of the sum $c_1 + c_2 + \cdots + c_n + \cdots$ even if the value of the sum is unknown to us.

Lemma 13. *If the sum of the series* $c_1 + c_2 + \cdots + c_n + \cdots$ *exists, then the sequence of numbers* $d_n = c_{n+1} + c_{n+2} + \cdots$ *approaches zero without limit.*

We use a simple property of a limit. Let a sequence $a_1, a_2, \ldots, a_n, \ldots$ have the limit α, that is, $a_n - \alpha \to 0$ as $n \to \infty$. Then for any number β, the sequence $\beta - a_1, \beta - a_2, \ldots, \beta - a_n, \ldots$ has the limit $\beta - \alpha$. Indeed, the difference $\beta - \alpha - (\beta - a_n) = a_n - \alpha$, and if $a_n - \alpha \to 0$, then $\beta - \alpha - (\beta - a_n) \to 0$ as $n \to \infty$.

Let α denote the sum of the series $c_1 + c_2 + \cdots + c_n + \cdots$ and a_m denote the number $c_1 + c_2 + \cdots + c_m$ for any m. By the definition of the sum of an infinite series, the sum α of the series $c_1 + c_2 + \cdots + c_n + \cdots$ is equal to the limit of the sequence $a_1, a_2, \ldots, a_m, \ldots$. In exactly the same way, the sum d_n of the series $c_{n+1} + c_{n+2} + \cdots$ is equal to the limit of the sequence $a_{n+1} - a_n, a_{n+2} - a_n, \ldots, a_{n+k} - a_n, \ldots$. According to the remark at the beginning of the proof, this last limit is equal to $\alpha' - a_n$, where α' is the limit of the sequence $a_{n+1}, a_{n+2}, \ldots, a_{n+k}, \ldots$ (for fixed n). But the limit of the sequence $a_{n+1}, a_{n+2}, \ldots$ is exactly the same as the limit of the sequence $a_1, a_2, \ldots$, that is, $\alpha' = \alpha$. We see that $d_n = \alpha - a_n$. And by the definition of limit, $\alpha - a_n \to 0$, that is, $d_n \to 0$ as $n \to \infty$.   □

For example, we set $d_n = 1/n^2 + 1/(n+1)^2 + \cdots$ and can see that $d_n \to 0$ as $n \to \infty$.

The consideration of limits and infinite sums leads us out of algebra, which by its spirit is connected with finite expressions. These questions are closer to the branch of mathematics that is called analysis. Therefore, we do not go deeper into these questions. However, we note that the more amazing results—such as equalities (3) and (4)—arise at the junction of these branches.

Problems:

1. Prove that if the sum of the series $c_1 + c_2 + \cdots + c_n + \cdots$ exists, then $c_n \to 0$ as $n \to \infty$.

2. Prove that if $a_n < C$ for any n and $a_n \to \alpha$ as $n \to \infty$, then $\alpha \le C$. Give an example where the equality is attained.

3. Given that $a_n \to \alpha$ as $n \to \infty$. We set $b_n = a_{2n}$. Does the sequence $b_1, b_2, \ldots$ have a limit, and what is it equal to? Can we conclude from the existence of the limit of such a sequence that the sequence $a_1, a_2, \ldots$ has a limit? If it has a limit, what is the limit?

4. Does the sequence $a_1, a_2, \ldots, a_n, \ldots$ in which $a_n = 1/2 - 1/3 + \cdots + (-1)^n/n$ have a limit?

5. Let $f(x)$ be a polynomial of degree d. Prove that if $a_n = f(n)/n^{d+1}$, then $a_n \to 0$ as $n \to \infty$.

6. Find the sum of the series $b + ba + ba^2 + \cdots + ba^n + \cdots$ for $|a| < 1$ and any number b. Usually, the sequence $b, ba, ba^2, \ldots$ is also called an infinite geometric progression.

7. In a square with the side a, the midpoints of adjacent sides are joined by line segments. Then the same operation is performed with the resulting square, and so on. Find the sum of the areas of all resulting squares.

8. Find the sum of the series

$$\frac{1}{1 \cdot 2} + \frac{1}{2 \cdot 3} + \cdots + \frac{1}{n(n+1)} + \cdots.$$

   Hint: Use the fact that $1/(n(n+1)) = 1/n - 1/(n+1)$ as in Problem 4.

9. Construct a sequence of positive rational numbers less than 1 in which a_n has the denominator n and which does not have a limit.

10. Prove that if the sequence $a_1, a_2, \ldots$ has the limit α and the sequence $b_1, b_2, \ldots$ has the limit β, then the sequence $a_1 + b_1, a_2 + b_2, \ldots$ has the limit $\alpha + \beta$.

11. Prove that if $0 \leq \alpha_i \leq \beta_i$ and the sum $\beta_1 + \beta_2 + \cdots + \beta_n + \cdots$ exists, then the sum $\alpha_1 + \alpha_2 + \cdots + \alpha_n + \cdots$ exists, and $\alpha_1 + \alpha_2 + \cdots + \alpha_n \cdots \leq \beta_1 + \beta_2 + \cdots + \beta_n + \cdots$.

16. Representation of Real Numbers as Decimal Fractions

In Sec. 14, we used a system of axioms to describe real numbers. We now show how it is possible to represent a real number concretely. We do not say anything new here—we speak about the basics of the well-known written representation of real numbers using infinite decimal fractions. But we now show how the existence of such a representation is deduced from the axioms in Sec. 14.

We have in mind the habitual written representation in which the negative real numbers are written with the same digits as the corresponding positive number (its absolute value) and with a minus sign in front. Clearly, we can use Axiom I_4 to establish a one-to-one correspondence between the subset of negative real numbers and the subset of positive real numbers. For simplicity (to avoid the continual use of the absolute value notation for example), we therefore treat the subset of positive real numbers in what follows.

Let A be an arbitrary nonnegative integer and $a_1, a_2, \ldots, a_n, \ldots$ be an infinite sequence of integers each of which has one of the ten values 0, 1, 2, 3, 4, 5, 6, 7, 8, 9. We write all this together as $A.a_1a_2a_3 \ldots$ and call it an infinite decimal fraction. Then it is just another way to write an infinite sequence. We now show how it can be associated with a real number. For this, we define the number

$$\alpha_n = A + \frac{a_1}{10} + \cdots + \frac{a_n}{10^n} \tag{5}$$

for any index n. Obviously, the sequence $\alpha_1, \alpha_2, \ldots, \alpha_n, \ldots$ is increasing. We prove that it is bounded. Indeed, because all $a_i \leq 9$,

$$\frac{a_1}{10} + \frac{a_2}{100} + \cdots + \frac{a_n}{10^n} \leq \frac{9}{10} \left(1 + \frac{1}{10} + \cdots + \frac{1}{10^{n-1}} \right).$$

We can apply the formula for the sum of a geometric progression (formula (12) in Chap. 1) to the sum in the parentheses:

$$1 + \frac{1}{10} + \cdots + \frac{1}{10^{n-1}} = \frac{1 - 1/10^n}{1 - 1/10} < \frac{1}{9/10}.$$

As a result, we obtain

$$\frac{a_1}{10} + \frac{a_2}{100} + \cdots + \frac{a_n}{10^n} < 1, \tag{6}$$

and therefore $\alpha_n < A + 1$.

According to Theorem 29, the sequence $\alpha_1, \alpha_2, \ldots, \alpha_n, \ldots$ has a limit α. We call the real number α the number corresponding to the infinite decimal fraction and write it as

$$\alpha = A.a_1 a_2 \ldots a_n \ldots . \tag{7}$$

We sometimes say that α is equal to the decimal fraction $A.a_1 \ldots a_n \ldots$. This simply means that α is equal to the sum of the infinite series $A + a_1/10 + a_2/10^2 + \ldots$.

Our further goal is to investigate this correspondence between infinite decimal fractions and real numbers. Is it a one-to-one correspondence? This presupposes two questions. Can two different decimal fractions correspond to the same real number? And does every real number correspond to some decimal fraction?

We consider the first question. We first note that the answer to it is sometimes affirmative. For example, we take the infinite decimal fraction $0.9999\ldots$, in which only nines occur after the decimal point. What real number does it correspond to? In accordance with the general definition, we must consider the sequence $\alpha_n = 9/10 + 9/10^2 + \cdots + 9/10^n$. It is simple to calculate this sum: according to the formula for the sum of a geometric progression (see formula (12) in Chap. 1), it is equal to

$$\frac{9}{10} \left(1 + \frac{1}{10} + \cdots + \frac{1}{10^{n-1}} \right) = \frac{9}{10} \frac{1 - 1/10^n}{1 - 1/10}$$

$$= \frac{9}{10} \frac{1 - 1/10^n}{9/10} = 1 - \frac{1}{10^n}.$$

It is obvious that the limit of the sequence $\alpha_1, \alpha_2, \ldots, \alpha_n, \ldots$ is equal to 1, and therefore $1 = 0.9999\ldots$. But on the other hand, we of course also have $1 = 1.0000\ldots$, where only zeros occur after the decimal point. Two different infinite decimal fractions (defined as a *written representation*) correspond to the same real number 1.

Clearly, we can construct many such examples. In the most general form, such an example can be constructed as follows. Let an infinite decimal fraction have the form $A.a_1 \ldots a_k 99 \ldots$, that is, we suppose that all the digits are nines after a certain position (for us, the kth). Moreover, we can suppose that $a_k \neq 9$, that is, the kth position is the first position after which there are only nines. We can then repeat the preceding argument literally and verify that this fraction is equal to the same number as the fraction $A.a_1 \ldots a_{k-1}(a_k + 1)00 \ldots$, where only zeros occur after the kth position and we have $a_k + 1$ in the kth position. Regarding a fraction in which only nines occur after some position, we say that it has a repeating nine. We saw that the one-to-one correspondence between infinite decimal fractions and real numbers is violated for such fractions. Our argument shows that every decimal fraction that has a repeating nine determines the same real number as some fraction that does not have a repeating nine.

It is somewhat unexpected that the violation is produced in only the cases considered.

Theorem 30. *Different real numbers correspond to two different infinite decimal fractions neither of which has a repeating nine.*

The proof of the theorem is obtained by itself if we connect our construction of a real number determined by a decimal fraction with measurement of a number to the accuracy $1/10^m$ with a deficit or surplus. We must divide a straight line into segments of the length $1/10^m$ at the end of which are rational numbers with the denominator 10^m. Then each point on the line, that is, each real number, falls in one segment. Its beginning and end yield the measure of the real number with a deficit and a surplus to the accuracy $1/10^m$. However, an ambiguity arises with the very ends of the segments, that is, with the rational numbers with the denominator 10^m. They participate in two segments. Which are they connected with—with the left-hand or the right-hand segment? This is the same difficulty that arose in connection with the repeating nine. We show that our choice (without a repeating nine) corresponds to the dividing point being connected with the right-hand segment. In other words, we prove that the numbers α_m we construct (without a repeating nine) and the number α they determine are connected by the relation

$$\alpha_m \le \alpha < \alpha_m + \frac{1}{10^m}. \tag{8}$$

That the number α_m has a denominator of the form 10^m is evident from representation (5).

We recall that the number α is defined as the limit of the sequence $\alpha_1, \alpha_2, \ldots, \alpha_n, \ldots$. All the numbers α_n with $n \ge m$ obviously satisfy the condition $\alpha_n \ge \alpha_m$. Hence, the same inequality $\alpha \ge \alpha_m$ follows for their limit α. Indeed, from the supposition that $\alpha < \alpha_m$, we could conclude that $\alpha_n - \alpha = (\alpha_n - \alpha_m) + (\alpha_m - \alpha) > \alpha_m - \alpha$ for all $n \ge m$. But by the definition of limit, the absolute value of the number $\alpha_n - \alpha$ becomes less than any positive number given in advance if n is sufficiently large. This contradicts our deduction that it is greater than the fixed positive number $\alpha_m - \alpha$. (See Problem 2 in Sec. 15.)

This also proves the left inequality in (8). The right inequality could be proved in exactly the same way if we agreed to use the sign $\le$ instead of $<$ in it. Namely, for any $n > m$, we have

$$\alpha_n = \alpha_m + \frac{a_{m+1}}{10^{m+1}} + \cdots + \frac{a_n}{10^n} \le \frac{1}{10^m}\left(\frac{a_{m+1}}{10} + \cdots + \frac{a_n}{10^{n-m}}\right). \tag{9}$$

Applying inequality (6), we can conclude that $\alpha_n < \alpha_m + 1/10^m$. Hence, repeating the preceding argument, we obtain $\alpha \le \alpha_m + 1/10^m$.

But if we want to obtain the right inequality in relation (8) with the sign $<$, we must use the fact that the fraction $A.a_1 a_2 \ldots$ does not have a repeating nine. The proof becomes a bit more complicated. We prove the right inequality in (8) for some fixed index m. We use the fact that the decimal fraction α does not have a repeating nine. Therefore, somewhere after a_m, we must have a digit a_k that is not 9. For any $n > k$, we write

$$\alpha_n = \alpha_m + \left(\frac{a_{m+1}}{10^{m+1}} + \cdots + \frac{a_k}{10^k}\right) + \left(\frac{a_{k+1}}{10^{k+1}} + \cdots + \frac{a_n}{10^n}\right).$$

As we saw above, we have

$$\frac{a_{k+1}}{10^{k+1}} + \cdots + \frac{a_n}{10^n} \le \frac{1}{10^k},$$

and therefore

$$\alpha_n \le \alpha_m + \left(\frac{a_{m+1}}{10^{m+1}} + \cdots + \frac{a_k + 1}{10^k}\right).$$

Because $a_k \neq 9$, $a_k + 1$ is one of the digits $0, 1, \ldots, 9$.

We set

$$c = \frac{a_{m+1}}{10} + \cdots + \frac{a_k + 1}{10^{k-m}}.$$

We can repeat the previous argument and conclude that $c < 1$. The number c depends only on the choice of m and k, not on n. Therefore, replacing α_n with its limit α, we obtain $\alpha \leq \alpha_m + c/10^m < \alpha_m + 1/10^m$ as before. Inequality (8) is thus proved.

It immediately follows from relation (8) that two different decimal fractions that do not have a repeating nine cannot correspond to the same real number. Let the fractions $A.a_1 a_2 \ldots$ and $A'.a'_1 a'_2 \ldots$ correspond to the same real number α. Then we have the relation

$$\alpha'_m \leq \alpha < \alpha'_m + \frac{1}{10^m},$$

where $\alpha'_m = A' + a'_1/10 + \cdots + a'_m/10^m$. Let $\alpha'_m \neq \alpha_m$, for example, $\alpha'_m > \alpha_m$. It follows from these relations that $\alpha'_m < \alpha_m + 1/10^m$, that is, $\alpha'_m - \alpha_m < 1/10^m$. But this contradicts the fact that α_m and α'_m are different rational numbers with the denominator 10^m. Therefore, $\alpha'_m = \alpha_m$ for all m. But the numbers a_m are uniquely determined by the numbers α_m because $\alpha_{m+1} - \alpha_m = a_m/10^m$. Therefore, the numbers a_m and a'_m must coincide in both fractions. □

We now turn to the second question of whether every real number corresponds to some infinite decimal fraction. You are undoubtedly already familiar with both the answer to this question and the method of its proof. We only want to verify that the usual arguments are based on the axioms we formulated.

We first note that every real number α is located between two consecutive integers, that is, there exists an integer A such that $A \leq \alpha < A+1$. First let α be positive. According to the axiom of Archimedes, there exists a natural number n such that $\alpha < n$. And because there is only a finite number of natural numbers not exceeding n, there exists the last (smallest) number with this property. Let m denote this number. Because we supposed that α is positive, $m > 0$. Then $\alpha < m$, but $m - 1$ already does not have this property. This means that $m - 1 \leq \alpha < m$, and $A = m - 1$ has the property we need. And if α is negative, then we set $\alpha' = -\alpha$ and apply the already proved assertion to it: there exists n such that $n \leq \alpha' < n + 1$. It easily follows from Axiom IV$_3$ that then $-(n + 1) < \alpha \leq -n$. If α is not a natural number, then this proves the desired property with $A = -(n+1)$. If $\alpha = -n$, then we can set $A = -n$ and obtain $A \leq \alpha < A + 1$. Thus, for any real number α, there exists an integer A such that $A \leq \alpha < A + 1$, and this means that α can be represented in the form $\alpha = A + \varepsilon$, where $0 \leq \varepsilon < 1$.

We now note that if $a_1 < a_2$ and $a_2 < a_3$ for three numbers a_1, a_2, and a_3, then any α satisfying the condition $a_1 \leq \alpha < a_3$ satisfies one of the conditions $a_1 \leq \alpha < a_2$ or $a_2 \leq \alpha < a_3$. This simply means that the segment

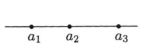

Fig. 22

$[a_1, a_3]$ is composed of the segments $[a_1, a_2]$ and $[a_2, a_3]$ in the drawing in Fig. 22. Formally, this is a consequence of the fact that any given α satisfies exactly one of the relations $\alpha < a_2$ or $a_2 < \alpha$ or $a_2 = \alpha$.

We consider a more general case: let n numbers $\alpha_1, \ldots, \alpha_n$ satisfy the conditions $\alpha_1 < \alpha_2$, $\alpha_2 < \alpha_3, \ldots$, $\alpha_{n-1} < \alpha_n$. Then any number α satisfying the condition $\alpha_1 \leq \alpha < \alpha_n$ satisfies the condition $\alpha_{i-1} \leq \alpha < \alpha_i$ for some $i = 2, 3, \ldots, n$. To prove this, we apply the preceding assertion to the three numbers $\alpha_1, \alpha_2, \alpha_n$. Then either $\alpha_1 \leq \alpha < \alpha_2$ (and our proposition holds for $i = 2$) or $\alpha_2 \leq \alpha < \alpha_n$. In the latter case, we must consider the numbers $\alpha_2, \alpha_3, \ldots, \alpha_n$ and repeat the same argument. For some i, we obtain the needed relation $a_{i+1} \leq \alpha < a_i$.

After this small diversion, we return to our basic question. We already proved that any real number α can be represented in the form $A + \varepsilon$, where A is an integer and $0 \leq \varepsilon < 1$. We now consider the numbers $k/10$, $k = 0, 1, \ldots, 9$. According to the preceding result, we can assert that $k/10 \leq \varepsilon < (k+1)/10$ for some k, $0 \leq k < 10$. Letting a_1 denote that number k, we can write $\varepsilon = a_1/10 + \varepsilon_1$, where $0 \leq \varepsilon_1 < 1/10$. Hence, $\alpha = A + a_1/10 + \varepsilon_1$. Continuing this process, we obtain the numbers $a_1, \ldots, a_n, \ldots$, where always $0 \leq a_i \leq 9$ and $\alpha = A + a_1/10 + \cdots + a_n/10^n + \varepsilon_n$ with $0 \leq \varepsilon_n < 1/10^n$ for any n. And this means that the sequence $\alpha_1, \alpha_2, \ldots, \alpha_n, \ldots$, where $\alpha_n = A + a_1/10 + \cdots + a_n/10^n$, has the limit α, that is, the infinite decimal fraction $A.a_1 a_2 \ldots a_n \ldots$ corresponds to the number α.

Summarizing all that has been proved, we can assert that *the association of infinite decimal fractions with real numbers does not constitute a one-to-one correspondence between infinite decimal fractions and real numbers, but this correspondence becomes one-to-one if the fractions with a repeating nine are excluded from the set of infinite decimal fractions.*

Problems:

1. Prove that the real number α corresponds to an infinite decimal fraction with a repeating zero if and only if α is a rational number a/b, where a and b are integers and moreover only the prime numbers 2 and 5 divide b. (If r is one of the digits $0, 1, \ldots, 9$, then we say that an infinite decimal fraction $\alpha = A.a_1 a_2 \ldots a_n \ldots$ has a repeating r if all a_n beginning with $n \geq n_0$ are equal to r.)

2. For associating an infinite decimal fraction with the rational number a/b, it is sufficient to calculate the fractional part, and we can therefore consider that $0 < a < b$. Let $\alpha_n = a_1/10 + a_2/10^2 + \cdots + a_n/10^n$, where $0.a_1a_2\ldots$ is an infinite decimal fraction, correspond to the number a/b. Prove that $a/b - \alpha_n = r_n/(10^n b)$, where $0 \leq r_n < b$ and the numbers r_n are connected by the relation $10r_{n-1} = ba_n + r_n$, that is, a_n is the quotient and r_n is the remainder from dividing $10r_{n-1}$ by b. Verify that this way of calculating the digits a_n of the decimal fraction coincides with the usual method of "long division."

3. Prove that an infinite decimal fraction corresponding to a rational number is periodic, that is, has the form $(* * \ldots)(P)(P)\ldots$, where $(* * \ldots)$ denotes some finite string of digits after which the same string of digits (P), called the period, is repeated endlessly. *Hint:* Use Problem 2 (that is, "long division") and note that the number of remainders from dividing $10r_{n-1}$ by b is finite (not greater than b).

4. Prove that if the denominator b of the fraction a/b is relatively prime to 10, then the period begins immediately after the decimal point.

5. With the condition in Problem 4, prove that the number of digits in the period is the smallest number k for which $10^k - 1$ is divisible by b.

6. With the condition in Problem 4, prove that the number of digits in the period does not exceed the number of natural numbers not exceeding b and relatively prime to b. This number is given by formula (25) in Chap. 3.

7. Prove that any periodic infinite decimal fraction corresponds to a rational number. Namely, if $A.a_1a_2\ldots a_n$ precedes the period, if the period is $(p_0p_1\ldots p_{m-1})$, if $A + a_1/10 + \cdots + a_n/10^n = Q$, and if $p_0 + p_1 10 + \cdots + p_{m-1}10^{m-1} = P$, then the fraction corresponds to the rational number $Q + P/(10^{n+1}(10^m - 1))$.

8. Prove that the infinite decimal fraction $0.1001000100001\ldots$, where the number of zeros between two ones increases by 1 each time, corresponds to an irrational number.

17. Real Roots of Polynomials

Having established a firmer foundation for the theory of real numbers, we can now obtain some new results concerning real roots of polynomials with real coefficients. For this, we must first study the behavior of a polynomial $f(x)$ in the neighborhood of one value $x = a$.

Theorem 31. *For any polynomial $f(x)$ and any number a, there exists a positive constant M such that the inequality*

$$|f(x) - f(a)| \leq M |x - a| \qquad (10)$$

is satisfied for all x for which $|x - a| \leq 1$.

We recall that $|A|$ (read "the absolute value of A") by definition is equal to A if $A \geq 0$ and to $-A$ if $A < 0$. Therefore, $|A|$ is always a nonnegative number. The properties of $|A|$ are known from a school course:

$$|A + B| \leq |A| + |B|, \tag{11}$$
$$|A + B| \geq |A| - |B|, \tag{12}$$
$$|AB| = |A| \cdot |B|. \tag{13}$$

Theorem 31 states that $f(x)$ becomes ever closer to $f(a)$ when x is chosen closer to a. We can therefore consider $f(a)$ an estimate of the value of $f(x)$, and formula (10) estimates the error of the estimate. To prove the theorem, we set $y = x - a$, that is, $x = a + y$, and substitute this value in the polynomial $f(x)$. Each term $a_k x^k$ of the polynomial then gives the expression $a_k(a+y)^k$, which we can expand in powers of y and then collect like terms in $f(a+y)$. As a result, we find that $f(a+y)$ is a polynomial in y, which can be denoted by $g(y) = c_0 + c_1 y + \cdots + c_n y^n$. Then $f(x) = f(a+y) = g(y)$ and $f(a) = f(a+0) = g(0)$. Inequality (10), which we want to prove, becomes

$$|g(y) - g(0)| \leq M |y| \tag{14}$$

for all y such that $|y| \leq 1$.

In the changed form, the expression $g(y) - g(0)$ takes the simple form $c_1 y + \cdots + c_n y^n$ because $g(0) = c_0$. Inequality (11) can be applied to the sum of any number of terms (which can be immediately verified by induction) and, in particular, to our sum $c_1 y + \cdots + c_n y^n$. We obtain

$$|g(y) - g(0)| = |c_1 y + \cdots + c_n y^n| \leq |c_1 y| + \cdots + |c_n y^n|.$$

By inequality (13) (also applicable to an arbitrary number of factors), $|c_k y^k| \leq |c_k| \cdot |y|^k$, and therefore

$$|g(y) - g(0)| \leq |c_1| \cdot |y| + \cdots + |c_n| \cdot |y|^n.$$

Because $|x - a| \leq 1$ by hypothesis, that is, $|y| \leq 1$, we have $|y|^k \leq |y|$ and consequently

$$|g(y) - g(0)| \leq (|c_1| + \cdots + |c_n|) |y|$$

for $|y| \leq 1$. It is now sufficient to set $M = |c_1| + \cdots + |c_n|$, and we obtain inequality (14) and therefore (10). $\qquad\Box$

We can now prove an important property of polynomials, formulated as Theorem 32.

Theorem 32 (Bolzano Theorem). *If the values of a polynomial have different signs for $x = a$ and $x = b$, then the polynomial becomes zero between those values.*

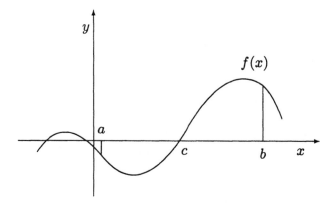

Fig. 23

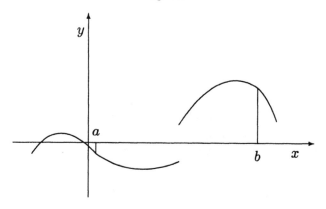

Fig. 24

That is, if the values $f(a)$ and $f(b)$ have different signs for some polynomial $f(x)$ and $a < b$, then there exists a number c such $a < c < b$ and $f(c) = 0$.

Theorem 32 seems obvious at a glance, especially looking at the graph of a polynomial $f(x)$ (see Fig. 23). The theorem asserts that the graph cannot "jump over" the x axis without intersecting it. But we could indeed *draw* such a graph (see Fig. 24).

We must prove that such a graph cannot be the graph of a polynomial. For more general functions, this is connected with a rather refined property called *continuity*. But in the case of polynomials, simple inequality (10), established in Theorem 31, is sufficient.

The proof is based on the same principle of "capturing a lion in the desert" that we used to prove Theorem 29. For definiteness, we can take $f(a) > 0$ and $f(b) < 0$. We consider the interval $[a, b]$ (that is, the set of real numbers x for which $a \leq x$ and $x \leq b$). We let I_1 denote this interval and divide it into two parts of equal length at the midpoint $r = (a - b)/2$. If $f(r) = 0$, then the theorem is proved $(c = r)$. If $f(r) \neq 0$ and $f(r) > 0$, then the polynomial $f(x)$ takes values with different signs for $x = r$ and $x = b$. We then let I_2 denote the interval $[r, b]$. And if $f(r) < 0$, then we let I_2 denote the interval $[a, r]$. In both cases, we obtain an interval I_2 that is contained in I_1, is half the length of I_1, and has endpoints at which the function $f(x)$ takes values with opposite sign—specifically, positive at the left endpoint and negative at the right endpoint.

This process can be continued further. Either we meet a root of the function at some point (and the theorem is proved with that value of c) or the process continues endlessly. It remains to consider the latter case. We obtain an infinite sequence of intervals with each containing the next: $I_1 \supset I_2 \supset \cdots \supset I_n \supset \ldots$, where $I_n = [a_n, b_n]$. Moreover, the length of a successive interval is half the length of the interval it follows, and $f(x)$ takes values with opposite signs at the endpoints of each interval, specifically, $f(a_n) > 0$ and $f(b_n) < 0$. And here we use the more rigorous analysis of the concept of a real number laid out in Sec. 14. The intervals I_n satisfy the conditions of Axiom VII (the axiom of included intervals) and Lemma 13, proved in Sec. 14. Indeed, the intervals I_n are included, each in the preceding, by construction, and because I_{n+1} is half the length of the interval I_n, the length of I_{n+1} is equal to $(b - a)/r^n$, that is, the length becomes arbitrarily small as n increases. Therefore, according to Axiom VII and Lemma 13, there exists a unique number c that belongs to all intervals I_n, that is, such that

$$a_n \leq c \leq b_n. \tag{15}$$

The construction of the number we need is finished. Namely, we can confirm that $f(c) = 0$ in accordance with the assertion in the theorem. But it remains to prove this.

We consider the values $f(a_n)$ of the polynomial $f(x)$ at the left-hand endpoints of the intervals I_n. All $f(a_n) > 0$ by construction. It follows from inequality (15) that the sequence $a_1, a_2, \ldots$ approaches arbitrarily close to the number c: indeed, $a_n \leq c \leq b_n$ and $0 \leq c - a_n \leq b_n - a_n$, and we moreover have the condition $b_n - a_n = (b - a)/2^{n-1}$. Therefore, we can satisfy the inequality $|a_n - c| < \varepsilon$ if $(b - a)/2^{n-1} < \varepsilon$, and this is true for any $\varepsilon > 0$ if n is chosen sufficiently large. We use this to prove that the values $f(a_n)$ approach arbitrarily close to the value

$f(c)$. To prove that $|f(a_m) - f(c)| \leq \varepsilon$ for sufficiently large m, we use inequality (10) in Theorem 31. Because a_m approaches arbitrarily close to c, we have $|a_m - c| < 1$ for sufficiently large m, and we can apply inequality (10). We see that $|f(a_m) - f(c)| < M\,|a_m - c|$. This means that $|f(a_m) - f(c)| < \varepsilon$ if $M\,|a_m - c| < \varepsilon$, that is, $|a_m - c| < \varepsilon/M$. But we already verified that such an inequality is satisfied for sufficiently large m (ε/M can again be denoted by ε).

What can we say about the number $f(c)$ if we know that a sequence of positive numbers approaches arbitrarily close to it? It is clear that $f(c) \geq 0$. Indeed, if $f(c)$ were negative, then we would have $f(a_n) - f(c) > -f(c)$ for positive $f(a_n)$ and therefore $|f(a_n) - f(c)| > -f(c)$. This would contradict the fact that $|f(a_n) - f(c)| < \varepsilon$ if $\varepsilon < -f(c)$.

We have thus proved that $f(c) \geq 0$. In exactly the same way, we consider the numbers b_n, for which $f(b_n) < 0$, and prove that $f(c) \leq 0$. Therefore, only one possibility remains for $f(c)$: $f(c) = 0$. The theorem is thus proved.   □

We note a completely new type of argument that we used to prove the theorem. We, in fact, proved the existence of a root c of a polynomial $f(x)$ (under certain conditions). But we did this not using a formula (such as the formula for solving a quadratic equation) but using the axiom of included intervals. At the same time, it is far from a pure "existence theorem," when we know only that a certain quantity exists and nothing more. For instance, we can actually find the root c with a deficit and surplus, and with desired accuracy, constructing the numbers a_n and b_n between which c is contained (see inequality (15)) and which become ever closer to one another.

The Bolzano theorem enables us to find out quite a bit about a concrete polynomial. For example, we consider the polynomial $f(x) = x^3 - 7x + 5$ and construct a table of its values for small integer values of x:

x	-3	-2	-1	0	1	2	3
$f(x)$	-1	11	11	5	-1	-1	11

It is evident from the table that the polynomial $f(x)$ takes values with opposite signs at the ends of the intervals $[2, 3]$, $[0, 1]$, and $[-3, -2]$. According to the Bolzano theorem, it has roots in each of these intervals. The polynomial $f(x)$ therefore has at least three roots. But its degree is equal to three, and according to Theorem 14, it cannot have more than three roots. Consequently, we have proved that the polynomial $f(x)$ has exactly three roots and they are contained in the intervals $[2, 3]$, $[0, 1]$, and $[-3, -2]$.

For certain other polynomials, the Bolzano theorem gives an exact answer. An important case is the polynomials $x^n - a$, whose roots are called the nth roots of a (denoted $\sqrt[n]{a}$). We first consider the case where $a > 0$. Then the polynomial $f(x) = x^n - a$ takes the negative value $-a$ for $x = 0$. However, it is easy to find a value $x = c$ for which $f(c) > 0$. For example, we can set $c = a + 1$. It then follows immediately from the axioms in group IV that $c^n > a$ and $f(c) > 0$. According to the Bolzano theorem, we can say that the polynomial has a root in the interval $[0, c]$. If $a < 0$ and n is even, the polynomial obviously has no root: $x^n \geq 0$ for an even power of a real number, and $x^n - a > 0$. And if n is odd, then setting $x = -y$, we obtain $x^n - a = -y^n - a = -(y^n + a)$. The polynomial $y^n + a$ (with $a < 0$), as we showed, has a root, and therefore $x^n - a$ does too. And we already dealt with the case where $a > 0$. All these arguments are usually omitted in school (for lack of a reliable theory of real numbers), although they prove (very simply) that a polynomial $x^n - a$ with odd n has no more than one root (as we saw, exactly one) and with even n and $a > 0$, not more than two roots differing in sign (and this means exactly two). In particular, we have proved the existence of the real number $\sqrt{2}$ for the first time.

But in the case of other polynomials, the Bolzano theorem might not give anything. For example, we consider the polynomial $x^2 - x + 2$. From analysis of the formula for solving a quadratic equation, we can conclude that this polynomial has no root. But if we give x the values $0, \pm 1, \pm 2, \ldots$, then we obtain positive values for the polynomial, and the Bolzano theorem gives us nothing. In view of this, we investigate polynomials further.

Theorem 31 clarifies the values of a polynomial for values of x close to some value a. We now prove a similar statement regarding the values of a polynomial for large values of x (in absolute value).

Theorem 33. *For the polynomial* $f(x) = a_0 + a_1 x + \cdots + a_n x^n$, *there exists a constant* $N > 0$ *such that*

$$|a_0 + a_1 x + \cdots + a_{n-1} x^{n-1}| < |a_n x^n| \qquad (16)$$

for values of x *for which* $|x| > N$.

We divide inequality (16) by $|x|^n$ and set $y = 1/x$. The inequality then becomes

$$|a_{n-1} y + \cdots + a_0 y^n| < |a_n|. \qquad (17)$$

We set $g(y) = a_{n-1}y + \cdots + a_0 y^n$ and apply Theorem 31 to it. There exists a constant M such that $|g(y)| < M\,|y|$ for $|y| < 1$. Choosing y such that $M\,|y| < |a_n|$, we find that inequality (17) is satisfied for $|y| < |a_n|/M$ and $|y| < 1$, that is, inequality (16) is satisfied for $|x| > M/|a_n|$ and $|x| > 1$. $\qquad\qquad\qquad\qquad\qquad\qquad\qquad\qquad\qquad\qquad\qquad\qquad\Box$

A series of useful consequences follow from Theorem 33. We note that under the conditions of the theorem (that is, for $|x| > N$), we necessarily have $|f(x)| > 0$. This follows immediately from inequality (12):

$$|f(x)| = |a_0 + a_1 x + \cdots + a_n x^n|$$
$$\geq |a_n x^n| - |a_0 + a_1 x + \cdots + a_{n-1} x^{n-1}|, \qquad (18)$$

that is, in accordance with Theorem 33, $|f(x)| > 0$.

But this means that the polynomial $f(x)$ cannot have any roots x with $|x| > N$. In other words, all roots of $f(x)$ (if they exist) must be contained in the interval $|x| \leq N$. Moreover, as we showed in the proof of Theorem 33, we can take the larger of the two numbers $|a_0| + \cdots + |a_{n-1}|$ and 1 as N (for $a_n = 1$). We say that such an N is a *bound of the roots* of the polynomial. For the polynomial $x^3 - 7x + 5$, for example, we can take $N = 12$. This means that the roots of the polynomial lie between -12 and 12. Earlier, we verified that they are in fact contained between -3 and 3 (see the table on page 162).

We can deduce even more from Theorem 33 than just the affirmation that $f(x) \neq 0$ if $|x| > N$ for the value N found in that theorem. Calculating the value $a_0 + a_1 x + \cdots + a_{n-1} x^{n-1} + a_n x^n$, we add the two real numbers $a_0 + a_1 x + \cdots + a_{n-1} x^{n-1}$ and $a_n x^n$, of which the absolute value of the first is less than that of the second (for $|x| > N$). But then the sign of the sum is determined by the sign of the second summand. We obtain the statement in Corollary 1.

Corollary 1. *For $|x| > N$, where N is a bound of the roots defined in Theorem 33, the value of the polynomial $f(x)$ has the same sign as the leading term $a_n x^n$.*

We now consider a polynomial of an odd degree n. Then the sign of the leading term $a_n x^n$ is the sign of the coefficient a_n for $x > 0$ and is opposite to the sign of that coefficient for $x < 0$. Corollary 1 indicates that for $x > N$ and $x < -N$, the polynomial itself then takes values with opposite signs (namely, the sign of a_n and the sign of $-a_n$). The Bolzano theorem states that the polynomial has at least one root contained between these values. We have thus obtained Corollary 2.

Corollary 2. *A polynomial of odd degree has at least one root.*

This is a very unexpected statement. As you know, a second-degree polynomial might have no root (for example, the polynomial $x^2 + 1$). It would seem that this phenomenon would be even more likely for higher-degree polynomials, of degree three, four, etc. But here, for example, according to the corollary, a third-degree polynomial must have a root. The situation turns out to be more complicated: it depends not only on the degree of the polynomial but also on the parity (even or odd) of the degree.

Finally, we consider one more property of polynomials that significantly facilitates investigating concrete cases. Theorem 31 gives us information about the absolute value of the difference $f(x) - f(a)$ when the difference $x - a$ is small. We now investigate the *sign* of the difference $f(x) - f(a)$. For this, we exclude those cases where the value $x = a$ is a root of the derivative $f'(x)$ of the polynomial $f(x)$. The excluded values of a could be easily investigated with the same approach, but we have no need for it now.

Theorem 34. *Let a polynomial $f(x)$ and a value $x = a$ that is not a root of its derivative $f'(x)$ (that is, $f'(a) \neq 0$) be given. If $f'(a) > 0$, then values of $f(x)$ close to a on the left are less than $f(a)$ and close to a on the right are greater. If $f'(a) < 0$, then values of $f(x)$ close to a on the left are greater than $f(a)$ and close to a on the right are less.*

This means that there exists a sufficiently small number $\varepsilon > 0$ (depending on $f(x)$ and a) such that for $f'(a) > 0$, we have $f(x) < f(a)$ for $a - \varepsilon < x < a$ and $f(x) > f(a)$ for $a < x < a + \varepsilon$. And if $f'(a) < 0$, then we have $f(x) > f(a)$ for $a - \varepsilon < x < a$ and $f(x) < f(a)$ for $a < x < a + \varepsilon$ (see the graphs in Figs. 25 and 26).

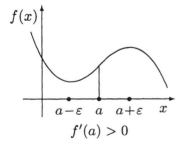

Fig. 25

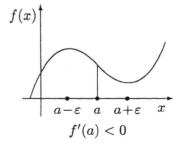

Fig. 26

The proof is very simple. According to the Bézout theorem, we know that the polynomial $f(x) - f(a)$ is divisible by $x - a$. Therefore,

$$f(x) - f(a) = (x - a)g(x, a), \qquad (19)$$

where the coefficients of the polynomial $g(x, a)$ depend on a. For $x = a$, the polynomial $g(x, a)$ takes the value $f'(a)$—this was our definition of the derivative (see formula (14) in Chap. 2). Because $f'(a) \neq 0$ by hypothesis, we have $g(a, a) = f'(a) \neq 0$. Let ε denote any positive number less than the distance from a to the nearest root of the polynomial $g(x, a)$ (we have a fixed here, and x is the variable), and then the polynomial $g(x, a)$ does not become zero in the interval $[a - \varepsilon, a + \varepsilon]$. Then it retains the same sign in this interval as for $x = a$: if it did take two values with different signs, then it would become zero in this interval according to the Bolzano theorem, which would contradict our choice of the number ε. This in essence already contains the assertion in Theorem 34. For example, let $f'(a) > 0$. Then $g(a, a) = f'(a) > 0$, and by what we just showed, $g(x, a) > 0$ for $a - \varepsilon < x < a + \varepsilon$. The other factor $x - a$ in formula (19) also behaves in a known way: $x - a < 0$ for $a - \varepsilon < x < a$ and $x - a > 0$ for $a < x < a + \varepsilon$. Multiplying in formula (19), we find that $f(x) - f(a) < 0$ for $a - \varepsilon < x < a$ and $f(x) - f(a) > 0$ for $a < x < a + \varepsilon$. And this is the assertion in the theorem. The case where $f'(a) < 0$ is treated in exactly the same way.   □

An interesting consequence follows from the proved theorem.

Theorem 35 (Rolle Theorem). *Between two neighboring roots of a polynomial without multiple roots is found a root of the derivative.*

As previously, we assume that a polynomial does not have multiple roots—just to shorten the argument. We have only such cases in what follows.

Let α and β, $\alpha < \beta$, be two neighboring roots of the polynomial $f(x)$, that is, it does not have a root lying between them. Because we assume that the polynomial does not have multiple roots, α and β are not multiple roots, and it follows from Theorem 16 that $f'(\alpha) \neq 0$ and $f'(\beta) \neq 0$. For instance, let $f'(\alpha) > 0$. We prove that then $f'(\beta) < 0$. Indeed, if $f'(\beta) > 0$, then according to the preceding theorem, we would have $f(x) > f(\alpha) = 0$ for $\alpha + \varepsilon > x > \alpha$ and $f(y) < f(\beta) = 0$ for $\beta - \varepsilon < y < \beta$. We would thus have $f(x) > 0$ for any x for which $\alpha + \varepsilon > x > \alpha$ and $f(y) < 0$ for any y for which $\beta - \varepsilon < y < \beta$. It then follows from the Bolzano theorem that the polynomial $f(x)$ would have a root located between x and y, that is, within the interval $[\alpha, \beta]$. And this would contradict our given condition that α and β are neighboring roots of the polynomial $f(x)$. We see that only the possibility $f'(\beta) < 0$ remains, but then the polynomial $f'(x)$ would have a root between α and β according to the Bolzano theorem. The case where $f'(\alpha) < 0$ is treated in exactly the same way.   □

The impossible and possible cases of signs of $f'(\alpha)$ and $f'(\beta)$ (for $f'(\alpha) > 0$) are shown in Figs. 27 and 28.

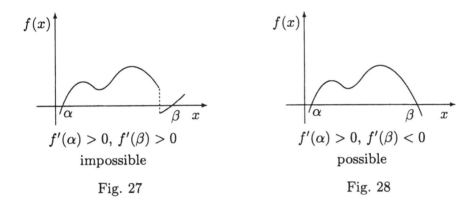

$f'(\alpha) > 0, \; f'(\beta) > 0$ $f'(\alpha) > 0, \; f'(\beta) < 0$

impossible   possible

Fig. 27   Fig. 28

In concluding this section, we show that the theorems we have proved are already sufficient to completely answer the question of the number of roots of a third-degree polynomial. We saw in Sec. 6 that any third-degree equation can be replaced with an equivalent equation of the form $x^3 + ax + b = 0$. We consider equations of this form in what follows.

We first solve the problem of multiple roots. We proved in Sec. 5 that multiple roots are common roots of the polynomial and its derivative. According to formula (16) in Chap. 2, the derivative of the polynomial $f(x) = x^3 + ax + b$ is equal to $f'(x) = 3x^2 + a$. If $a > 0$, then the derivative has no root, and this means that the polynomial $f(x)$ does not have multiple roots. And if $a < 0$, then we let δ denote the positive root of the polynomial $3x^2 + a$ (that is, $\delta = +\sqrt{-a/3}$). Then the polynomial $f(x)$ can have multiple roots only equal to δ or $-\delta$. Because we can write the poylnomial in the form $f(x) = (x^2 + a)x + b$ and for $x = \pm\delta$, we have $x^2 = -a/3$ and $x^2 + a = 2a/3$, the condition for $f(x)$ to have multiple roots takes the form $\pm\delta(2a/3) = -b$ or $\delta^2(4a^2/9) = b^2$. Because $\delta^2 = -a/3$, we obtain the condition $-4a^3/27 = b^2$ or $4a^3 + 27b^2 = 0$. If this condition is satisfied by the coefficients a and b, then the polynomial $f(x)$ has a multiple root α and can be represented in the form $(x - \alpha)^2 g(x)$.

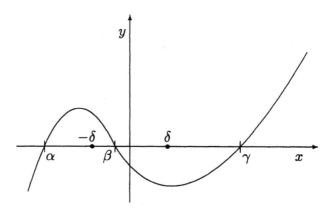

Fig. 29

Here, the polynomial $g(x)$ must be a first-degree polynomial, and this means it has one root β. Therefore, the polynomial has two roots equal to α and one root equal to β.

We now consider the remaining case, where the polynomial $f(x)$ does not have multiple roots, that is, $4a^3 + 27b^2 \neq 0$. According to Corollary 2 to Theorem 33, the polynomial $f(x)$ has at least one root α. If it has another root β, then it must be divisible by $(x-\alpha)(x-\beta)$, that is, it can be represented in the form $(x-\alpha)(x-\beta)g(x)$, where $g(x)$ is a first-degree polynomial and consequently has a root γ. Therefore, the polynomial $f(x)$ has three roots: α, β, and γ. A third-degree polynomial cannot have more than three roots. We see that there can only be two cases: the polynomial $f(x)$ has a single root, or the polynomial $f(x)$ has three roots. Our task is to find out which of these cases occurs (for given coefficients a and b).

We suppose that the polynomial $f(x)$ has three roots α, β, and γ and, moreover, $\alpha < \beta < \gamma$. This means that the polynomial does not have a root less than α or a root greater than γ. But according to Corollary 1 to Theorem 33, there exists a number N such that for sufficiently large x (more precisely, $x \geq N$), the value of the polynomial has the same sign as the value of the leading term x^3—that is, positive. And for $x \leq -N$, it has a negative sign for the same reason. This means that always $f(x) < 0$ for $x < \alpha$, and always $f(x) > 0$ for $x > \delta$ (Fig. 29).

Because $f(x) < 0$ for $\alpha - \varepsilon < x < \alpha$ for any $\varepsilon > 0$, we have $f'(\alpha) > 0$ according to the theorem, and this means that $f(x) > 0$ for $\alpha < x < \alpha + \varepsilon$ with some $\varepsilon > 0$. Because $f(x)$ does not have a root between α and β, its values have the same sign according to the Bolzano theorem, and this means that $f(x) > 0$ for $\alpha < x < \beta$. Analogously, we find that $f(x) < 0$

for $\beta < x < \gamma$. According to Theorem 35, roots of the derivative $f'(x)$ of the polynomial $f(x)$ lie between the roots α and β and between the roots β and γ. Because $f'(x) = 3x^2 + a$, the derivative does not have roots for $a > 0$, and the case where the polynomial $f(x)$ has three roots is impossible. For $a = 0$, we have $f(x) = x^3 + b$. As we saw previously, such a polynomial has only one root. Finally, for $a < 0$, the derivative $f'(x) = 3x^2 + a$ has two roots: $\delta > 0$ and $-\delta < 0$ (here $\delta = +\sqrt{-a/3}$). Obviously, $\alpha < -\delta < \beta < \delta < \gamma$. Because the polynomial takes positive values on the interval from α to β and negative values on the interval from β to γ, we have

$$f(-\delta) > 0, \qquad f(\delta) < 0 \qquad (20)$$

(assuming that the polynomial $f(x)$ has three roots).

Conversely, if relations (20) are satisfied, then the polynomial $f(x)$ has a root located between $-\delta$ and δ according to the Bolzano theorem. Let β denote this root. Moreover, according to Corollary 1 to Theorem 33, the polynomial takes positive values for sufficiently large x and negative values for sufficiently small x. It then follows from the Bolzano theorem that the polynomial has a root less than $-\delta$ and also a root greater than δ. Let α and γ respectively denote these roots. It thus follows from relations (20) that the polynomial has three roots: α, β, and γ. In other words, relations (20) are *necessary and sufficient* for the polynomial $f(x)$ to have three roots. In the remaining cases, it has one root.

The proved assertion accomplishes our task. But we further transform conditions (20) to an even simpler form. Because $f(x) = (x^2 + a)x + b$ and $3\delta^2 + a = 0$, $\delta^2 = -a/3$, we have

$$f(\pm\delta) = (\delta^2 + a)(\pm\delta) + b = \pm\delta\frac{2a}{3} + b.$$

Therefore, relations (20) become

$$-\frac{2a}{3}\delta + b > 0, \qquad \frac{2a}{3}\delta + b < 0$$

or $2a\delta/3 < b < -2a\delta/3$. These two inequalities are equivalent to one: $b^2 < 4a^2\delta^2/3^2$. Because $4a^2\delta^2/3^2 = -4a^3/(27b^2)$, inequalities (20) are equivalent to the inequality $4a^3 + 27b^2 < 0$. And this is the final answer: if $4a^3 + 27b^2 < 0$, then the polynomial $x^3 + ax + b$ has three roots; if $4a^3 + 27b^2 = 0$, then it has two equal roots and one more root; and if $4a^3 + 27b^2 > 0$, then it has only one root.

Understandably, everything said here relates only to third-degree polynomials. It is possible to investigate polynomials of an arbitrary degree analogously, but the arguments are somewhat more complicated, and we relegate them to the supplement.

Problems:

1. At the end of Chap. 1, we proved that the polynomial $x^3 - 7x^2 + 14x - 7$ does not have rational roots; therefore, its roots, if they exist, are irrational numbers. Determine the number of roots of this polynomial and their signs. Show which consecutive integers they are located between.

2. Prove that the polynomial $x^4 + ax + b$ (exclude the case where $a = b = 0$) either does not have a root or has two roots. Deduce the condition (on the coefficients a and b) for which case occurs.

3. Prove that the number of roots of an even-degree polynomial is even and the number of roots of an odd-degree polynomial is odd (counting each root with its multiplicity).

4. Prove that the polynomial $x^n + ax + b$ has either zero or two roots for even n and either one or three roots for odd n. Deduce the condition (on the coefficients a and b) for which case occurs.

5. Determine the number of roots of the polynomial $x^n + ax^{n-1} + b$ (depending on n, a, and b).

6. Prove that the value of any polynomial $f(x)$ becomes arbitrarily large in absolute value for sufficiently large absolute values of x.

7. Prove that the number $1 + M/|a_n|$, where M is the largest of the numbers $|a_0|, \ldots, |a_{n-1}|$, can be taken as a bound N of the roots. *Hint:* Use the inequality

$$|a_0 + \cdots + a_{n-1}z^{n-1}| \le M\left(1 + |z| + \cdots + |z|^n\right).$$

8. Prove that the polynomial $a_0 + a_1x + \cdots + a_{n-1}x^{n-1} + a_nx^n$ for which $a_n > 0$, $a_i \le 0$ for $i = 1, \ldots, n - 1$, and $a_0 < 0$ has exactly one positive root. *Hint:* Write $f(x)$ in the form $a_nx^n(1 + a_{n-1}/(a_nx) + \cdots + a_0/(a_nx^n))$, and study the expressions $a_{n-k}/(a_nx^k)$ for increase or decrease as x increases while remaining positive.

9. For the polynomial $f(x)$, let all coefficients of even powers be equal to zero and all coefficients of odd powers be positive. Prove that the polynomial has exactly one root.

10. Deduce the conditions (known to you) for the polynomial $x^2 + px + q$ to have zero, one, or two roots, repeating the arguments applied at the end of this section.

11. Calculate $\sqrt{\pi}$ to two digits after the decimal point.

Supplement: Sturm's Theorem

Here, we describe a method that allows determining the number of roots located in a given interval $[a, b]$ for any polynomial $f(x)$. The idea of the method is based on the fact that although there is no method for relating the properties of one polynomial $f(x)$ to the properties of a polynomial of lower degree, such a method is known for a pair of polynomials $f(x)$ and $g(x)$. It consists of dividing $f(x)$ by $g(x)$ with remainder, $f(x) = g(x)q(x) + r(x)$, and passing from the pair of polynomials (f, g) to the pair of polynomials of lower degree (g, r). Repeating this process, we obtain the Euclidean algorithm for finding the greatest common divisor of the polynomials f and g.

For example, the question of the existence of common roots of the polynomials f and g can be reduced to the question of the existence of common roots of the lower-degree polynomials g and r and, as a result, to the question of the existence of roots of the lower-degree polynomial $GCD(f, g)$. This method can be applied to a pair consisting of a polynomial and its derivative. We then obtain an answer to the question of the existence of multiple roots of the polynomial. We used this method in Chap. 2. We now proceed thus: we first consider a certain property of roots of a pair of polynomials (f, g) to which division with remainder can be applied. Applying this property to the pair consisting of the polynomial and its derivative, we then find an answer to the question we are interested in.

We begin with a perfectly simple remark concerning one polynomial $F(x)$. Let the polynomial have the root $x = \alpha$, and let the root have the multiplicity k. By the definition of the multiplicity of a root (given in Sec. 5), we can write

$$F(x) = (x - \alpha)^k G(x), \tag{21}$$

where $G(\alpha) \neq 0$. Therefore, if a positive number ε is less than the distance from α to the nearest root of $G(x)$, then $G(x)$ takes values with the same sign in the interval $[\alpha - \varepsilon, \alpha + \varepsilon]$. Indeed, if the polynomial G had values $G(x)$ and $G(y)$ with different signs for two numbers x and y in that interval, then there would be a root between x and y according to the Bolzano theorem. But we choose ε such that the polynomial G does not have a root in the interval $[\alpha - \varepsilon, \alpha + \varepsilon]$. In particular, all values of $G(x)$ for x in the interval $[\alpha - \varepsilon, \alpha + \varepsilon]$ have the same sign as $G(\alpha)$. It is now evident from formula (21) that for an even multiplicity k, the values of the polynomial $F(x)$ for x in the interval $[\alpha - \varepsilon, \alpha + \varepsilon]$ also have the same sign as $G(\alpha)$. Possible arrangements of the graph are shown in Fig. 30.

We now consider the case where the multiplicity k is odd. Formula (21) indicates that for $G(\alpha) > 0$, $F(x) < 0$ for $\alpha - \varepsilon \leq x < \alpha$ and $F(x) > 0$ for $\alpha < x \leq \alpha + \varepsilon$, and conversely for $G(\alpha) < 0$, $F(x) > 0$ for $\alpha - \varepsilon \leq x < \alpha$ and $F(x) < 0$ for $\alpha < x \leq \alpha + \varepsilon$. In the first case (that is, for $G(\alpha) > 0$), we call α a *root with increase*, and in the second case (for

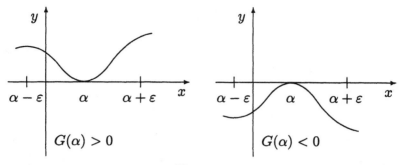

Fig. 30

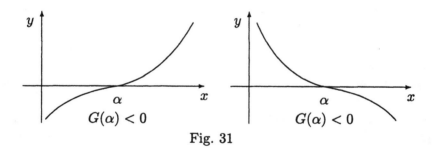

Fig. 31

$G(\alpha) < 0$), we call it a *root with decrease*. The possible arrangements of the graph of the polynomial $F(x)$ are shown in Fig. 31.

Definition. Let $F(x)$ be a polynomial that has neither a nor b as a root. The difference between the number of its roots with increase and the number of its roots with decrease on the interval $[a, b]$ (where roots with an even multiplicity are not counted) is called the *characteristic of the polynomial $F(x)$ on the interval $[a, b]$.*

The characteristic is denoted by $[F(x)]_a^b$. For example, the polynomial $F(x)$ depicted in Fig. 32 has three roots with increase and two roots with decrease; therefore, $[F(x)]_a^b = 1$.

In brief, we can say that near a root with increase, the polynomial passes (as x increases) from negative values to positive values, and near a root with decrease, from positive to negative. It follows from the definition that a root following a root with increase must be a root with decrease (roots with an even multiplicity do not count). The characteristic is therefore determined by the signs of the numbers $F(a)$ and $F(b)$:

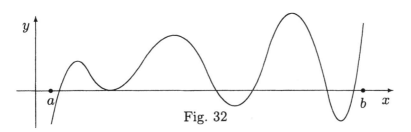

Fig. 32

$$[F(x)]_a^b = \begin{cases} 0 & \text{if } F(a) \text{ and } F(b) \text{ have the same sign,} \\ 1 & \text{if } F(a) < 0 \text{ and } F(b) > 0, \\ -1 & \text{if } F(a) > 0 \text{ and } F(b) < 0. \end{cases}$$

We can picture our situation by imagining a traveler who crosses the border between France and Germany several times. What is the difference between the numbers of border crossings from France to Germany and from Germany to France? Obviously, the difference is zero if the trip began and ended in the same country, one if the trip began in France and ended in Germany, and minus one if the trip began in Germany and ended in France. The route plan is represented by a line similar to the graph in Fig. 32, where France is represented by the territory below the x axis and Germany above.

The characteristic of the polynomial $F(x)$ on a given interval is thus determined by its values at the ends of the interval. This means that it can always be easily calculated even though it is connected by definition with the roots of the polynomial, which are difficult to find as a rule.

We now consider two polynomials f and g and assume that, first, they do not have common roots and, second, the first of them (that is, $f(x)$) is not zero for $x = a$ and for $x = b$. The *characteristic of the polynomial $f(x)$ with respect to $g(x)$ on the interval $[a, b]$ is* the difference between the number of roots of $f(x)$ in the interval $[a, b]$ that are roots with increase for the polynomial fg and the number of its roots in that interval that are roots with decrease for fg. This characteristic is denoted by $(f, g)_a^b$.

The main reason for introducing the concept of a characteristic is given by the assertion in Theorem 36.

Theorem 36. *If a polynomial $f(x)$ does not have multiple roots in and is not zero at the ends a and b of the interval $[a, b]$, then the characteristic $(f, f')_a^b$ is equal to the number of roots contained in the interval $[a, b]$.*

This theorem is a simple consequence of Theorem 34. We simply confirm that all roots of the polynomial $f(x)$ are roots with increase for the polynomial ff', and the theorem then follows from the definition of the characteristic. Indeed, according to Theorem 16, the polynomials f and f' do not have common roots. If α is a root of the polynomial $f(x)$ and $f'(\alpha) > 0$, then α is a root with increase for the polynomial $f(x)$ by Theorem 34, and because $f'(x) > 0$ close to α, it is also a root with increase for $f(x)f'(x)$. And if $f'(\alpha) < 0$, then α is a root with decrease for $f(x)$, and because $f'(x) < 0$ close to α, it is again a root with increase for $f(x)f'(x)$.   □

The characteristic $(f,g)_a^b$ is just the kind of expression that can be calculated using division with remainder. We first note some simple properties of this characteristic.

1. $(f, -g)_a^b = -(f, g)_a^b.$

This is obvious because when the polynomial g is multiplied by -1, roots with increase for fg become roots with decrease and vice versa.

2. If $g(a) \neq 0$ and $g(b) \neq 0$, then $(f, g)_a^b + (g, f)_a^b = [fg]_a^b.$

This is also obvious because the polynomials f and g do not have common roots by hypothesis. The roots of fg can therefore be segregated into the roots of the polynomial f and the roots of the polynomial g. The number of roots with increase (and exactly the same for decrease) of the polynomial fg is equal to the sum of the numbers of such roots of the polynomial f and of the polynomial g, which proves the given property.

3. If the polynomials g and h take identical values at roots of the polynomial f (that is, $g(\alpha) = h(\alpha)$ for $f(\alpha) = 0$), then

$$(f, g)_a^b = (f, h)_a^b.$$

Indeed, if $g(\alpha) = h(\alpha)$, then a root α of the polynomial $f(x)$ is either a root with increase or a root with decrease for the two polynomials fg and fh simultaneously.

4. Let the polynomial f be divisible by the polynomial g. Then roots of the polynomial g would be common roots of the polynomials f and g. Because f and g do not have common roots by condition, this means that the polynomial g cannot have roots. It follows that the polynomials f and fg have exactly the same roots, and this means that

$$(f, g)_a^b = [fg]_a^b.$$

We now describe the process of calculating the characteristic $(f, g)_a^b$. We divide f by g with a remainder:

$$f = gq + r. \tag{22}$$

According to property 2, we have $(f, g)_a^b = -(g, f)_a^b + [fg]_a^b$. On the other hand, it follows from relation (22) that $f(\alpha) = r(\alpha)$ for $g(\alpha) = 0$. Therefore, according to property 3, we obtain $(g, f)_a^b = (g, r)_a^b$. Together, the two equalities obtained show that

$$(f, g)_a^b = -(g, r)_a^b + [fg]_a^b. \tag{23}$$

Relation (23) itself solves our problem because it reduces calculating the characteristic $(f, g)_a^b$ to calculating the characteristic $(g, r)_a^b$ of the lower-degree polynomials g and r insofar as the expression $[fg]_a^b$ is determined by the values of the polynomials f and g at the ends a and b of the interval $[a, b]$ (see the expression on page 173).

Our process of passing from the pair (f, g) to a pair of polynomials of lower degree is the same as when finding the greatest common divisor of the polynomials f and g. In the end, we therefore obtain two polynomials u and v, where u is divisible by v. In that case, the characteristic is determined by property 4.

We further perfect our result in two directions. First, in a more harmonious and elegant form, we present the final result obtained by passing from (f, g) to (g, r), then taking the next step, and so on for all steps of the Euclidean algorithm. Second, our inductive type of argument requires that the conditions placed on f and g must then be applied to g and r, and so on; we show how this additional limitation can be easily eliminated.

There are two such conditions: (1) the polynomials f and g do not have a common root, and (2) both $f(a) \neq 0$ and $f(b) \neq 0$. Satisfaction of the first condition is retained in passing from the pair of polynomials (f, g) to the pair (g, r). This is immediately evident from relation (22): a common root of the polynomials g and r would also be a root of the polynomial f, but f and g have no common root by condition 1. However, the second condition ($f(a) \neq 0$, $f(b) \neq 0$) might not be satisfied by g or r or one of the other resulting polynomials; we must therefore additionally assume that condition 2 is satisfied by all these polynomials.

We first transform the answer obtained (formula (23)) somewhat. We begin by changing the notation. Let f_1 denote f, f_2 denote g, and f_3 denote $-r$. By property 1, formula (23) then becomes

$$(f_1, f_2)_a^b = (f_2, f_3)_a^b + [f_1 f_2]_a^b, \qquad (24)$$

and the formula for division with a remainder (formula (22)) is

$$f_1 = f_2 q_1 - f_3$$

(where q is now denoted by q_1). How to consecutively apply formula (24), lowering the degree of the polynomials considered, is now clear. Beginning with f_1 and f_2, we define the polynomial f_i by induction, setting

$$f_{i-1} = f_i q_{i-1} - f_{i+1}, \qquad (25)$$

where the degree of f_{i+1} is less than the degree of f_i (considering that f_{i-1} and f_i are already defined). Clearly, f_{i+1} is the polynomial obtained as the remainder in the Euclidean algorithm, only with a change of the sign. After some steps, we reach the polynomial f_k that coincides with $\mathrm{GCD}(f_1, f_2)$ up to the sign.

Applying formula (24) to f_2 and f_3 instead of f_1 and f_2, we obtain $(f_2, f_3)_a^b = (f_3, f_4)_a^b + [f_2 f_3]_a^b$. Substituting this value for $(f_2, f_3)_a^b$ in formula (24), we obtain

$$(f_1, f_2)_a^b = (f_3, f_4)_a^b + [f_1 f_2]_a^b + [f_2 f_3]_a^b.$$

Completely analogously, we obtain the formula

$$(f_{i-1}, f_i)_a^b = (f_i, f_{i+1})_a^b + [f_{i-1} f_i]_a^b.$$

The sequence of polynomials $f_1, f_2, \ldots, f_k$ is called the *Sturm series* for the polynomials f_1 and f_2. To find the Sturm series, it is necessary to perform almost the same operations as for finding the greatest common divisor of the polynomials f_1 and f_2 (only changing the sign of the remainder). Moreover, the Sturm series allows easily calculating the characteristic $(f_1, f_2)_a^b$.

We perform the basic process $k-2$ times and note that $(f_{k-1}, f_k)_a^b = [f_{k-1} f_k]_a^b$ according to property 4. As a result, we obtain

$$(f_1, f_2)_a^b = [f_1 f_2]_a^b + [f_2 f_3]_a^b + \cdots + [f_{k-1} f_k]_a^b. \qquad (26)$$

We note, however, that we must now assume that $f_i(a) \neq 0$ and $f_i(b) \neq 0$ for all $i = 1, 2, \ldots, k$ in order to have the right to apply formula (24).

We examine the expression $[fg]_a^b$, which can be calculated on the basis of the expression on page 173 for $F = fg$, more carefully. In our case, we can rewrite it as

$$[fg]_a^b = \begin{cases} 0 & \text{if either } f(a) \text{ and } g(a) \text{ have the same sign and} \\ & \quad f(b) \text{ and } g(b) \text{ have the same sign or} \\ & \quad f(a) \text{ and } g(a) \text{ have different signs and} \\ & \quad f(b) \text{ and } g(b) \text{ have different signs,} \\ 1 & \text{if } f(a) \text{ and } g(a) \text{ have different signs and} \\ & \quad f(b) \text{ and } g(b) \text{ have the same sign,} \\ -1 & \text{if } f(a) \text{ and } g(a) \text{ have the same sign and} \\ & \quad f(b) \text{ and } g(b) \text{ have different signs.} \end{cases}$$

Given two nonzero numbers A and B, we say that the pair (A, B) has one change of sign if the two numbers have different signs and that the pair has no change of sign if they have the same sign. We can use this terminology to construct a table of values of $[fg]_a^b$, letting m denote the changes of sign in the pair $(f(a), g(a))$ (it is equal to 0 or 1) and n denote the changes of sign in the pair $(f(b), g(b))$. Our table has the following form:

$[fg]_a^b$	m	n
0	0	0
0	1	1
1	1	0
-1	0	1

We see that $[fg]_a^b = m - n$ in all cases. We now apply this equality to formula (26), letting m_i denote the changes of sign in the pair $(f_i(a), f_{i+1}(a))$ and n_i denote the changes of sign in the pair $(f_i(b), f_{i+1}(b))$. Formula (26) then becomes

$$(f_1, f_2)_a^b = m_1 - n_1 + m_2 - n_2 + \cdots + m_{k-1} - n_{k-1}$$
$$= (m_1 + \cdots + m_{k-1}) - (n_i + \cdots + n_{k-1}). \tag{27}$$

What is the sense of the number $m_1 + m_2 + \cdots + m_{k-1}$? We simply need to write the numbers $f_1(a), f_2(a), \ldots, f_k(a)$ in order and see how many times numbers with different signs occur next to each other—this is equal to $m_1 + \cdots + m_{k-1}$. In general, if a sequence of nonzero numbers $A_1, \ldots, A_r$ is given, the number of positions where the succeeding number has a different sign is called the *number of sign changes* (it takes values from 0 to $r - 1$). For example, the sequence $1, -1, 2, 1, 3, -2$ has three changes of sign. We can say that $m_1 + m_2 + \cdots + m_{k-1}$ is the number of sign changes in $f_1(a), f_2(a), \ldots, f_k(a)$ and $n_1 + n_2 + \cdots + n_{k-1}$ is the number of sign changes in $f_1(b), f_2(b), \ldots, f_k(b)$. We can now state formula (27) in the form of Theorem 37.

Theorem 37. *If all the polynomials in the Sturm series* $f_1, \ldots, f_k$ *for* f_1, f_2 *are nonzero both at* a *and at* b *and if the polynomials* f_1 *and* f_2 *do not have a common root, then the characteristic* $(f,g)_a^b$ *is equal to the difference between the numbers of sign changes in the Sturm series at* a *and at* b.

It remains to eliminate the limitation $f_i(a) \neq 0$, $f_i(b) \neq 0$ for $i = 1, \ldots, k$, which might be inconvenient in applications: we assume only that $f_1(a) \neq 0$ and $f_1(b) \neq 0$. For this, we generalize the concept of the number of sign changes somewhat. If some of the numbers in a sequence $A_1, \ldots, A_r$ are zero, then the number of sign changes in the shortened sequence obtained by omitting the zeros is called the number of sign changes in the original sequence. For example, omitting the zeros from the sequence $1, 0, 2, -1, 0, 3, 1$, we obtain the sequence $1, 2, -1, 3, 1$, which has two sign changes. Consequently, the original sequence has two sign changes (by definition).

We let ε denote the distance from a to the nearest root (different from a) for whatever $f_i(x)$. Therefore, $f_i(x) \neq 0$ for $a < x < a+\varepsilon$. We choose an arbitrary value a' such that $a < a' < a+\varepsilon$. We choose b' analogously such that $f_i(b') \neq 0$ for $b - \eta < b' < b$. We now prove Lemma 14.

Lemma 14. *The number of sign changes in the sequence* $f_1(a), \ldots, f_k(a)$ *is equal to the number of sign changes in the sequence* $f_1(a'), \ldots, f_k(a')$. *The same holds if* a *is replaced with* b *and* a' *with* b'.

We first verify that the lemma allows applying Theorem 37 to arbitrary polynomials f_1 and f_2 with the only conditions that $f_1(a) \neq 0$, $f_1(b) \neq 0$, and f_1 and f_2 do not have a common root. Indeed, by hypothesis, f_1 does not have a root in the interval $[a, a']$ nor in $[b', b]$. Therefore, all its roots in the interval $[a, b]$ must be contained in the interval $[a', b']$. Hence, $(f_1, f_2)_a^b = (f_1, f_2)_{a'}^{b'}$. We can apply Theorem 37 to the characteristic $(f_1, f_2)_{a'}^{b'}$ because of the choice of a' and b'. The number of sign changes in the sequence $f_1(a'), \ldots, f_k(a')$ and in the sequence $f_1(b'), \ldots, f_k(b')$ is determined by Lemma 14. We hence obtain the desired result in the form of Theorem 38.

Theorem 38. *If the polynomials* f_1 *and* f_2 *do not have a common root and* $f_1(a) \neq 0$ *and* $f_1(b) \neq 0$, *then the characteristic* $(f_1, f_2)_a^b$ *is equal to the difference between the numbers of sign changes in the sequence* $f_1(a), \ldots, f_k(a)$ *and in the the sequence* $f_1(b), \ldots, f_k(b)$, *where* $f_1(x), \ldots, f_k(x)$ *is the Sturm series corresponding to the pair of polynomials* f_1 *and* f_2.

We now verify the correctness of Lemma 14. We consider the value $x = a$ for example. We suppose that $f_i(a) = 0$ for some $i = 1, \ldots, k$. Because $f_1(a) \neq 0$ by hypothesis, we have $i \neq 1$. Further, $i \neq k$ because the polynomial $f_k(x)$ coincides with $\mathrm{GCD}(f_1, f_2)$ up to the sign and therefore has no root. We note that then $f_{i-1}(a) \neq 0$ and $f_{i+1}(a) \neq 0$. Indeed, if we had $f_i(a) = 0$ and $f_{i+1}(a) = 0$, then it would follow from formula 25 that $f_{i-1}(a) = 0$. Similarly, we would then have $f_{i-2}(a) = 0$, and so on, all the way to $f_1(a) = 0$, which would contradict the given condition. But we can state more than this: the numbers $f_{i-1}(a)$ and $f_{i+1}(a)$ not only differ from zero but also have different signs—this follows immediately from substituting $x = a$ in equality (25) with our supposition that $f_i(a) = 0$.

We now compare the sequences $f_1(a), \ldots, f_k(a)$ and $f_1(a'), \ldots, f_k(a')$. Let $f_i(a) = 0$. Then, as we saw, $f_{i-1}(a) \neq 0$ and $f_{i+1}(a) \neq 0$; moreover, $f_{i-1}(a)$ and $f_{i+1}(a)$ have different signs. But then $f_{i-1}(a') \neq 0$ and $f_{i+1}(a') \neq 0$. Moreover, $f_{i-1}(a')$ has the same sign as $f_{i-1}(a)$, and $f_{i+1}(a')$ has the same sign as $f_{i+1}(a)$. This follows because the polynomials f_{i-1} and f_{i+1} have no roots in the interval $[a, a']$ and therefore (by the Bolzano theorem) cannot take values with different signs. We write the corresponding parts of our sequences, supposing that $f_{i-1}(a) > 0$. Based on our arguments, we can construct the following table:

	$f_{i-1}(x)$	$f_i(x)$	$f_{i+1}(x)$
$x = a$	$+$	0	$-$
$x = a'$	$+$	$?$	$-$

The characteristic $(f_1, f_2)_{a'}^{b'}$ depends on the number of sign changes in the lower row. But we can see that it coincides with the number of sign changes in the row for $x = a$: no matter what the unknown sign (denoted by ?) is, we always have one sign change. The case where $f_{i-1}(a) < 0$ is treated in exactly the same way. Lemma 14 is proved. $\square$

Combining Theorem 38 and Theorem 36, we obtain the basic result in the form of Theorem 39.

Theorem 39 (Sturm Theorem). *If the polynomial $f(x)$ does not have multiple roots and is nonzero for $x = a$ and $x = b$, then the number of its roots in the interval $[a, b]$ is equal to the difference between the numbers of sign changes in the Sturm series of polynomials for $f(x)$ and $f'(x)$ at $x = a$ and at $x = b$.*

We need only note that the absence of multiple roots of the polynomial $f(x)$ is equivalent to the absence of common roots of the polynomial $f(x)$ and its derivative $f'(x)$ according to Theorem 16. We can therefore apply Theorem 36 to the polynomial $f(x)$ and then apply Theorem 38 to the pair of polynomials $f(x)$ and $f'(x)$.   □

The Sturm theorem allows answering basic questions about the distribution of the roots of a polynomial. First, we can use it to determine the number of roots of a polynomial. For this, we need only recall Corollary 1 to Theorem 33, which provides a number N such that the roots of the polynomial lie between $-N$ and N. After this, it is sufficient to apply the Sturm theorem to the interval $[-N, N]$. However, it is remarkable that there is no need to calculate the value N itself (using Corollary 1 to Theorem 33) nor to calculate the values of the polynomials in the Sturm series for $x = -N$ and $x = N$ in order to determine the number of roots. Indeed, we do not need to know the actual values of $f_i(\pm N)$, but only their signs. For this, we choose the number N so large that the interval $[-N, N]$ includes not only all roots of $f_1(x)$ but also all roots of each polynomial $f_i(x)$ in the Sturm series (that is, we choose a corresponding N_i for each $f_i(x)$ and take the largest of them as N). According to Corollary 1 to Theorem 33, the sign of the value $f_i(N)$ or $f_i(-N)$ coincides with the sign of the value of the leading term of the polynomial $f_i(x)$ for $x = N$ and $x = -N$. These signs are determined by the sign of the coefficient of the leading term and the parity of its power. Therefore, we do not need to calculate N and the values $f_i(N)$ and $f_i(-N)$.

When the number of roots is determined, then we can find the intervals that each contain exactly one root. For this, we must actually calculate the value of N indicated in Theorem 33. We then divide the inteval $[-N, N]$ in half and use the Sturm theorem to determine how many roots there are in each half. We then do the same with the intervals $[-N, 0]$ and $[0, N]$ and continue until we have intervals that each contain exactly one root.

If we know that the interval $[a, b]$ contains exactly one root of the polynomial $f(x)$ and the polynomial has no multiple roots, then the values $f(a)$ and $f(b)$ must have opposite signs. Indeed, if the root is equal to α, then the values $f(\alpha - \varepsilon)$ and $f(\alpha + \varepsilon)$ have different signs for sufficiently small ε according to Theorem 34. But $f(\alpha - \varepsilon)$ and $f(a)$ must have the same sign—otherwise the polynomial would have a root in the interval $[a, \alpha - \varepsilon]$. The same holds for the values $f(\alpha + \varepsilon)$ and $f(b)$. Hence, $f(a)$ has the same sign as $f(\alpha - \varepsilon)$, $f(b)$ has the same sign as $f(\alpha + \varepsilon)$, and $f(\alpha - \varepsilon)$ and $f(\alpha + \varepsilon)$ have opposite signs. Therefore, $f(a)$ and $f(b)$ have opposite signs. Knowing this, we can calculate the root α to any degree of accuracy. For this, we divide the interval into two parts (for instance, in half) with the value c. We calculate the sign of the value of $f(c)$. If $f(c) = 0$, then the root is found. (We assume that c is not a root and therefore continue.) Either $f(a)$ and $f(c)$ or $f(c)$ and $f(b)$ have opposite signs. In the first case, α is contained in the interval $[a, c]$, and in the second case, in the interval $[c, b]$. We then do the same with the interval that contains α. We continue thus until α is included in an interval with a length less than the desired accuracy. This means that we have calculated the value of α to whatever accuracy we desired.

As an example, we consider the polynomial $f(x) = x^3 + 3x - 1$. To apply the condition deduced in Sec. 16, we must set up the expression $4a^3 + 27b^2 = 4 \cdot 27 + 27$. Because it is positive, the polynomial has one root. Applying Theorem 33, we determine the value $N = 3$ from the formula given there. Therefore, the root is located between -3 and 3; moreover, $f(-3) < 0$ and $f(3) > 0$. Because $f(0) < 0$, the root lies between 0 and 3. Because $f(1) = 3$, the root lies between 0 and 1. To find its first digit after the decimal point, we must determine which of the ten intervals (from 0 to $1/10$, $1/10$ to $2/10$,..., $9/10$ to 1) contains it. We first set $x = 1/2$ and obtain $f(1/2) = 5/8$. Because $f(0)$ and $f(1/2)$ have different signs, the root is contained between 0 and $1/2$. We then set $x = 3/10$. Because $f(3/10) = 27/1000 + 9/10 - 1 = 27/1000 - 1/10 < 0$, the root is contained between $3/10$ and $5/10$. Finally, $f(4/10) = 64/1000 + 12/10 - 1 > 0$. This means the root lies between $3/10$ and $4/10$ and has the form $\alpha = 0.3 \ldots$.

In view of the large number of its applications and the elegance of its formulation, the Sturm theorem became widely known immediately after it was proved. When the French mathematician Sturm, who proved it, spoke about it in his lectures, he said, "I now prove the theorem whose name I have the honor to bear."

Problems:

1. Construct the Sturm series for the polynomials $f(x)$ and $f'(x)$ in the cases where $f(x) = x^2 + ax + b$ and $f(x) = x^3 + ax + b$. Using the Sturm theorem, once again deduce the results concerning the number of roots of these polynomials obtained at the end of Sec. 16. *Hint:* For the case $f(x) = x^3 + ax + b$, separately consider the cases where the numbers a and $D = 4a^3 + 27b^2$ have one or another sign.

2. Use the Sturm theorem to determine the number of roots of the polynomial $x^n + ax + b$ depending on n (whose parity plays a role), a, and b.

3. Find the number of roots of the polynomial $x^5 - 5ax^3 + 5a^2x + 2b$. *Hint:* The answer depends on the sign of the expression $a^5 - b^9$.

4. Let a be a root of the derivative $f'(x)$ of the polynomial $f(x)$. We set $f_1(x) = f(x)$ and $f_2(x) = f'(x)/(x - a)$. Let $f(x)$ have no multiple roots and $f_1(x), \ldots, f_k(x)$ be the Sturm series for $f_1(x)$ and $f_2(x)$. Express the number of roots of the polynomial $f(x)$ in terms of the number of sign changes in the sequences $f_i(N)$, $f_i(a)$, and $f_i(-N)$, where $i = 1, \ldots, k$ and N is a sufficiently large number.

5. Let two polynomials $f_1(x)$ and $f_2(x)$ of the degrees n and $n - 1$ be given. Assume that the degree of the polynomial $f_i(x)$ in the Sturm series for them is equal to $n - i + 1$ and that the coefficient of x^{n-i+1} in it is positive. Prove that the polynomial $f_1(x)$ then has n roots. Moreover, each polynomial $f_i(x)$ has $n-i+1$ roots, and the interval between two neighboring roots of the polynomial $f_i(x)$ contains a root of the polynomial $f_{i+1}(x)$.

6. Let the nth-degree polynomial $f(x)$ have n roots. Prove that each successive polynomial in the Sturm series (for f and f') then has a degree exactly one less than its predecessor and that the coefficient of its leading term is positive.

6
Infinite Sets

Topic: Sets

18.   Equipotence

In this chapter, we study infinite sets. Some "infinite" generalizations of certain concepts introduced in Chap. 3 naturally appear. For example, let $S_1, S_2, S_3, \ldots$ be an infinite system of subsets of some set S. The subset S' of the set S that contains the elements that belong to at least one of the S_i is called the *union* of these subsets, as before, and is denoted by the same symbol $S' = S_1 \cup S_2 \cup S_3 \cup \ldots$. This can be written briefly as

$$S' = \bigcup_{n \geq 1} S_n.$$

For example, if S is the set of all natural numbers and the subset S_n consists of the numbers k satisfying $k \leq 2^n$, then $\bigcup_{n \geq 1} S_n = S$.

In exactly the same way, if $S_1, S_2, \ldots, S_n, \ldots$ is an infinite system of subsets of some set S, then the elements of S that belong to all the subsets S_n, $n = 1, 2, \ldots$, form their *intersection*. It is denoted by $S_1 \cap S_2 \cap \ldots$ or $\bigcap_{n \geq 1} S_n$. For example, if S is the set of all natural numbers and S_n is the subset of numbers divisible by n, then $\bigcap_{n \geq 1} S_n$ is empty.

If the sets S_i are pairwise disjoint (do not have common elements, that is, $S_i \cap S_j = \varnothing$ for $i \neq j$), then their union is called the *sum* and is denoted by $S_1 + S_2 + \cdots + S_n + \ldots$. For example, if S is the set of all natural numbers and S_n consists of the numbers k satisfying $2^{n-1} \leq k < 2^n$, then $S = S_1 + S_2 + \cdots + S_n + \ldots$.

In considering infinite sets, we immediately encounter the absence of the basic concept used to formulate all questions and assertions in

Chap. 3: the concept of the number of elements of a set. In Chap. 3, however, we established a principle expressing the *equipotence* of two finite sets (that is, that they have the same number of elements) differently: *two finite sets are equipotent if and only if a one-to-one correspondence can be established between them.* Of the two concepts "number of elements of a set" and "one-to-one correspondence," the first is not applicable to infinite sets, but the second is. We can therefore transform the principle expressed above into a *definition* of equipotence that has meaning for completely arbitrary sets: *two sets are said to be* equipotent *if a one-to-one correspondence can be established between them.*

Among finite sets, we can thus distinguish sets with one, two, three, etc., elements, and among infinite sets, we can distinguish those that are equipotent to each other. Such an approach to investigating infinite sets attracted thinkers long ago. But on that path, they found manifestations that seemed paradoxical to them: a set can be equipotent to its parts. For example, the set of natural numbers is equipotent to the set of even natural numbers. We need only associate the natural number n with the number $2n$, and this is obviously a one-to-one correspondence between the sets of natural and even natural numbers. The "paradoxicalness" of such a situation evidently consists in the fact that it cannot hold for finite sets. This is why the concept of the equipotence of infinite sets seemed senseless. For instance, Galileo in his *Discourses* gives the example of the one-to-one correspondence between natural numbers and square numbers in which n and n^2 correspond to each other. One participant in the discussion summarizes, "The concept of equality and also of greater and lesser magnitude does not have a place in matters concerning infinity."

Only quite later, in the second half of the 19th century, Dedekind, in contrast, introduced the concept of equipotence as the foundation of the study of infinite sets. He even propose taking the property that was previously considered "paradoxical" as the *definition* of an infinite set: a set S that is equipotent to a subset $S' \subset S$ that differs from S. We soon prove that this property is indeed equivalent to the infiniteness of a set.

We constantly use the following fact: *if the sets A and B are equipotent and the sets B and C are equipotent, then the sets A and C are equipotent.* Indeed, that A and B are equipotent means that a one-to-one correspondence can be established between them. Let the correspondence associate the pair (a, b) of elements from those sets. Analogously, the equipotence of B and C allows uniting pairs of elements from these two sets. We take some element $a \in A$. By the one-to-one correspondence between A and B, it is combined in a pair (a, b) with some element

$b \in B$. By the one-to-one correpsondence between B and C, that element is associated with some element $c \in C$. We now pair the elements a and c. A one-to-one correspondence is thus established between A and C (verify this yourself). Therefore, A and C are equipotent.

Based on this property, when we want to prove that the sets A and B are equipotent, we can replace A with a set A' that is equipotent to it and then show that A' and B are equipotent. Analogously, the set B can be replaced with a set B' that is equipotent to it. We often use this technique in what follows, not justifying it each time. It is analogous to the reasoning that we can prove $a = b$ by showing that $a = a'$ for some a' and that $a' = b$.

We first focus on some simpler infinite sets. The simplest example of an infinite set is the "sequence of natural numbers," that is, the set of all natural numbers. A set that is equipotent to the set of natural numbers is said to be *countable*.

For a countable set S, there must therefore exist a one-to-one correspondence between S and the set N of natural numbers. If the correspondence combines an element $a \in S$ with the natural number n, then the index n can be assigned to it, and the elements of the set S can be thus *enumerated*. In other words, a set is countable if it can be written in the form of an infinite sequence: $S = \{a_1, a_2, \dots \}$.

In a certain sense, countable sets are the "smallest" infinite sets. At least, this is indicated by the fact that *any subset of a countable set is either finite or countable*. Indeed, a countable set S can be enumerated: $S = \{a_1, a_2, \dots, a_n, \dots \}$. Let S' be a subset of S. We enumerate S', calling the element a_k with the least index in S' the first element, the element a_l with the least index of those remaining in S' the second, and so on. Either this process terminates (and the subset is finite) or it continues endlessly, giving a complete enumeration of S' (each element of S' is encountered among the elements of $S = \{a_1, a_2, \dots, a_n, \dots \}$ sooner or later).

The same property of countable sets as the "smallest" of infinite sets is characterized differently in Theorem 40.

Theorem 40. *Each infinite set has a countable subset.*

Proof. Let the set S be infinite. We choose an arbitrary element of S and call it a_1. Because S is infinite, it must contain elements different from a_1. We choose an arbitrary element of S different from a_1 and call it a_2. Because S is infinite, it must contain elements different from a_1 and a_2. From them, we choose our a_3. We can continue thus endlessly. If we have chosen n elements $a_1, \dots, a_n$ from the set S, then because S is infinite and must therefore contain elements different from the elements

$a_1, \ldots, a_n$, we can arbitrarily chose one of them as a_{n+1}. The subset $T = \{a_1, a_2, a_3, \ldots\}$ consists of different elements. Their enumeration establishes a one-to-one correspondence between T and the set of natural numbers.   □

Corollary 3. *Any infinite set S is equipotent to some subset of S that is different from S.*

We prove this in a more explicit form: no matter which element a of S is chosen, the set $\overline{\{a\}}$, obtained by excluding a from S, is equipotent to S. Here, $\overline{\{a\}}$ is the complement of the set $\{a\}$ containing one element of S. We first consider the case where S is the set N of natural numbers and $N' = \overline{\{a\}}$, where a is a natural number. We associate a natural number $n < a$ with n itself and a natural number $n \geq a$ with $n + 1$. Obviously, we thus establish a one-to-one correspondence between the set N of all natural numbers and the set N' of all natural numbers that are different from a. Therefore, the same assertion is true for any countable set.

For an arbitrary infinite set S, we use Theorem 40. As we saw, we can construct a countable subset N such that $a \in N$. As we proved, there exists a one-to-one correspondence between N and N', where N' is obtained by excluding a from N. Let $\overline{N}$ be the complement of the set N in S, and let S' be obtained by excluding a from S. Then $S = N + \overline{N}$ and $S' = N' + \overline{N}$. We establish a one-to-one correspondence between S and S' by combining elements from N and N' into pairs in accordance with the previously constructed one-to-one correspondence and by combining each element $b \in \overline{N}$ into a pair with itself. We thus obviously obtain a one-to-one correspondence between S and S', and they are therefore equipotent.   □

We present a few examples of countable sets.

1. *The set of integers is countable.*

We pair the integer 0 with the number 1, positive integers n with the numbers $2n$, and negative integers $-m$ with the numbers $2m + 1$. We obviously obtain a one-to-one correspondence between the set of integers and the set of natural numbers.

Corollary 4. *A set that is the sum of two countable subsets is countable.*

Let $A = B + C$. If B and C are countable, then B is equipotent to the set of positive integers, and C is equipotent to the set of negative integers. Therefore, A is equipotent to the set of all integers and is hence countable. (In this argument, only nonzero integers were obtained. Find a way around this difficulty yourself.)

2. *The set of all positive rational numbers is countable.*

Given a positive rational number m/n (with m and n relatively prime), we call the number $m + n$ the *height* of m/n. Obviously, the number of rational numbers of a given height is finite. We first write the rational numbers of height 2, then of height 3, and so on. We obtain an infinite sequence in which we can find any positive rational number sooner or later. Pairing each rational number and the number of its position in this sequence, we obtain a one-to-one correspondence between the set of all positive rational numbers and the set of natural numbers. The beginning of our sequence is

$$1, \quad \frac{1}{2}, \quad 2, \quad \frac{1}{3}, \quad 3, \quad \frac{1}{4}, \quad \frac{2}{3}, \quad \frac{3}{2}, \quad 4, \quad \dots .$$

Here, we assume that $1 = 1/1$, $2 = 2/1$, $3 = 3/1$, $4 = 4/1, \dots$.

Corollary 5. *The set of all rational numbers is countable.*

Indeed, we established a one-to-one correspondence between the set of positive rational numbers and the set of natural numbers. This means that there exists a one-to-one correspondence between the set of negative rational numbers and the set of negative integers. Pairing zero with zero in addition, we obtain a one-to-one correspondence between the set of all rational numbers and the set of all integers. Because the second set is countable according to Example 1, the first set must be countable.

In Example 2, we represented the set S of rational numbers as a sum of a countable number of finite subsets S_k, where S_k is the set of rational numbers of height k. Therefore, the claim in Example 2 follows from a general result: the sum of a countable number of finite sets is countable. We prove an even more general statement.

3. *The sum of a countable number of finite or countable sets is countable.*

The proof is based on the same principle we used in Example 2. Let $S = S_1 + S_2 + S_3 + \dots$. Because the set S_i is either finite or countable, its elements can be enumerated. We assume that each set S_i is in fact enumerated. We define the height of the element a of the set S. If it belongs to the set S_i and has the index j in its enumeration, then we call $i + j$ its height. Indeed, if $i + j = n$ is given, then $i < n$, and an element of height n can only belong to one of the sets $S_1, \dots, S_{n-1}$. If it belongs to the set S_i, then it has the index $j = n - i < n$. There is only a finite number of such elements. We can therefore first enumerate elements of height 2, then of height 3, and so on. As a result, we enumerate all elements of the set S. This means that it is countable. A procedure for enumerating S is shown schematically in Fig. 33, where the set S_1 is written in the first line, the set S_2 in the second, and so on. The order of enumerating all elements of S is represented by the zigzag line. It is assumed in Fig. 33 that all the sets S_i are countable. Draw a similar scheme if, for example, the set S_1 contains three elements, the set S_2 contains two elements, and the sets S_i for $i = 3, 4, \ldots$ are countable.

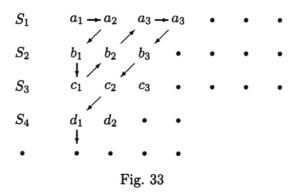

Fig. 33

We have given a series of examples of countable sets. Obviously, they are all equipotent to each other. We now give a few examples of other sets that are equipotent to each other.

4. *Any two line segments are equipotent.*

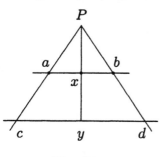

Fig. 34

Does P always exist?)

We can place our two line segments $[a, b]$ and $[c, d]$ on two parallel lines and establish a one-to-one correspondence as in Fig. 34. That is, connecting any point x on the segment $[a, b]$ with the intersection P of the lines ac and bd, we associate x with the point y that is the intersection of the line Px with the line cd on which the segment $[c, d]$ is placed. (Verify that y cannot lie outside the segment $[c, d]$.

5. *Any line segment is equipotent to the entire line, that is, to the set of all real numbers.*

Recalling Example 4, it is sufficient to consider one such line segment, for example, $[0, 1]$. Let I denote this segment. We divide the segment into two equal parts by the point $1/2$ and place it on the plane, bending it at the midpoint, as shown in Fig. 35: in the form of two equal segments $[A, B]$ and $[B, C]$ of length $1/2$ making equal angles with the x axis.

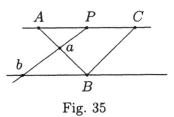

Fig. 35

Letting P denote the midpoint of the segment AC, we project the bent segment I on the x axis as in Example 4, that is, we associate the point a with the point b that is the intersection of the line Pa and the x axis. Obviously, the bent segment I is equipotent to the original segment. The described correspondence between the bent segment and the straight line x is one-to-one with the exclusion of the endpoints of the segment—the points A and C are not associated with any point on the x axis, because the line $PA = PC$ is parallel to the axis and does not intersect it. We therefore change our construction slightly. We recall that in proving the corollaries to Theorem 40, we established that any infinite set is equipotent to a set obtained by excluding one element from it. In particular, the segment $[0, 1]$ is equipotent to the set obtained by excluding one of its endpoints, the number 0. Repeating this argument, we verify that the segment is equipotent to the set obtained by excluding both its endpoints. In Fig. 35, this set corresponds to the set obtained by excluding the endpoints A and C from the bent segment I. Let J denote this set. As we saw, the set J is equipotent to our original segment. On the other hand, the correspondence shown in Fig. 35 is a one-to-one correspondence between J and the entire x axis. Our assertion is thus proved.

Problems:

1. Prove that the product of two countable sets is countable. The concept of the product of two sets is defined in Sec. 7 and is as suitable for infinite sets as for finite sets.
2. Prove that any circle is equipotent to a line segment.
3. We divide the segment $[0, 1]$ into two equal parts by the number $1/2$, then the segment $[1/2, 1]$ into two equal parts by the number $1/2 + 1/4$, and so on. We obtain the numbers $\alpha_n = 1/2 + 1/4 + \cdots + 1/2^n$. Prove that the segment $[0, 1)$ is the sum of a countable number of segments $[\alpha_{n-1}, \alpha_n)$, where $\alpha_0 = 0$ and the notation $[a, b)$ means that a is included in the segment and b is not. Analogously, prove that the straight line is the sum of a countable number of segment $[n-1, n)$, where n ranges all integers. Prove that the segments $[\alpha_{n-1}, \alpha_n)$ and $[n-1, n)$ are equipotent. From this, obtain a new deduction of the statement that any line segment is equipotent to the entire straight line (the statement in Example 5).
4. Give a formula for the one-to-one correspondence between the segments $[a, b]$ and $[c, d]$ constructed in Example 4.
5. Prove that there does not exist a one-to-one correspondence between the set of all integers and the set of all natural numbers that preserves addition, that is, such that if m corresponds to m' and n corresponds to n', then $m+n$ corresponds to $m' + n'$.
6. Prove that there does not exist a one-to-one correspondence between the set of all integers and the set of all natural numbers that preserves the order relation,

that is, such that if m corresponds to m' and n corresponds to n' and $m < n$, then $m' < n'$.

7. For two points on a circle, we call the length of the shortest arc connecting those points the distance between those points. As usual, we define the distance between two points A and B on a line segment as the length of the segment AB. Prove that there does not exist a one-to-one correspondence between a circle and a line segment that preserves distance, that is, such that if A corresponds to A' and B corresponds to B', then the distance between A and B is equal to the distance between A' and B'. Does this statement still hold if one point is excluded from the circle?

8. Given a set of line segments on a straight line that are pairwise disjoint, prove that this set is finite or countable. *Hint*: Prove that a set of line segments contained in a given line segment must be finite or countable. For this, consider the set of segments with a length greater than 1, greater than 1/2, greater than 1/3, and so on.

9. Some crosses are located on the plane. Each cross consists of horizontal and vertical line segments of length 1 that intersect in the middle. The crosses are pairwise disjoint. Prove that the set of these crosses is finite or countable.

10. Let $S_1, S_2, \ldots, S_n, \ldots$ be a countable system of subsets of S. Prove that if all sets S_n are countable, then the set $\bigcup_{n \geq 1} S_n$ is also countable.

19.   Continuum

At the end of the preceding section, we presented several examples of infinite sets, which can be segregated in two groups: A being countable sets (all equipotent to each other by definition) and B being sets that are equipotent to a line segment (and this also means equipotent to each other). An open question remains: are sets in the first group equipotent to sets in the second group? The principle of classifying sets by their equipotency would probably have remained a way to present already existing mathematical theories more distinctly and smoothly if it were not proved that sets from the groups A and B are *not equipotent*. This fact has a fundamental importance for all mathematics.

Theorem 41. *The set of points in a line segment is not countable.*

We present two different proofs of this theorem using different properties of a line segment. As we saw in Sec. 18, all line segments are equipotent to one line segment, for example, to $[0,1]$. We therefore consider this segment.

First proof. If the segment $[0,1]$ were countable, then the set obtained by excluding the right endpoint, the number 1, would also be countable. We write each number in this set in the form of a decimal fraction $0.a_1 a_2 \ldots$, where a_i takes the values $0, 1, \ldots, 9$. We suppose that all these numbers can be enumerated $x_1, x_2, \ldots, x_n, \ldots$ and write their representations as decimal fractions in a list, one after another:

$$x_1 : 0.a_1 a_2 a_3 \ldots$$
$$x_2 : 0.b_1 b_2 b_3 \ldots$$
$$x_3 : 0.c_1 c_2 c_3 \ldots \tag{1}$$
$$\vdots$$

We obtain a contradiction of the supposition that *all* numbers in the segment are enumerated by constructing a decimal fraction that is not in list (1). We construct it in the form $y = 0.k_1 k_2 k_3 \ldots$, choosing the digits $k_1, k_2, k_3, \ldots$ such that $k_1 \neq a_1$. Then for any choice of the following digits, $y \neq x_1$. We then choose k_2 such that $k_2 \neq b_2$. Then for any choice of the following digits, $y \neq x_2$. We then choose $k_3 \neq c_3$, and so on: k_n is chosen such that it is not equal to the digit in the nth position of the nth line in list (1). Then $y \neq x_n$. Choosing all the digits k_n in this way, we find that y does not coincide with any x_n.

The following objection can be made to this procedure. We saw in Sec. 15 that the correspondence between the numbers in the segment and all decimal fractions is *not* one-to-one—it is violated by the infinite decimal fractions that contain a repeating nine. But if we exclude such fractions, then the correspondence is one-to-one. Therefore, we must use only fractions without a repeating nine. As a result, all fractions in list (1) do not have a repeating nine. But we must also construct our fraction y such that it does not have a repeating nine. This is entirely possible. In choosing each digit k_n, we had the only restriction that it must be different from the digit in the nth position in the nth line in list (1). But because we then have nine of the ten digits $0, 1, 2, \ldots, 9$ to choose from, we can choose k_n to be also different from 9. As a result, our constructed fraction y, which is *not* in list (1), does not contain any digit 9. The proof is complete. This procedure for constructing the infinite decimal fraction y is called the *diagonal procedure*. $\square$

Second proof. In this proof, we use segments determined by two numbers a and b, $a \neq b$, whose order is not specified, that is, either $a < b$ or $b < a$ is possible. In either case, the segment is denoted by $[a, b]$ such that if $b < a$, then $[a, b]$ means the segment $[b, a]$ in the usual notation. We note that any segment contains a point different from its endpoints, for example, the segment $[a, b]$ contains the midpoint $(a+b)/2$. Applying the same reasoning to the segment $[a, (a+b)/2]$ and repeating it, we verify that a segment contains an infinite set of points.

We now turn directly to the proof and suppose that the points of the segment $[0, 1]$ are enumerated:

$$x_1, \quad x_2, \quad x_3, \quad x_4, \quad \ldots \ . \tag{2}$$

Because the enumeration of the points in the segment is a one-to-one correspondence by definition, the numbers x_m and x_n with $m \neq n$ are different. We prove that the supposition of the countability of the points in the segment contradicts the axiom of included intervals. We construct a certain system of included intervals. We start with the segment $[x_1, x_2]$ and select only those points from sequence (2) that are contained in this segment. As we saw, there is an infinite number of such points, and they comprise the sequence

$$x_1, \quad x_2, \quad x_p, \quad x_q, \quad x_r, \quad \ldots, \tag{3}$$

where $1 < 2 < p < q < r \ldots$.

We consider the segment $[x_p, x_q]$ and select only those points from sequence (3) that are contained in this segment. We obtain one more infinite sequence

$$x_p, \quad x_q, \quad x_s, \quad x_t, \quad x_u, \quad \ldots,$$

where $p < q < s < t < u \ldots$.

This process can be repeated endlessly: at each step, we obtain the numbers for a certain segment, of which there are infinitely many and which are contained in sequence (2). We thus obtain a countable number of sequences:

$$x_1, \quad x_2, \quad x_p, \quad \ldots$$
$$x_p, \quad x_q, \quad x_s, \quad \ldots$$
$$x_s, \quad x_t, \quad x_v, \quad \ldots$$
$$\vdots$$

with the first containing all points of the segment $[x_1, x_2]$, the second containing all points of the segment $[x_p, x_q]$, and so on. By construction, each new sequence begins with a larger number than the preceding sequence. Therefore, no single number in sequence (2) can belong to all such sequences. But sequences (2), (3), etc., are simply all numbers in the intervals $[x_1, x_2] \supset [x_p, x_q] \supset [x_s, x_t] \supset \ldots$. According to the axiom of included intervals, there exists a number common to all the intervals $[x_1, x_2], [x_p, x_q], [x_s, x_t], \ldots$. In other words, it belongs to all the segments, and this, as we just saw, is impossible. $\square$

The second proof is a bit more complicated than the first, but it has the advantage that it does not use writing the numbers in any system of representation but proves Theorem 41 directly from the axioms of real numbers.

Both of these proofs were found by Cantor in the 1870s: he first found the second proof and then the first (he was bothered by the difficulty connected with fractions with a repeating nine). As we saw, the proofs are quite simple, but the posing of the problem was then completely novel. In a later work, Cantor wrote that he required eight years to find the proof that the segment is not countable. Some very interesting correspondence between Cantor and Dedekind has been preserved, and it indicates the effort with which these new ideas were developed. Cantor wrote that he could not answer the question whether a segment is countable and asked whether Dedekind knew the answer. Dedekind replied that he did not know the proof (both had correctly guessed the right answer), but said that the question, in his opinion, was not worth any great effort, because it was not likely to have any interesting consequences.

It is amazing that Dedekind did not immediately sense the importance of the question—if only because the existence of irrational numbers follows immediately from the uncountability of the segment and the countability of the set of rational numbers, moreover, on the basis of completely new reasoning (before then, the existence of irrational numbers was established analogously to our arguments in Chap. 1). It is even more amazing when we consider that Dedekind himself proved an assertion that together with the theorem of the uncountability of the segment led to a significantly more remarkable conclusion.

We now discuss a concept, new to us, that was well known at the time of Dedekind and Cantor. The number α is said to be *algebraic* if it is the root of a polynomial $a_0 + a_1 x + \cdots + a_n x^n$ with rational coefficients a_i. Because the number of roots does not change when it is multiplied by a constant, we can multiply the polynomial $a_0 + a_1 x + \cdots + a_n x^n$ by a common denominator of all the numbers $a_0, \ldots, a_n$ and therefore immediately suppose in the definition of an algebraic number that the coefficients $a_0, \ldots, a_n$ are integers. Those numbers that are not algebraic are said to be *transcendental*.

Theorem 42. *The set of all algebraic numbers is countable.*

Proof. We define the *height* of a polynomial $a_0 + a_1 x + \cdots + a_n x^n$ with integer coefficients as the number $n + |a_0| + |a_1| + \cdots + |a_n|$. The height is obviously a natural number. It is just as obvious that there is only a finite number of polynomials whose height does not exceed a given number m. Indeed, if $n + |a_0| + |a_1| + \cdots + |a_n| \leq m$, then $n \leq m$ and $|a_i| \leq m$ for all $i = 0, 1, \ldots, n$. Therefore, there are at most $2m+1$ possibilities $(-m, -m+1, \ldots, 1, 0, 1, \ldots, m)$ for each coefficient a_i, and the number of all such polynomials is finite.

We now consider all algebraic numbers that are roots of polynomials with integer coefficients having a height that does not exceed a given natural number m. Let A_m denote this set. It is finite: indeed, the number of polynomials whose height does not exceed m is, as we just saw, finite, and each polynomial has a finite number of roots (according to Theorem 13 in Chap. 2). Finally, the union of all sets A_m for $m = 1, 2, \ldots$ coincides with the set of all algebraic numbers. From Example 3 in Sec. 18, it consequently follows that the set of all algebraic numbers is countable.   □

Because the set of all real numbers is equipotent to a segment (Example 5 in Sec. 18), the existence of transcendental numbers follows from Theorem 42. And this fact is far from simple. Although it follows from Theorems 41 and 42 that the transcendental numbers are "incomparably more" than the algebraic numbers, the construction of a specific transcendental number or, still worse, the proof of the transcendality of some known number is a difficult task. Only in the middle of the 19th century was a specific number constructed that could be proved to be transcendental. The most widely known transcendental number is π (the ratio of the circumference of a circle to it diameter). Its transcendality was proved in the 1880s.

The assertion that we labeled Theorem 42 was proved by Dedekind in a letter to Cantor. Evidently, just to emphasize concrete applications of the new ideas, Cantor published an article with the title "On one property of the collection of all real algebraic numbers," in which proofs of the assertions we called Theorem 41 and Theorem 42 were given and the existence of transcendental numbers was deduced from them. After this, Dedekind admitted that his objection to the question of the countability or uncountability of the segment on the basis of little interest had been "decisively refuted."

But Cantor's next discovery was a shock even to Cantor himself. It is our next theorem.

Theorem 43. *The set of all points in a square is equipotent to the set of points in a segment.*

We compare a square with side 1 and a segment of length 1. We start with a simple technical change in the task. We let S denote the square and separate two adjacent sides AB and BC from it (Fig. 36). We let P denote the remainder of the square and L denote the union of the sides AB and AC. Then $S = P + L$. It is obvious that L is equipotent to a segment (it is a bent segment). If we prove that P is equipotent to a segment, then it will be proved that S is also equipotent to a segment. Indeed, as we saw, a segment is equipotent to the set obtained by excluding one of its endpoints.

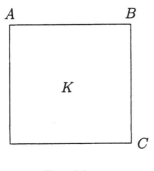

We can therefore consider that P is equipotent to the interval $[0, 1)$ with the number 1 excluded and L is equipotent to the interval $[0, 2]$. And then S is equipotent to the interval $[0, 2]$. In just the same way, it is sufficient for us to prove that the set P is equipotent to the segment $[0, 1)$ with the point 1 excluded. These obvious remarks are needed to make it convenient to pass from points to numbers. The coordinate method allows assigning each point of the square P a pair of numbers (x, y), where $0 \leq x < 1$ and $0 \leq y < 1$. In exactly the same way, the points in the segment $[0, 1)$ with the point 1 excluded correspond to the numbers t such that $0 \leq t < 1$. Consequently, we must show a one-to-one correspondence between the pairs (x, y) and the numbers t.

Fig. 36

For this, we write all the numbers we encounter in the form of decimal fractions

$$x = 0.a_1 a_2 a_3 \ldots, \tag{4}$$
$$y = 0.b_1 b_2 b_3 \ldots, \tag{5}$$
$$t = 0.c_1 c_2 c_3 \ldots, \tag{6}$$

where a_i, b_j, and c_k are some digits $0, 1, \ldots, 9$. We thus write all points (x, y) of the set P and all numbers t of the segment with an excluded endpoint. We now combine a pair (x, y) and a number t by "mixing" fractions (4) and (5), that is, we set $c_1 = a_1$, $c_3 = a_2$, $c_5 = a_3, \ldots$, $c_2 = b_1$, $c_4 = b_2$, $c_6 = b_3, \ldots$. The correspondence is one-to-one because we can reconstruct fractions (4) and (5) from fraction (6): the digits c_j with odd indices give all digits of the first fractions, and the digits c_j with even indices give all digits of the second fraction.

This argument provokes the same objection as the first proof we offered for Theorem 41, specifically, in connection with decimal fractions having a repeating nine. Indeed, for the correspondence between numbers in the interval and decimal fractions to be one-to-one, we must prohibit fractions with a repeating nine. But if fraction (6) does not have a repeating nine, one of the reconstructed fractions (4) or (5) could have a repeating nine. For example, if $t = 0.909090\ldots$, then $x = 0.999\ldots$. Dedekind made this objection when Cantor communicated his proof in a letter. Cantor was unable to eliminate this defect in the chosen approach, and he soon proposed a different proof that did not use decimal fractions.

Actually, it became clear later that the proof is easily corrected. For this, it is necessary to complicate the process of "mixing" fractions (4) and (5) somewhat, focusing special attention on the appearance of the digit 9. Namely, we mix them as previously, taking in turn one digit from the first and another from the second as long as all the digits differ from 9. If the digit a_k or b_k is equal to 9, then we take it together with all 9s that immediately follow it and the first following digit that differs from 9, and we place the entire group of digits in fraction (6). For example, from the fractions $x = 0.12(995)76\ldots$ and $y = 0.4(93)51\ldots$, we obtain $t = 0.142(93)(995)5716\ldots$, where we show digits taken as an entire group in parentheses. The process of reconstructing fractions (4) and (5) from fraction (6) also changes. As previously, we in turn place one digit from (6) in (4) and then in (5) as long as they are not 9. If we reach a digit 9, then we take it together with all immediately following 9s and the first following digit that is not 9, and we place the entire group of digits directly in the corresponding fraction. Thus, each group of transferred digits contains a digit that is not 9, and a repeating nine cannot occur.   □

The result obtained with this simple argument seems to contradict our geometric intuition: figures of different dimensionalities—a square and a line segment—are equipotent! It is possible to prove that even a cube is equipotent to them (Problem 1). This result stunned even Cantor himself. He writes to Dedekind, "What I recently revealed to you is so unexpected, so new to me that I cannot calm my mind until I receive, my respected friend, your judgment regarding it. Until you endorse it, I can only say"—here, in a letter written in German, Cantor unexpectedly changes to French—"I see, but I don't believe" (probably an allusion to the evangelical saying "you believed because you saw Me: blessed are they that did not see but believed"). It seemed to Cantor that the mathematical description of our intuition of dimensionality required revision: "The difference possessed between two formations with different numbers of dimensions most likely must be sought in a completely different cause than the number of independent coordinates...."

In answer, Dedekind confirmed the validity of the new proof, but he expressed his conviction that it does not contradict our belief that dimensionality coincides with the number of coordinates ("although it might seem devastated by your theory"): "Why, all authors, obviously, took account of the tacit, completely natural assumption that for a new definition of points (- - -) using new coordinates, the latter must be *continuous* functions of the old."

By that time, the concept of a continuous function had been precisely formulated and was well known. We do not delve into this question in all its generality, but consider examples of what curious phenomena arise from the one-to-one correpondence we have constructed between the points of a square and of a segment. Moreover, they are connected namely with the difficulty given by a repeating nine and our way of overcoming it. We consider two points (x, y) and (x', y), where $x = 0.10 \ldots 0 \ldots$, $y = 0.0 \ldots 0 \ldots$, and $x' = 0.09 \ldots 90 \ldots$ (there are n consecutive 9s in x'). It is obvious that

$$y = 0, \qquad x = \frac{1}{10}, \qquad x' = \frac{9}{10^2} + \cdots + \frac{9}{10^{n+1}}.$$

Then

$$x' = \frac{9}{10^2}\left(1 + \frac{1}{10} + \cdots + \frac{1}{10^{n-1}}\right) = \frac{9}{10^2}\frac{1 - 1/10^n}{1 - 1/10}$$

(by the formula for the sum of a geometric progression). This number is equal to

$$\frac{1}{10}\left(1 - \frac{1}{10^n}\right) = \frac{1}{10} - \frac{1}{10^{n+1}}.$$

Therefore, x and x' become arbitrarily close as n increases.

We now consider which points t and t' in the segment are associated with our points by the rule given in the proof of Theorem 43. To obtain t, we simply "mix" the fractions corresponding to x and y. After the decimal point, we take the digit 1 from x, then 0 from y, and further there are only 0s; therefore, $t = 0.10 \cdots = 1/10$. We now attentively follow the process of constructing t'. After the decimal point, we must take the digit 0 from x', then 0 from y, and then we encounter the digit 9 and must take the entire group of 9s and the digit 0 that follows. Then we again take alternately one digit from y and from x', but they are all 0s. As a result, we obtain $t' = 0.009 \ldots 90 \ldots$. In other words, $t' = 9/10^3 + \cdots + 9/10^{n+2}$. As previously, we calculate

$$t' = \frac{9}{10^3}\left(1 + \frac{1}{10} + \cdots + \frac{1}{10^{n-1}}\right) = \frac{9}{10^3}\frac{1 - 1/10^n}{9/10}$$

$$= \frac{1}{100}\left(1 - \frac{1}{10^n}\right) = \frac{1}{100} - \frac{1}{10^{n+2}}.$$

We see that x and x' become arbitrarily close as n increases. Because y does not change, the points (x, y) and (x', y) also become arbitrarily close. But the corresponding points t and t' are always separated from each other by more than a certain constant amount:

$$t = \frac{1}{10}, \qquad t' = \frac{1}{100} - \frac{1}{10^{n+2}},$$

and

$$t - t' = \frac{1}{10} - \frac{1}{100} + \frac{1}{10^{n+2}} > \frac{1}{10} - \frac{1}{100} = \frac{9}{100}.$$

With our correspondence, the square bursts apart in a sense: all ever closer points of the square correspond to points separated from each other by a distance greater than some fixed positive quantity. This process is similar to tearing a sheet of paper into ever smaller pieces.

In a letter to Cantor, Dedekind expressed the proposition that if the requirement for continuity (whose formulation we omit) is included in the definition of one-to-one correspondence, then geometric forms with different dimensionalities (for example, a square and a segment, or a cube and a square) could never be set in one-to-one correspondence with each other. He wrote that he lacked the time to even try proving it, but he advanced the hypothesis "as my conviction and belief." Cantor later agreed with Dedekind's view and even published a proof of this hypothesis, but the proof turned out to be invalid. Dedekind's hypothesis was finally proved only in 1910.

A set that is equipotent to a segment is called a *continuum*. Theorem 41 confirms that *a continuum is uncountable*. Practically all infinite sets encountered in mathematics are of one of these two types: either a continuum or countable. It is possible to construct a set that is not a continuum, not countable, and not finite (Problem 9) and even to construct an infinite family of infinite sets that are pairwise nonequipotent to each other. But their importance for the rest of mathematics is not great. Similarly, the equipotence of a square and a segment (Theorem 43) does not have the applications that it would seem we could count on. We already explained part of the reason. Usually, a set in mathematics arises more concretely, its elements connected by some kind of relations, for example, definite operations on them or inequalities or (in geometry) distance. Or we assert that a certain sequence a_n approaches an element a arbitrarily closely. Then we are only interested in one-to-one correspondences that preserve these relations between the elements—they can turn out to be significantly fewer (see Problems 5, 6, and 7 in Sec. 18). Therefore, Theorem 43, despite its mind-blowing formulation, is not a "working" mathematical result. But Theorem 41 is one of the most significant results in mathematics.

Problems:

1. Prove that the set of points in the unit cube, consisting of (x, y, z), $0 \leq x < 1$, $0 \leq y < 1$, $0 \leq z < 1$, is equipotent to the segment t, $0 \leq t < 1$.
2. Prove that if S_1 and S_2 are nonintersecting subsets of the set S and S_1 and S_2 are equipotent, then their complements $\overline{S_1}$ and $\overline{S_2}$ are also equipotent.
3. Prove that if N is a countable subset of X and $\overline{N}$ is infinite, then X and $\overline{N}$ are equipotent. *Hint*: Use Theorem 40.
4. Prove that the set of irrational numbers in a segment and the set of transcendental numbers in a segment are continuums.
5. Prove that the set of all infinite sequences of the form $a_1, a_2, a_3, \ldots, a_n, \ldots$, where a_k can take the values $0, 1, \ldots, 9$, is a continuum.
6. Given a natural number $k > 0$, prove that the set of infinite sequences $a_1, a_2, \ldots, a_n, \ldots$, where a_i is an arbitrary integer satisfying $0 \leq a_i \leq k$, is a continuum.
7. Let U be the set of all subsets of the set S (we encountered it in Chap. 3 for the case where S is finite). Prove that the sets U and S are not equipotent. Prove that if S is countable, then U is a continuum. *Hint*: Suppose that a one-to-one correspondence $a \leftrightarrow A$ can be established between elements and subsets of the set S. Consider the set B of all elements a such that a is not contained in the subset to which it corresponds. Let $b \leftrightarrow B$ in our correspondence. Treat both hypotheses: $b \in B$ and $b \notin B$.
8. Construct an infinite set that is neither countable nor a continuum.
9. Prove that the set of decimal fractions in the interval $[0, 9]$ with a repeating nine is a countable set. Using this, give a different end to the proof of Theorem 41 (the difficulty with a repeating nine).

20. Thin Sets

In this section, we show which specific properties are connected with the countability of a set. We consider sets S contained in the interval $[0, 1]$. We discuss whether the "length" of such a set can be measured somehow. Clearly, the length of the interval $[0, 1]$ can be naturally considered equal to 1. The length of the interval $[a, b]$ contained in $[0, 1]$ is just as naturally set equal to $b - a$. If the set consists of several nonintersecting intervals, then its length can be set equal to the sum of those intervals. For example, the length of the set $S = [0, 1/2] \cup [3/4, 1]$ is equal to $1/2 + 1/4 = 3/4$. But an arbitrary set in general is not required to be composed of intervals, and it is not so easy to find a natural definition of length for an arbitrary set. For example, how can the "length" of the set of rational numbers or of irrational numbers in the interval $[0, 1]$ be defined?

There is a theory that allows defining the length for, if not an arbitrary set, at least a very wide class of sets contained in an interval, and the definition, moreover, has all the properties that our intuition suggests. This theory is called *measure theory*. We do not go deeply into it and only give the definition of when a set is considered to have the "length 0." In measure theory, these sets are called *sets of measure* 0. We call them *thin sets*.

We first justify the definition, appealing to geometric intuition, and give a formal definition at the end of our discussion. We already specified what to call the length of an interval $[a, b]$ contained in the interval $[0, 1]$ and also the sum of a finite number of nonintersecting intervals. It is one step from this to sets that are sums of a countable number of intervals. Namely, if $S = I_1 + I_2 + \cdots + I_n + \ldots$, where the I_k are nonintersecting intervals and the length of I_k is equal to α_k, then the intervals $I_1, I_2, \ldots, I_n$ are contained within the interval $[a, b]$ and do not intersect, and the sum of their lengths is therefore not greater than $b - a$. In other words, $\alpha_1 + \alpha_2 + \cdots + \alpha_n \leq a - b$ for all $n = 1, 2, \ldots$. It follows from Theorem 29 that the infinite sum $\alpha_1 + \alpha_2 + \cdots + \alpha_n + \cdots$ exists and does not exceed $b - a$. We call this sum the length of the set S. For example, if

$$I_1 = \left[0, \frac{1}{2}\right], \qquad I_k = \left[1 - \frac{1}{2^{2k-2}}, 1 - \frac{1}{2^{2k-1}}\right]$$

(Fig. 37), then

$$\alpha_k = \left(1 - \frac{1}{2^{2k-1}}\right) - \left(1 - \frac{1}{2^{2k-2}}\right) = \frac{1}{2^{2k-2}} - \frac{1}{2^{2k-1}},$$

and hence

$$\alpha_k = \frac{1}{2^{k-1}}.$$

Therefore,

$$\alpha_1 + \alpha_2 + \cdots = \frac{1}{2} + \frac{1}{8} + \frac{1}{32} + \cdots$$
$$= \frac{1}{2}\left(1 + \frac{1}{4} + \left(\frac{1}{4}\right)^2 + \cdots\right) = \frac{1}{2}\frac{1}{1 - 1/4}$$

(by the formula for the sum of an infinite geometric progression). For the length of our set, we obtain the value 2/3.

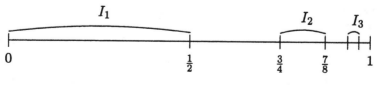

Fig. 37

We now introduce two assumptions that allow formulating the desired definition. First, we consider a changed situation in which the set S is the union of intervals $I_1, I_2, \ldots, I_n, \ldots$, which possibly intersect. It is natural to assume that no matter how we define the length of the set S, it *cannot exceed* the sum of the lengths of the intervals I_k. Now, of course, we can no longer assert that the sums $\alpha_1 + \alpha_2 + \cdots + \alpha_n$ are bounded, where, as before, α_k denotes the length of the interval I_k. Even the case where all the intervals I_k coincide is now possible. Therefore, our assumption is applicable only in those cases where the sum $\alpha_1 + \alpha_2 + \cdots + \alpha_n + \cdots$ exists. The second assumption is even more obviously intuitive: we assume that if the lengths of the sets S_1 and S_2 are defined somehow and if $S_1 \subset S_2$, then the length of S_1 cannot exceed the length of S_2.

The assumptions made still do not give the possibility to *measure* the length of an arbitrary set, because it is impossible, generally speaking, to represent it as the union of a countable number of intervals. For example, it is impossible to thus represent the set of irrational numbers in the interval $[0, 1]$ (see Problem 1). But we can use the second assumption to estimate the length of a set, no matter how the length is defined, if only the definition of length satisfies the given assumption. For example, we suppose that the set S is contained in the union of the intervals $I_1, I_2, \ldots, I_n, \ldots$. Then its length must not exceed the sum of the lengths of the intervals I_k. We measure the set S, in a sense, including it in different sets that are unions of countable sequences of intervals. If we obtain ever lower (closer to zero) estimates for the length of S as a result of such "measurements," then nothing remains but to assume that the length of the set S is equal to zero. This leads to the definition of a thin set.

In what follows, for the brevity of formulations, if a set S is contained in the union of the sets $S_1, S_2, \ldots, S_n, \ldots$, then we say that it is *covered* by those sets, and we call the inclusion

$$S \subset S_1 \cup S_2 \cup \cdots \cup S_n \cup \cdots$$

itself a *covering* of the set S by the sets S_k.

Definition. A set S on the line is called a *thin set* if for any arbitrarily small positive number ε, there exists a covering of the set S by the intervals $I_1, I_2, \ldots, I_n, \ldots$ (a finite or countable set of intervals) such that the sum of the lengths of the intervals does not exceed ε.

We emphasize once more that all the preceding discussion in this section proved nothing, but was only an *explanation* of this definition.

We now consider examples of thin sets. Of course, a set consisting of a single number x is a thin set because for any $\varepsilon > 0$, it has the covering

$x \in I_\varepsilon \cap [0, 1]$, where I_ε is the interval $[x - \varepsilon/2, x + \varepsilon/2]$ (and we include the intersection to account for the case where I_ε is not totally contained in $[0, 1]$), with a length not exceeding ε. We can similarly prove that a set consisting of a finite number of numbers is a thin set.

We can now clarify how the concept of a thin set is connected with countability.

Theorem 44. *A countable set of points in the interval* $[0, 1]$ *is a thin set.*

Let S be a countable set somehow enumerated: $S = \{a_1, a_2, \ldots, a_n, \ldots\}$. For each positive $\varepsilon > 0$, we construct a covering of this set by intervals $I_1, I_2, \ldots, I_n, \ldots$ such that the sum of the lengths of these intervals does not exceed ε, as is required by the definition of a thin set. For this, we take I_1 to be the interval of length $\varepsilon/2$ centered at a_1, $I_1 = [a_1 - \varepsilon/4, a_1 + \varepsilon/4]$. And for any k, we take the interval I_k to be the interval $[a_k - \varepsilon/2^{k+1}, a_k + \varepsilon/2^{k+1}]$. Clearly, the length of the interval I_k is $\varepsilon/2^k$, and the sum of the lengths of all intervals I_k cannot exceed the sum $\varepsilon/2 + \varepsilon/4 + \varepsilon/8 + \cdots = \varepsilon$ according to the formula for the sum of an infinite geometric progression (also see Problem 11 in Sec. 15). Because $a_k \in I_k$, the set S is covered by the intervals $I_1, I_2, \ldots$, the sum of whose lengths does not exceed ε. And this proves that S is a thin set. $\square$

Therefore, the set of rational numbers, for example, is a thin set, which is not obvious at first glance.

We pass to the next property.

Theorem 45. *The union of two thin sets is a thin set.*

Let S_1 and S_2 be thin sets and $S = S_1 \cup S_2$. To prove that S is a thin set, for any given $\varepsilon > 0$, we must construct a covering of S by intervals $I_1, I_2, \ldots, I_n, \ldots$ such that the sum of the lengths of the intervals I_k does not exceed ε. We use the fact that the sets S_1 and S_2 are thin sets. This means that for any $\eta > 0$, both S_1 and S_2 have a covering by intervals such that the sum of the lengths of the intervals in each covering do not exceed η. This is true for any η and, in particular, for $\eta = \varepsilon/2$, where ε is the positive number initially given. We write the coverings obtained:

$$S_1 \subset J_1 \cup J_2 \cup \cdots \cup J_n \cup \ldots,$$
$$S_2 \subset K_1 \cup K_2 \cup \cdots \cup K_n \cup \ldots.$$

We consider the sequence $J_1, K_1, J_2, K_2, \ldots, J_n, K_n, \ldots$. The union contains both S_1 and S_2, and this means it contains $S = S_1 \cup S_2$, that is, it is a covering of S by intervals. We prove that the sum of the lengths of the intervals in this covering does not exceed ε. For this, we let a_m

denote the length of the interval J_m and b_m denote the length of the interval K_m. The the sum of the lengths of the intervals in the sequence $J_1, K_1, J_2, K_2, J_3, K_3, \ldots$ is equal to

$$a_1 + b_1 + a_2 + b_2 + \cdots = (a_1 + a_2 + \cdots) + (b_1 + b_2 + \cdots). \qquad (7)$$

Because $a_1 + a_2 + \cdots \leq \varepsilon/2$ and $b_1 + b_2 + \cdots \leq \varepsilon/2$ by construction, the sum in the left-hand side of equality (7) does not exceed $\varepsilon/2 + \varepsilon/2$, that is, ε, as was necessary to prove.

This argument requires one refinement, which concerns formula (7). If the matter concerned finite sums, then we would say that parentheses can be placed arbitrarily when calculating the sum. This follows from the commutative and associative laws for addition (Axioms I_1 and I_2 in Chap. 5). But we have not introduced such axioms for infinite sums, and equality (7) therefore requires proof. Actually, it is sufficient for us to prove here that the left-hand side of equality (7) does not exceed the right-hand side. Let $a_1 + a_2 + \cdots = \alpha$ and $b_1 + b_2 + \cdots = \beta$. Any finite sum of k terms in the left-hand side of equality (7) can be segregated into the series of terms $a_1, a_2, \ldots, a_n$ and $b_1, b_2, \ldots, b_m$ ($m = n$ if k is even and $m = n - 1$ if k is odd). But (by the rules for operating with finite sums) this sum is equal to $a_1 + \cdots + a_n + b_1 + \cdots + b_m$. Because $a_1 + \cdots + a_n \leq \alpha$ and $b_1 + \cdots + b_m \leq \beta$, all our finite sums cannot exceed $\alpha + \beta$. This means that the infinite sum in the left-hand side of equality (7) cannot exceed $\alpha + \beta$. $\qquad \square$

It obviously follows immediately from Theorem 45 by induction that the union of a finite number of thin sets is a thin set. But an even stronger statement holds, containing Theorem 44 as a particular case.

Theorem 46. *The union of a countable number of thin sets is a thin set.*

The proof is completely similar to the proof of the preceding theorem, and we therefore present it more briefly. Let $S_1, S_2, \ldots, S_n, \ldots$ be a countable set of thin sets and S be their union. Let a positive number ε be given. We construct a covering of the set S by a countable number of intervals, the sum of whose lengths does not exceed ε—we thus prove that S is a thin set. Because each of the sets S_n is a thin set, it (by definition) has a covering by a countable set of intervals, the sum of whose lengths does not exceed $\varepsilon/2^n$. We now consider all intervals in all the coverings considered above. Let the covering of the set S_1 have the form $S_1 \subset I_1 \cup I_2 \cup \ldots$; for S_2, the covering $S_2 \subset J_1 \cup J_2 \cup \ldots$; for S_3, the covering $S_3 \subset K_1 \cup K_2 \cup \ldots$; and so on for all S_n for $n = 1, 2, 3, \ldots$. We now consider the set of all intervals I_r, all intervals J_s, all intervals K_t, and so on. The set of intervals obtained is a union of a countable number of countable sets (because each S_n is covered by a countable set of intervals). We obtain a countable set of intervals (see Problem 10 in Sec. 18). The sum of their lengths does not exceed $\varepsilon/2 + \varepsilon/2^2 + \cdots + \varepsilon/2^n + \cdots = \varepsilon$. This proves the theorem. $\square$

We note that we encounter the same question here as in the proof of Theorem 45 in connection with equality (7), only in a more complicated situation. We now deal with a countable number of equalities:

$$a_1 + a_2 + \cdots = \alpha, \qquad b_1 + b_2 + \cdots = \beta, \qquad c_1 + c_2 + \cdots = \gamma,$$

and so on. We then mix all the numbers a_i, b_j, c_k, etc., somehow renumber them, and then consider their sums. It is sufficient to verify that the sum does not exceed $\alpha + \beta + \gamma + \ldots$. The proof is literally the same as in Theorem 45, and we omit it.

Finally, we consider one more, very important property of thin sets. Namely, we must still verify that the presuppositions used to formulate the definition of this concept are not mutually contradictory. We used the notion that the length of an interval $[a, b]$ contained in the interval $[0, 1]$ is equal to $b - a$. A thin set has "length zero." If it turned out that the interval $[a, b]$ with $b \neq a$ is a thin set, then our theory would be self-contradictory. We prove that this is not so. Simultaneously, we establish the first example of a set that is not a thin set—until now, we only proved that one or another set is a thin set. If thin sets play the same role in our theory as countable sets in Sec. 19, then the interval plays the role of a continuum, and the theorem we now prove is analogous to Theorem 42 that the continuum is not countable.

Theorem 47. *An interval is not a thin set.*

Let an interval $I = [a, b]$, and contrary to the assertion of the theorem, for any $\varepsilon > 0$, let it have a covering by intervals, the sum of whose lengths does not exceed ε. Let a covering denoted by

$$I \subset I_1 \cup I_2 \cup \ldots,$$

where α_k denotes the length of I_k, be such that the sum $\alpha_1 + \alpha_2 + \cdots$ exists and does not exceed ε. In what follows, we see that it is convenient to deal with intervals (a_k, b_k) with the endpoints a_k and b_k excluded (it is the set of the numbers x satisfying $a_k < x < b_k$). Such intervals are called *open intervals* and are denoted by the symbol (a_k, b_k). Here, we let them be denoted by the letters $J_1, J_2, \ldots$. As before, the length of an interval $J = (c, d)$, where $c < d$, is equal to $d - c$. We prove that under the given suppositions, the interval I has a covering by open intervals, the sum of whose lengths can be made arbitrarily small. For this, we take an arbitrarily small positive number η and include the interval $I_k = [a_k, b_k]$ in the open interval $J_k = (a_k - \eta/2^{k+1}, b_k + \eta/2^{k+1})$. Its length differs from the length of I_k by $\eta/2^k$. Therefore, the sum of their lengths does not exceed

$$\left(\alpha_1 + \frac{\eta}{2}\right) + \left(\alpha_2 + \frac{\eta}{4}\right) + \cdots + \left(\alpha_n + \frac{\eta}{2^n}\right) + \cdots$$

$$\leq \varepsilon + \eta\left(\frac{1}{2} + \frac{1}{4} + \cdots\right) = \varepsilon + \eta.$$

On the other hand, $I_k \subset J_k$ by construction, and therefore

$$I \subset I_1 \cup I_2 \cup \cdots \subset J_1 \cup J_2 \cup \ldots, \tag{8}$$

and $J_1, J_2, \ldots$ is therefore a covering of the interval I. Because we can take arbitrarily small numbers for ε and η, the open intervals J_k can be chosen such that the sum of their lengths is arbitrarily small.

We first prove the theorem in the simple case where the number of intervals in covering (8) is finite. We let $l(I)$ and $l(J)$ denote the respective lengths of the interval $I = [a, b]$ and the open interval $J = (a, b)$. Let the number of intervals in formula (8) be finite and equal to n. We prove that $l(I) \leq l(J_1) + \cdots + l(J_n)$ always holds. This then leads to a contradiction of the existence of a covering such that $l(J_1) + \cdots + l(J_n) < \varepsilon$ for any $\varepsilon < l(I)$.

We choose an enumeration of the intervals J_k such that $b \in J_n$, that is, $a_n < b < b_n$ (we recall that $I = [a, b]$ and $J_n = (a_n, b_n)$). We consider two cases.

1. The first case is the case where the interval J_n does not intersect any interval J_k for $k < n$. This means that if $J_k = (a_k, b_k)$, then either $b_k < a_n$ or $a_k > b_n$. In either case, a_n does not belong to the interval J_k for $k < n$. At the same time, a_n does not belong to the interval J_n (because J_n is an open interval). But $I \subset \bigcup J_k$, and therefore a_n does not belong to I, that is, either $a_n < a$ or $a_n > b$. This means that $a_n < a$. Similarly, we have $b < b_n$. Therefore, $l(J_n) = b_n - a_n > b - a = l(I)$. The necessary inequality is proved (and even $l(I) \leq l(J_n)$ in this case).

2. The interval J_n intersects some other interval, which can be denoted by J_{n-1}. Then $J_{n-1} \cup J_n$ is an open interval (verify this!), which can be denoted by J'_{n-1}. Then relation (8) can be rewritten in the form

$$I \subset J_1 \cup J_2 \cup \cdots \cup J_{n-2} \cup J'_{n-1}.$$

We can apply the principle of induction (on the number n) and can therefore assume that our assertion is proved in the case of this covering, that is,

$$l(I) \le l(J_1) + \cdots + l(J_{n-2}) + l(J'_{n-1}).$$

(The basis of induction is the case $n = 1$, which is obvious; see the end of the proof in case 1.)

It remains to verify that $l(J'_{n-1}) \le l(J_{n-1}) + l(J_n)$. Then we can obtain the necessary inequality. That the intervals J_{n-1} and J_n intersect means that there exists a number x with the property that

$$a_n < x < b_n,$$
$$a_{n-1} < x < b_{n-1}.$$

If $J'_{n-1} = J_{n-1} \cup J_n = (c, d)$, then c is the lesser of the numbers a_n and a_{n-1}, and d is the larger of the numbers b_n and $b_n - 1$. Therefore,

$$l(J_{n-1}) = d - c = d - x + x - c$$
$$\le b_n - a_n + b_{n-1} - a_{n-1} = l(J_{n-1}) + l(J_n).$$

The theorem is proved in this case.

We now turn to the consideration of a finer case, the case where the number of open intervals is infinite. We consider the open interval J_k as a subset of the set of all real numbers. We let $\overline{J_k}$ denote its complement and let $I'_k = \overline{J_k} \cap I$. In other words, I'_k is the set of numbers in the interval I that do not belong to the open interval J_k. Our given relation (8) is equivalent to

$$I'_1 \cap I'_2 \cap I'_3 \cap \cdots = \varnothing. \tag{9}$$

Relation (9) is just another way of writing the assertion that every number in the interval I belongs to some open interval J_k, that is, relation (8).

The set I'_k is similar to an interval, but it is not necessarily an interval: it might be an interval or two intervals. Moreover, the ends are now included in the interval, and we must consider the case where it reduces to a point: $[a, a] = a$ (Fig. 38).

We let $A_n = I'_1 \cap I'_2 \cap \cdots \cap I'_n$. It is obvious that $A_n \subset A_{n-1}$. Relation (9) implies that

$$A_1 \cap A_2 \cap A_3 \cap \cdots = \varnothing. \tag{10}$$

We have a picture similar to that in the axiom of included intervals. If the sets A_n were intervals, then we could apply that axiom and deduce

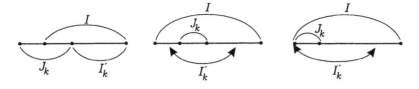

Fig. 38

from it that there exists a number x that belongs to all of them. This contradiction with relation (10) would prove the theorem. However, A_n is in general not an interval, but a somewhat more complicated set. It remains to consider this difficulty.

Because the intersection of two open intervals is an interval (possibly reduced to a point), A_n must consist of nonintersecting intervals (possibly degenerated to points). Let these intervals (or points) be denoted by $A_n^{(1)}, A_n^{(2)}, \ldots, A_n^{(k)}$. Because $A_n = A_{n-1} \cap I'_{n-1}$, the number of different intervals $A_n^{(i)}$ with a given n is finite (cf. Problem 9). We take the entire interval I as A_0. We represent this system of intervals with each interval $A_n^{(i)}$ shown as a small circle placed in rows below each other in order of increasing n (Fig. 39). Moreover, we connect the circle representing $A_n^{(i)}$ to the circle representing $A_{n+1}^{(j)}$ with a path if $A_n^{(i)} \subset A_{n+1}^{(j)}$. We say that the circle $A_n^{(j)}$ follows the circle $A_m^{(i)}$ ($m < n$) if we can pass from $A_m^{(i)}$ to $A_n^{(j)}$ by going down paths. This implies that the interval $A_n^{(j)}$ is contained in the interval $A_m^{(i)}$. Because $A_n \subset A_{n-1}$, each circle $A_{n+1}^{(i)}$ follows some circle $A_{n-1}^{(j)}$. No circle follows the circle $A_n^{(i)}$ if the intersection of the interval $A_n^{(i)}$ and the set I'_{n+1} is empty.

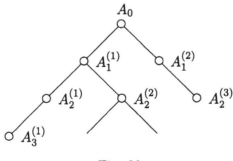

Fig. 39

In Fig. 39, we have a picture that is reminiscent of a branching root system. It can be one of two types.

1.   There is a finite number of circles in the picture. If they all have the form $A_m^{(i)}$ with $m \le n$, then no circle follows any circle $A_n^{(i)}$. Consequently, $A_n^{(i)} \cap I'_{n+1} = \varnothing$, and because $A_n = I'_1 \cap I'_2 \cap \cdots \cap I'_n$ and $A_n = \bigcap A_n^{(i)}$, we have $I'_1 \cap \cdots \cap I'_{n+1} = \varnothing$. This implies that relation (9) holds for the system of intervals $I'_1, \ldots, I'_{n+1}$ and relation (8) holds for the system of intervals $J_1, \ldots, J_{n+1}$. But this is the case with a finite number of intervals, which we already treated.

2.   The picture represented in Fig. 39 contains an infinite number of circles. If we find a connected sequence of paths going endlessly down in it (that is, a sequence $A_1^{(i)}, A_2^{(j)}, A_3^{(k)}, \ldots$, where $A_2^{(j)}$ follows $A_1^{(i)}$, $A_3^{(k)}$ follows $A_2^{(j)}$, and so on), then we obtain a system of included intervals $A_1^{(i)} \supset A_2^{(j)} \supset A_3^{(k)} \supset \ldots$. According to the axiom of included intervals (see Axiom VII in Sec. 14), this system of intervals must contain at least one number x (it makes a difference here that the intervals $A_n^{(i)}$ are considered together with their endpoints).

True, we have a situation that is somewhat more general´because there might be a degenerate interval $[a_k, b_k]$ with $a_k = b_k$ among the intervals $A_n^{(i)}$, that is, a point. But the axiom of included intervals also holds in this case: if one interval in the chain of included intervals is reduced a single point x, then all the following intervals must coincide with x. Therefore, there always exists a number x that belongs to all the sets A_n, and this means that $x \in A_1 \cap A_2 \cap A_3 \cap \ldots$, which contradicts condition (10).

We verify that we can always find such a sequence of paths going down endlessly in our case. This is very simple. By hypothesis, the number of circles in the representation given by Fig. 39 is infinite in our case. This means that an infinite number of circles follow A_0. It follows from this that an infinite number of circles follow $A_1^{(i)}$ for some i. It then follows that one of the $A_2^{(j)}$ following that $A_1^{(i)}$ must be followed by an infinite number of circles. Continuing, we obtain an infinite sequence $A_0, A_1^{(i)}, A_2^{(j)}, \ldots$ going down endlessly. This completes the proof of Theorem 47. $\square$

In view of Theorem 44, the uncountability of the continuum follows from Theorem 47. We thus obtain a third proof of Theorem 42.

Now, for the first time, we can assert that there exists a set that is not a thin set. For example, an entire interval is such a set. But we have proved more than this, a thin set is so "small" that it cannot contain an interval, no matter how small the length of the interval is. This justifies the representation of thin sets as sets "infinitely smaller" than an interval. In connection with this, we say that a certain property holds for *almost all numbers in an interval* if the set of numbers for which the property does not hold is a thin set. For example, almost all numbers are irrational, and almost all numbers are transcendental. (This is a combination of Example 2 in Sec. 18, Theorem 42, and Theorem 44.)

Until now, we discussed general properties of thin sets, but we still have only a few examples of such sets. Specifically, we only know that finite and countable sets are thin sets. We now give one of the most interesting examples of a thin set, a set that is uncountable.

This set is connected with the representation of numbers in the interval $[0, 1]$ using decimal fractions. We choose some digit—for definiteness in what follows, we take the digit 2, but our reasoning is applicable to any of the digits $r = 0, 1, 2, \ldots, 9$. We let S denote the set of all numbers in the interval whose decimal fraction does not contain our chosen digit. We prove that the set S is not countable and is a thin set. That S is not countable is obvious. It contains the subset S' consisting of all fractions that have only two specified digits in their written representation, where both digits are different from 2, for example, 0 and 1. The set of such fractions is obviously equipotent to the set of sequences $a_1, a_2, a_3, \ldots$, where all a_i are equal to 0 or 1. We know that the set S' is not countable (see Problem 6 in Sec. 19). Because a subset of a countable set is countable, it hence follows that the set S is not countable. It is easy to prove that S is a continuum (see Problem 4 at the end of this section).

We now prove that the set S is a thin set. We let S_n denote the set of numbers in the interval $[0, 1]$ that do not have the digit 2 in the first n positions of their representations as decimal fractions. The representation of a number $x \in S_n$ thus has the form

$$x = 0.a_1 a_2 a_3 \ldots a_n \ldots, \tag{11}$$

where $a_1 \neq 2$, $a_2 \neq 2, \ldots, a_n \neq 2$, and further can be any decimal digit. It is obvious that $S_n \supset S$. For each n, we construct a covering of the set S_n by intervals such that the sum of the lengths of the intervals becomes arbitrarily close to zero as n increases. Because $S_n \supset S$, a covering of the set S_n is simultaneously a covering of the set S, and we thus prove that S is a thin set.

We arbitrarily fix n digits $a_1, a_2, \ldots, a_n$ and ask ourselves: what does the set of numbers whose representations as a decimal fractions begin with these n digits look like? In other words, numbers from our set have a representation as a decimal fraction in form (11), where $a_1, a_2, \ldots, a_n$ are fixed and the remaining digits are arbitrary. We set

$$\alpha = 0.a_1 a_2 a_3 \ldots a_n 00 \ldots, \qquad \beta = 0.00 \ldots 0 a_{n+1} a_{n+2} \ldots,$$

where β has 0 as the first n digits after the decimal point and the following digits are the same as in x in (11). Then $x = \alpha + \beta$. Moreover, α is the same for all numbers in our set, and β ranges all numbers with n zeros immediately after the decimal point. In other words,

$$\beta = \frac{a_{n+1}}{10^{n+1}} + \cdots = \frac{1}{10^n}\left(\frac{a_{n+1}}{10} + \frac{a_{n+2}}{10^2} + \cdots\right).$$

Because the digits $a_{n+1}, a_{n+2}, \ldots$ are arbitrary, the expression in the parentheses represents arbitrary numbers y, $0 \le y < 1$, and $\beta = y/10^n$, that is, $\alpha + \beta$ (and this means our entire set) is contained in the interval $[\alpha, \alpha + 1/10^n]$. We considered numbers with the first n digits fixed. The possible choices for these n digits $a_1, a_2, \ldots, a_n$, where all $a_i \ne 2$, is 9^n according to Theorem 17 because the set of choices of digits is equipotent to the product of n sets of nine elements $\{0, 1, 3, 4, 5, 6, 7, 8, 9\}$ (we omit 2). Our set S_n is thus covered by 9^n intervals of the form $[\alpha, \alpha + 1/10^n]$. Each of these intervals has the length $1/10^n$, and the sum of their lengths is equal to $9^n/10^n = (9/10)^n$. Because $9/10 < 1$, we have $(9/10)^n \to 0$ as $n \to \infty$ according to Lemma 12. This proves our statement.

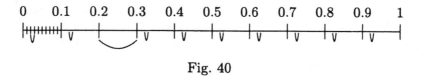

Fig. 40

The covering of the set S_n by 9^n intervals of length $1/10^n$ is shown in Fig. 40 for the cases $n = 1$ and 2. To construct the covering of the set S_1, it is sufficient to delete the interior of the interval $[0.2, 0.3]$ marked with the large arc, which we call the open interval $(0.2, 0.3)$. There remain nine intervals $[0, 0.1]$, $[0.1, 0.2]$, $[0.3, 0.4]$, $[0.4, 0.5]$, $[0.5, 0.6]$, $[0.6, 0.7]$, $[0.7, 0.8]$, $[0.8, 0.9]$, $[0.9, 1]$, each of length $1/10$. To construct the covering of the set S_2, we must delete the interior of an interval of length $1/100$ from each of those nine intervals. They are marked with small arcs in Fig. 40. For example, we delete the open interval $(0.02, 0.03)$ from the interval $[0, 0.1]$, the open interval $(0.12, 0.13)$ from the interval $[0.1, 0.2]$, and so on. In each of the nine large intervals, there remain nine smaller intervals of length $1/100$. We therefore have a total of 9^2 intervals, and the sum of their lengths is equal to $9^2/10^2$.

Let r be one of the digits $0, 1, \ldots, 9$. Let S_r denote the set of numbers in the interval $[0, 1]$ whose decimal representation does not contain the digit r. We proved that S_r is a thin set (for definiteness, we considered the set S_2). Hence, according to Theorem 45, their union $S_0 \cup S_1 \cup \cdots \cup S_9$

is a thin set. In other words, almost all decimal fractions (in the interval
$[0, 1]$) contain *all* digits. But this statement is only a very particular
case of a more general regularity.

It turns out that for almost all decimal fractions, each digit occurs
an identical number of times on the average. This has the following
precise meaning. We choose an arbitrary digit r for a given number x in
the interval $[0, 1]$. We let k_n denote the number of times that the digit
r occurs in the first n digits of the decimal fraction. For the number
$x = 0.12237152097$ and the digit $r = 2$, for example, we have $k_1 = 0$,
$k_2 = 1$, $k_3 = 2$, $k_4 = 2$, $k_5 = 2$, $k_6 = 2$, $k_7 = 2$, $k_8 = 3$, $k_9 = 3$,
$k_{10} = 3$, and $k_{11} = 3$. The number k_n depends on the choice of the digit
r. For example, for the same number x and $r = 1$, we have $k_1 = 1$,
$k_2 = 1$, $k_3 = 1$, $k_4 = 1$, $k_5 = 1$, $k_6 = 2$, $k_7 = 2$, $k_8 = 2$, $k_9 = 2$,
$k_{10} = 2$, and $k_{11} = 2$. If we add the numbers k_n calculated for all
$r = 0, 1, 2, \ldots, 9$ and a fixed n, then we obtain n, the total number of
digits in the first n digits. If each digit r occurs an identical number of
times in the first n digits of a number x, then the number k_n is the same
for all r, namely, $n/10$. In other words, we would have $k_n/n = 1/10$.
We say that all digits r occur an identical number of times in a decimal
representation *on the average* if the same relation is satisfied in the limit,
that is, $k_n/n \to 1/10$ as $n \to \infty$, and this is true for k_n calculated for
all the digits $r = 0, 1, 2, \ldots, 9$. Such numbers are said to be *normal*. A
remarkable theorem states that *almost all numbers are normal*, that is,
the set of numbers that are not normal is a thin set. This, of course, is
significantly more than we have proved. If some digit r does not occur
in the representation of a number, then $k_n = 0$ for all n, and $k_n/n \to 0$.
Such a number, of course, is not normal.

That almost all numbers are normal is an astounding fact. Indeed,
it is easy to construct numbers greatly different from normal, for example, such that for a given digit r and the k_n calculated for it, we
have $k_n/n \to p$, where p is any number between 0 and 1; or such that
$k_n/n \to 0$ for the digit 0 and $k_n/n \to 1$ for the digit 1; or such that
for a given digit r, the number k_n/n does not tend to any limit (see
Problems 6, 7, and 8). It would seem that complete chaos should rule
in the distribution of digits of arbitrary decimal fractions and a simple
regularity (that is, the number is normal) would occur only in certain
exceptional circumstances. In fact, the regularity holds almost always,
and "chaos" is only found for a set of numbers that is a thin set.

The proof of the theorem that almost all numbers are normal is based
on ideas the we have met in this chapter or elsewhere in this book. However, it is somewhat complicated, and we relegate it to the supplement.

Problems:

1. Prove that the set of irrational numbers does not contain any interval.
2. Prove that the set of irrational numbers is not a thin set.
3. Prove the equality

$$a_1 + b_1 + a_2 + b_2 + \cdots = (a_1 + a_2 + \cdots) + (b_1 + b_2 + \cdots)$$

   assuming that both sums in the right-hand side exist (in the text, we only proved that the left-hand side does not exceed the right-hand side).
4. Prove that the set of numbers in the interval $[0,1]$ whose decimal representation does not contain the digit 2 is a continuum. *Hint:* Use Problem 6 in Sec. 19.
5. It was proved in the text that the set S_n of those numbers in the interval $[0,1]$ whose decimal representation does not contain the digit 2 in the first n digits is covered by 9^n intervals of length $1/10^n$. Which numbers in these intervals are not contained in the set S_n?
6. For any real number p, $0 \leq p \leq 1$, construct a real number x for which the sequence of numbers k_n calculated for the digit 2 is such that $k_n/n \to p$ as $n \to \infty$.
7. Let two nonnegative real numbers p and q such that $p+q \leq 1$ be given. Construct a real number x such that the sequence of the numbers k_n calculated for the digit 2 is such that $k_n/n \to p$ as $n \to \infty$ and the sequence of numbers k_n' calculated for the digit 3 is such that $k_n'/n \to q$ as $n \to \infty$.
8. Construct a real number x such that the sequence of numbers k_n calculated for the digit 2 does not have a limit.
9. Prove that the number of circles $A_n^{(i)}$ in Fig. 39 is always less than or equal to 2^n.

Supplement: Normal Numbers

We consider real numbers x located between 0 and 1, that is, belonging to the interval $[0,1]$. We write such numbers in the form of an infinite decimal fraction

$$x = 0.a_1 a_2 \ldots a_n \ldots . \tag{12}$$

We recall that the number x is said to be *normal* if any digit r, $0 \leq r \leq 9$, occurs in the representation of x "equally often." This expression has the following meaning. Let k_n denote the number of times that the digit r occurs among the first n digits $a_1, a_2, \ldots, a_n$ of fraction (12). We require that the relation

$$\frac{k_n}{n} \to \frac{1}{10} \quad \text{as } n \to \infty \tag{13}$$

be satisfied. Therefore, a number x is said to be normal if condition (13) is satisfied for all $r = 0, 1, \ldots, 9$. The matter is that each such digit r has its own sequence of numbers k_n. We can let N_r denote the set of numbers for which condition (13) is satisfied for the given r (the sequence of number k_n constructed for that r). Then

$$N = N_0 \cap N_1 \cap \cdots \cap N_9 \tag{14}$$

is the set of all normal numbers.

This supplement is devoted to proving the following assertion.

Theorem 48. *The set of numbers in the interval* $[0,1]$ *that are not normal is a thin set.*

We first carefully analyze what in fact we must prove. The set of numbers that are not normal is obviously the set $\overline{N}$, where N is considered a subset of the interval $[0,1]$ and $\overline{N}$ is its complement. It follows from relation (14) that

$$\overline{N} = \overline{N_0} \cup \overline{N_1} \cup \cdots \cup \overline{N_9}.$$

Because the union of a finite number of thin sets is a thin set, it is sufficient to prove that each set $\overline{N_r}$ is a thin set (for $r = 0, 1, \ldots, 9$). Therefore, the digit r is taken as fixed in what follows, and we consider the set $\overline{N_r}$ of those numbers in the interval $[0,1]$ for which condition (13) is not satisfied, assuming that the sequence of numbers k_n is constructed for the digit r (that is, it shows how many times r occurs among the digits $a_1, a_2, \ldots, a_n$).

We let U denote the set of all numbers x in the interval $[0,1]$ for which condition (13) is not satisfied (for the given fixed r). We recall what condition (13) means: for any $\varepsilon > 0$, there exists $n(\varepsilon)$ such that

$$\left| \frac{k_n}{n} - \frac{1}{10} \right| < \varepsilon \quad \text{for } n > n(\varepsilon),$$

that is, for all n beginning from a certain number $n(\varepsilon)$, which depends on ε. If x does not have this property, then it means that there exists a value $\varepsilon > 0$ such that the inequality $|k_n/n - 1/10| < \varepsilon$ is not satisfied for some n that exceeds any fixed magnitude given in advance. In other words, we have x such that

$$\left| \frac{k_n}{n} - \frac{1}{10} \right| \geq \varepsilon \tag{15}$$

for an infinite number of values of n. We let $U(\varepsilon)$ denote the set of such numbers x (with a given ε). Then for each $x \in U$, there exists ε such that $x \in U(\varepsilon)$; in other words, U is the union of all $U(\varepsilon)$. We can simplify this description somewhat. Indeed, it follows from the very meaning of the definition that $U(\varepsilon_1) \supset U(\varepsilon_2)$ if $\varepsilon_1 < \varepsilon_2$. Therefore, each set $U(\varepsilon)$ is contained in some set $U(1/m)$ for sufficiently large m (such that $\varepsilon < 1/m$). Therefore, the union of all $U(\varepsilon)$ (for all ε) coincides with the union of all $U(1/m)$, that is, with $\bigcup_{m \geq 1} U(1/m)$. We could just as successfully take any sequence of numbers ε_n tending to zero instead of the sequence $1/m$. What is essential is that U is the union of sets $U(\varepsilon_m)$ for a countable sequence of numbers ε_m: $U = \bigcup_{m \geq 1} U(\varepsilon_m)$. Because the union of a countable number of thin sets is a thin set, it is sufficient to prove that each of the sets $U(\varepsilon_m)$ is a thin set. We prove that the set $U(\varepsilon)$ *is a thin set for any* $\varepsilon > 0$.

Let $V(n, \varepsilon)$ denote the set of numbers for which inequality (15) is satisfied for given n and ε. Then the fact that $x \in U(\varepsilon)$ means that $x \in V(n_i, \varepsilon)$ for some infinite sequence of natural numbers $n_1 < n_2 < n_3 < \ldots$, that is, no matter how large we choose the natural number N, $x \in V(n, \varepsilon)$ for some $n > N$.

We set

$$U_N(\varepsilon) = V(N, \varepsilon) \cup V(N+1, \varepsilon) \cup V(N+2, \varepsilon) \cup \ldots .$$

We can write this briefly as

$$U_N(\varepsilon) = \bigcup_{n \geq N} V(n, \varepsilon). \tag{16}$$

We can therefore say that if x is not normal, then $x \in U_N(\varepsilon)$, and this means that

$$U(\varepsilon) \subset U_N(\varepsilon) \tag{17}$$

for all N. We prove that for sufficiently large N, the set $U_N(\varepsilon)$ can be covered by intervals, the sum of whose lengths becomes arbitrarily small. In view of relation (17), this will prove that the set $U(\varepsilon)$ is a thin set.

All of this was a purely logical decoding of what the formulation of the theorem means in point of fact. We must now see which numbers are actually contained in our sets. The set $V(n, \varepsilon)$ is the key to the whole question. A careful examination shows that each of these sets is a union of a finite number of intervals, just as was the case for the sets of fractions with a missing digit in Sec. 20.

We first suppose that the first n decimal digits are fixed. Then

$$x = 0.a_1 a_2 \ldots a_n c_{n+1} c_{n+2} \ldots,$$

where $a_1, a_2, \ldots, a_n$ are fixed and the c_i, $i \geq n+1$, are arbitrary elements of $\{0, 1, \ldots, 9\}$. We set $\alpha = 0.a_1 a_2 \ldots a_n$ and $\gamma = 0.0 \ldots 0 c_{n+1} c_{n+2} \ldots$, where the first n digits in the decimal representation of γ are all zero. Then $x = \alpha + \gamma$, where α is fixed and γ ranges all numbers of the form $c_{n+1}/10^{n+1} + c_{n+2}/10^{n+2} + \ldots$. In other words, $\gamma = \beta/10^n$, where $\beta = 0.c_{n+1}c_{n+2}\ldots$, that is, β is an infinite decimal fraction representing an arbitrary number in the interval $[0,1]$ with $\beta = 1$ excluded. This means that $x = \alpha + \beta/10^n$, where α is fixed and β is any element of the interval $[0,1]$, $\beta \neq 1$. Clearly, such a number is contained in the interval $[\alpha, \alpha + 1/10^n]$ of length $1/10^n$. The set $V(n,\varepsilon)$ thus breaks into intervals of length $1/10^n$, and the number of such intervals is equal to the number of sequences $a_1, a_2, \ldots, a_n$ (of the digits $0, 1, \ldots, 9$) in which a fixed digit r occurs k times, where k satisfies inequality (15):

$$\left| \frac{k}{n} - \frac{1}{10} \right| \geq \varepsilon. \tag{18}$$

We must now calculate the number of sequences in which the digit r occurs a given number k times. If r is located in k fixed positions, then digits different from r occur in the remaining $n-k$ positions. This means there are a total of 9^{n-k} such sequences (we apply Theorem 17 in Chap. 3). Moreover, the k positions occupied by the digit r can be chosen from the n positions in C_n^k different ways, where C_n^k is the binomial coefficient, that is, the number of subsets of k elements from a set of n elements, as given in Theorem 19. The total number of sequences we consider is equal to

$$C_n^k 9^{n-k}.$$

The final answer is that the number of sequences $a_1, a_2, \ldots, a_n$ is equal to $T_n(\varepsilon)$, where

$$T_n(\varepsilon) \text{ is the sum of all expressions } C_n^k 9^{n-k} \tag{19}$$

for all numbers k satisfying inequality (18).

Amazingly, we have almost the same sum that arose in the Chebyshev theorem concerning the Bernoulli scheme in the supplement to Chap. 3. To see their connection, we divide the sum $T_n(\varepsilon)$ by 10^n. We find that $T_n(\varepsilon)/10^n$ is equal to the sum of all expressions $C_n^k(1/10)^k(9/10)^{n-k}$ for all numbers k satisfying inequality (18). Setting $p = 1/10$ and $q = 9/10$, we obtain the sum S_ε considered in the supplement to Chap. 3.

We could make it somewhat more understandable why consideration of two apparently totally different questions leads to exactly the same expression. Namely, the sequence $a_1, a_2, \ldots, a_n$ can be considered as a Bernoulli scheme I^n, where the probability scheme I consists of two events: the event "digit is r" with the probability $1/10$ and the event "digit is not r" with the probability $9/10$. But we do not clarify this connection because, in any event, we cannot apply the results proved in the supplement to Chap. 3. The matter is that instead of the theorem proved there, we need a more precise inequality, which we now prove. We formulate and prove it for the situation considered in Chap. 3, where the probability p is an arbitrary number between 0 and 1. We then apply it to the case where $p = 1/10$, but it is useful to have the more general case available.

Strengthened Chebyshev Inequality. *The sum S_ε for all expressions $C_n^k p^k q^{n-k}$ for all k for which $0 \le k \le n$ satisfies inequality (18) does not exceed $1/(4\varepsilon^4 n^2)$.*

We give the proof of this inequality later and now give the deduction of the theorem from it. We saw that the set $V(n, \varepsilon)$ is contained in the union of intervals of length $1/10^n$ and the number of intervals is equal to $T_n(\varepsilon)$. On the other hand, we just noted that $T_n(\varepsilon)/10^n = S_\varepsilon$, and therefore $T_n(\varepsilon) = 10^n S_\varepsilon$. Because the length of each interval is $1/10^n$, the sum of their lengths is exactly equal to S_ε. According to the strengthened Chebyshev inequality, $S_\varepsilon \le 1/(4\varepsilon^4 n^2)$. Therefore, the set $V(n, \varepsilon)$ is the union of a finite number of intervals, the sum of whose lengths does not exceed $1/(4\varepsilon^4 n^2)$.

We now recall that according to relation (16), $U_N(\varepsilon)$ is the union of all $V(n, \varepsilon)$ for $n \ge N$, and this means that the set $U_N(\varepsilon)$ is contained in the union of intervals, the sum of whose lengths does not exceed

$$\frac{1}{4\varepsilon^4} \left(\frac{1}{N^2} + \frac{1}{(N+1)^2} + \cdots \right). \tag{20}$$

We encountered such a sum in Sec. 15 (see Lemma 13). We saw there that it follows from the boundedness of the sums $1/1^2 + 1/2^2 + \cdots + 1/n^2$ for all natural numbers n that the sum

$$\frac{1}{N^2} + \frac{1}{(N+1)^2} + \cdots$$

will be less than any small positive number given in advance if N is chosen sufficiently large. We can therefore choose N so large that sum (20) is less than any number $\delta > 0$ given in advance. It follows from this that for a sufficiently large N, the set $U_N(\varepsilon)$ is contained in the union of a countable number of intervals, the sum of whose lengths is less than δ. We finally recall that according to relation (17), the set $U(\varepsilon)$ is contained in any $U_N(\varepsilon)$. It is therefore true for $U(\varepsilon)$ that for any arbitrarily small number $\delta > 0$, this set is contained in the union of a countable number of intervals, the sum of whose lengths is less than δ. In other words, $U(\varepsilon)$ is a thin set. We already showed previously that the assertion of the theroem follows from this: the complement $\overline{N}$ of the set of all normal numbers is a thin set.

It remains to prove the strengthened Chebyshev inequality formulated above. For those who solved Problem 5 in the supplement to Chap. 3, there is nothing more to be done—that problem required proving the strengthened Chebyshev inequality we need here. For those who did not solve that problem, we give the proof here. It requires writing some lengthy formulas, but the basic idea is the same as in the proof of the Chebyshev inequality in the supplement to Chap. 3. It is only necessary to open parentheses and combine like terms.

We consider the sum S_ε, consisting of the terms $C_n^k p^k q^{n-k}$ for all k for which $0 \le k \le n$ and $|k/n - p| \ge \varepsilon$. Following the idea of the proof of the Chebyshev inequality (see the supplement to Chap. 3), we multiply each term $C_n^k p^k q^{n-1}$ in this sum by $\big((k - np)/(n\varepsilon)\big)^4$. The sum does not decrease because the desired sum only contains terms $C_n^k p^k q^{n-1}$ for which $|k/n - p| \ge \varepsilon$, that is, $|(k - np)/(n\varepsilon)| \ge 1$. We then consider the sum for all terms

$$\left(\frac{k - np}{n\varepsilon}\right)^4 C_n^k p^k q^{n-k}$$

for all $k = 0, 1, \ldots, n$. The sum can only increase from this because we include new positive terms in it. The new sum obtained is denoted by $\overline{\overline{S_\varepsilon}}$. As we saw, $S_\varepsilon \le \overline{\overline{S_\varepsilon}}$. It turns that the sum $\overline{\overline{S_\varepsilon}}$ can be calculated explicitly, which yields the needed inequality.

In the sum $\overline{\overline{S_\varepsilon}}$, we can bring the common denominator $1/(n^4\varepsilon^4)$ of all terms outside the parentheses, that is, $\overline{\overline{S_\varepsilon}} = \big(1/(n^4\varepsilon^4)\big)P$, where P is the sum of all terms $(k - np)^4 C_n^k p^k q^{n-k}$ for $k = 0, 1, \ldots, n$. We now expand $(k - np)^4$ according to the binomial formula:

$$(k - np)^4 = k^4 - 4k^3 np + 6k^2 n^2 p^2 - 4k n^3 p^3 + n^4 p^4.$$

We obtain

$$P = \sigma_4 - 4np\sigma_3 + 6n^2 p^2 \sigma_2 - 4n^3 p^3 \sigma_1 + n^4 p^4 \sigma_0, \qquad (21)$$

where σ_r for any $r \ge 0$ denotes the sum of all terms $k^r C_n^k p^n q^{n-k}$ for $k = 0, 1, \ldots, n$. Just as in the proof of the Chebyshev inequality, we must find explicit expressions for σ_0, σ_1, σ_2, σ_3, and σ_4.

Lemma 15. Let σ_r denote the sum of the terms $k^r C_n^k p^k q^{n-k}$ for $k = 0, 1, \ldots, n$. Then

$$\sigma_0 = 1,$$
$$\sigma_1 = np,$$
$$\sigma_2 = n^2p^2 + npq,$$
$$\sigma_3 = n^3p^3 + 3n^2p^2q - npq(2p - 1),$$
$$\sigma_4 = n^4p^4 + 6n^3p^3q - n^2p^2q(11p - 7) + npq(1 - 6pq).$$

The proof consists of consecutive repetitions of the technique used to prove Lemma 7 in the supplement to Chap. 3. There, we introduced the sums σ_r consisting of the terms $k^r C_n^k p^k q^{n-k}$ for $k = 0, 1, \ldots, n$ and found their expression as $\sigma_r = q^n f_r(p/q)$ (formula (42) in the supplement to Chap. 3), where $f_r(t)$ is a polynomial that is the sum of the terms $k^r C_n^k t^k$ for $k = 0, 1, \ldots, n$. The task is thus reduced to finding the polynomial $f(t)$. They are found consecutively based on $f_0(t) = (1 + t)^n$ and

$$f_{r+1}(t) = f_r'(t)t \tag{22}$$

(see formula (42) in the supplement to Chap. 3). Moreover, we already found the polynomials $f_0(t)$, $f_1(t)$, and $f_2(t)$ in Chap. 3 (see formulas (46) and (48) in the supplement to Chap. 3). Therefore,

$$f_0(t) = (1 + t)^n, \qquad f_1(t) = n(1 + t)^{n-1}t,$$
$$f_2(t) = n(n - 1)(1 + t)^{n-2}t^2 + n(1 + t)^{n-1}t.$$

We can now use the last of these formulas and formula (22) to find the polynomial $f_3(t)$. We write $f_2(t)$ in the form $g(t) + h(t)$, where $g(t) = n(n - 1)(1 + t)^{n-2}t^2$ and $h(t) = n(1 + t)^{n-1}t$, and apply the rule for differentiating a sum in Sec. 5. We obtain

$$f_2' = g'(t) + h'(t). \tag{23}$$

To calculate the derivatives $g'(t)$ and $h'(t)$, we must apply formula (19) in Chap. 2 (as we already did for $f_2(t)$ in the supplement to Chap. 3). We obtain

$$g'(t) = n(n - 1)(n - 2)(1 + t)^{n-3}t^2 + 2n(n - 1)(1 + t)^{n-2}t,$$
$$h'(t) = n(n - 1)(1 + t)^{n-2}t + n(1 + t)^{n-1}.$$

Substituting these expressions in relation (23), substituting the result in formula (22), and combining like terms, we obtain

$$f_3(t) = n(n - 1)(n - 2)(1 + t)^{n-3}t^3$$
$$+ 3n(n - 1)(1 + t)^{n-2}t^2 + n(1 + t)^{n-1}t. \tag{24}$$

We now turn to calculating the polynomial $f_4(t)$, using formula (22) for $r = 3$. As before, we set $f_3(t) = u(t) + v(t) + w(t)$, where

$$u(t) = n(n-1)(n-2)(1+t)^{n-3}t^3,$$
$$v(t) = 3n(n-1)(1+t)^{n-2}t^2,$$
$$w(t) = n(1+t)^{n-1}t.$$

Then

$$f_3'(t) = u'(t) + v'(t) + w'(t). \tag{25}$$

To calculate the derivatives $u'(t)$, $v'(t)$, and $w'(t)$, we represent each of the polynomials $u(t)$, $v(t)$, and $w(t)$ in the form of a product of powers of $1 + t$ and powers of t and then apply formula (19) in Chap. 2. We thus obtain

$$u'(t) = n(n-1)(n-2)(n-3)(1+t)^{n-4}t^3$$
$$+ 3n(n-1)(n-2)(1+t)^{n-3}t^2,$$
$$v'(t) = 3n(n-1)(n-2)(1+t)^{n-3}t^2 + 6n(n-1)(1+t)^{n-2}t,$$
$$w'(t) = n(n-1)(1+t)^{n-2}t + n(1+t)^{n-1}.$$

It remains to substitute these expressions in formula (25) and then use formula (22) for $r = 3$. After combining like terms, we obtain

$$f_4(t) = n(n-1)(n-2)(n-3)(1+t)^{n-4}t^4$$
$$+ 6n(n-1)(n-2)(1+t)^{n-3}t^3$$
$$+ 7n(n-1)(1+t)^{n-2}t^2 + n(1+t)^{n-1}t. \tag{26}$$

We can now substitute the value $t = p/q$ in expressions (24) and (26) for $f_3(t)$ and $f_4(t)$. In doing this, we recall that $1+p/q = (p+q)/q = 1/q$ because $p + q = 1$. Using the relation $\sigma_r = q^n f_r(p/q)$, we obtain the expressions for σ_3 and σ_4. For σ_3, we obtain the expression

$$\sigma_3 = n(n-1)(n-2)p^3 + 3n(n-1)p^2 + np.$$

Here, we should substitute $n(n-1)(n-2) = n^3 - 3n^2 + 2n$ and $n(n-1) = n^2 - n$. Combining like terms and expressing the polynomial as a sum of terms in powers of n, we obtain

$$\sigma_3 = n^3 p^3 + 3n^2 p^2 (1-p) + n(2p^3 - 3p^2 + p).$$

Because $2p^3 - 3p + p = p(p-1)(2p-1)$, the expression for σ_3 given in the lemma follows from this expression.

We now turn to calculating the expression for σ_4. Similarly to the calculation of σ_3, from relation (26), we obtain

$$\sigma_4 = n(n-1)(n-2)(n-3)p^4 + 6n(n-1)(n-2)p^3 + 7n(n-1)p^2 + np.$$

We must open the parentheses in the product $n(n-1)(n-2)(n-3)$:

$$(n-1)(n-2)(n-3) = n^3 - 6n^2 + 11n - 6.$$

(The Viète formula in Chap. 3 can be used.) We already calculated the expressions $n(n-1)(n-2)$ and $n(n-1)$. Grouping terms with the same powers of n, we obtain

$$\sigma_4 = n^4 p^4 + (-6p^4 + 6p^3)n^2$$
$$+ (11p^4 - 18p^3 + 7p^2)n + (-6p^4 + 12p^3 - 7p^2 + p)n.$$

It remains to note that

$$-6p^4 + 6p^3 = 6p^3(1-p) = 6p^3 q,$$
$$11p^4 - 18p^3 + 7p^2 = -p^2(1-p)(11p-7) = -p^2 q(11p-7),$$
$$-6p^4 + 12p^3 - 7p^2 + p = -p(1-p)(6p^2 - 6p + 1) = pq(1 - 6pq),$$

and we obtain the expression for σ_4 given in the lemma. Lemma 15 is thus proved. $\qquad\qquad\square$

To prove the strengthened Chebyshev inequality, it remains to substitute the expressions for σ_0, σ_1, σ_2, σ_3, and σ_4 given in the lemma in formula (21) and then combine like terms. We write the coefficients for different powers of n, noting that only terms from σ_3 and σ_4 affect the coefficient of n^2 and only terms from σ_4 affect the coefficient of n:

for n^4 : $p^4 - 4p^4 + 6p^4 - 4p^4 + p^4 = 0,$

for n^3 : $6p^3 q - 12p^3 q + 6p^3 q = 0,$

for n^2 : $-p^2 q(11p - 7) + 4p^2 q(2p - 1) = p^2 q(-11p + 7 + 8p - 4)$
$$= 3p^2 q^2,$$

for n : $pq(1 - 6pq).$

As a result, we obtain

$$P = 3p^2 q^2 n^2 + pq(1 - 6pq)n.$$

Because $\overline{\overline{S_\varepsilon}} = P/(n^4 \varepsilon^4)$, we have

$$\overline{\overline{S_\varepsilon}} = \frac{1}{n^4 \varepsilon^4} \left(3p^2 q^2 n^2 + pq(1 - 6pq)n \right).$$

For the sum S_ε we are interested in, we proved that $S_\varepsilon \leq \overline{\overline{S_\varepsilon}}$. Therefore,

$$S_\varepsilon \leq \frac{1}{n^4 \varepsilon^4} \left(3p^2 q^2 n^2 + pq(1 - 6pq)n \right). \qquad (27)$$

For the expression in the parentheses, we have the inequality

$$3p^2 q^2 n^2 + pq(1 - 6pq)n = pq(1 + 3(n - 2)pq)n$$
$$\leq \frac{1}{4} \left(1 + \frac{3(n - 2)}{4} \right) n \leq \frac{1}{4} n^2$$

because always $pq \leq 1/4$, as we noted in the supplement to Chap. 3. It therefore follows from inequality (27) that $S_\varepsilon \leq 1/(4n^2 \varepsilon^4)$ as was to be proved. $\qquad\square$

The strengthened Chebyshev inequality is proved, and, as we saw, Theorem 48 follows from it. $\qquad\square$

Remark 1. We proved the strengthened Chebyshev inequality in which we have the denominator $(\varepsilon^2 n)^2$ instead of $\varepsilon^2 n$ (in the original inequality). The proof was completely parallel to the proof of the Chebyshev inequality; we only considered the multiplier $((k - np)/(\varepsilon n))^4$ instead of $((k - np)/(\varepsilon n))^2$. The question naturally arises whether the Chebyshev inequality can be improved by using the multiplier $((k - np)/(\varepsilon n))^{2m}$, where m is any natural number. We can indeed do this. For each specific case (for example, $m = 3$), we must perform the same transformation as in our proof. In doing this for each larger value of m, the transformation becomes more complicated. For example, it is necessary to calculate $2m+1$ sums $\sigma_0, \sigma_1, \ldots, \sigma_{2m}$. An even more complicated argument is needed for an arbitrary m. As a result, we obtain an inequality with $(\varepsilon^2 n)^m$ in the denominator. We have no need for this improvement of the Chebyshev inequality here and therefore limited ourselves to the case where $m = 2$.

Remark 2. Of course, our arguments are not connected with the specifics of the decimal system of representation and are suitable for written numbers in any system of representation. If the base of the system of representation is equal to g, then we say that a number x is normal if $k_n/n \to 1/g$ as $n \to \infty$, where k_n is the number of digits in the first n digits of the representation of x in the base-g system that are identical to a given digit r. When considering the set of numbers that are not normal, we obtain the same sum of terms $C_n^k (1/g)^k (1 - 1/g)^{n-k}$, where k satisfies the inequality $|k/n - 1/g| \geq \varepsilon$. This is exactly the sum we considered with $p = 1/g$ and $q = 1 - 1/g$. Our proof can therefore be used without change for numbers written in a base-g system.

We obtain an interesting application of this fact if we set $g = 100$. A "digit" in a base-100 system of representation is any two-digit combination in the decimal representation. We thus find that if k_n denotes the number of occurrences of a given pair of decimal digits (for example, 13 or 27 or ...) in the first $2n$ pairs of digits in the decimal representation of a number x, then $k_n/n \to 1/100$ for all x except those x belonging to a certain thin set. In other words, for almost all numbers, pairs of neighboring digits are encountered equally often—with a "frequency" of $1/100$. We can set $g = 10^l$ for any natural number l and then find that any sequence of l digits is found in the decimal representation of a number with the single "frequency" $1/10^l$—for almost all x.

Remark 3. We proved that almost all numbers x are normal. But to show that a specific given number is normal is usually an extremely difficult task. Of course, the number $0.0123456789\ldots$, where the sequence of digits $0, 1, \ldots, 9$ repeats, is normal (see Problem 1). It is already much more complicated to prove the normality of the number $0.123\ldots9101112\ldots$, in which all the natural numbers are written in order after the decimal point. This was proved only in the 1930s. Finally, it is still not proved whether the number $\sqrt{2}$ or π is normal. No approach to this question is yet visible (when speaking of the normality of a number greater than 1, we can disregard the integer part and consider the number normal if the representation of the fractional part is normal).

Problems:

1. Consider the number $x = 0.0123456789\ldots$, where the sequence of digits $0, 1, \ldots, 9$ repeats. Find the number k_n for any digit $r = 0, 1, \ldots, 9$ and prove that the number x is normal.
2. Prove that for any rational number x, the sequence of numbers k_n/n for any given $r = 0, 1, \ldots, 9$ tends to a definite limit. *Hint*: Recall that an infinite decimal fraction representing a rational number is periodic (see Problem 3 in Sec. 16).
3. What condition must the period of a periodic decimal fraction satisfy for the corresponding number to be normal?

Topic: Polynomials

21. Polynomials as Generating Functions

More than once already, we have encountered a situation in which it was especially convenient to express properties of a finite sequence of numbers $a_0, a_1, \ldots, a_n$ via the polynomial $f(x) = a_0 + a_1 x + \cdots + a_n x^n$. In this case, the polynomial $f(x)$ is called the *generating function* of the sequence $a_0, a_1, \ldots, a_n$. A striking example of this is the question, considered in Chap. 3, in which a finite set S is given and the a_k are the number of its subsets with k elements. If the generating function $f_S(x) = a_0 + a_1 x + \cdots + a_n x^n$ is introduced, then the value of a_k for the set $S_1 \times S_2$ is expressed by the particularly simple formula $f_{S_1 \times S_2}(x) = f_{S_1}(x) f_{S_2}(x)$ (see formula (9) in Chap. 3).

In exactly the same way, the binomial coefficients C_n^k, where $k = 0, 1, \ldots, n$, are conveniently investigated using their generating function, which is equal to $(1+x)^n$. Many identities between binomial coefficients easily follow from this (see formula (26) in Chap. 2 and problems in Sec. 6).

We now give several similar examples. The *first example* is related to certain properties of natural numbers. We consider representations of a natural number n as sums of natural numbers, $n = a_0 + a_1 + \cdots + a_k$. Such a representation is also called a *partition* of the number n. Two partitions are considered identical if the sets of their terms $\{a_0, a_1, \ldots, a_k\}$ are identical (that is, the order of terms in the sum is not considered). For example, we consider the representation $6 = 4 + 1 + 1 = 1 + 4 + 1 = 1 + 1 + 4$ to be one partition of the number 6.

We let $P_{k,l}(n)$ denote the number of partitions of n into no more than k terms each of which does not exceed l. We want to investigate these numbers. For this, we establish the corresponding generating function. We set $P_{0,0}(0) = 1$ by definition. We note that if there exists at least one partition of n satisfying our conditions, then $n \leq kl$. We can therefore compose the sum of all expressions $P_{k,l}(n)x^n$ for $n = 0, 1, 2, \ldots$, and only numbers with $n \leq kl$ contribute to it, which makes it a polynomial. Let $g_{k,l}(x)$ denote this polynomial:

$$g_{k,l}(x) = P_{k,l}(0) + P_{k,l}(1)x + \cdots + P_{k,l}(kl)x^{kl}. \tag{1}$$

It is obvious that $g_{0,l}(x) = 1$ and $g_{k,0}(x) = 1$. We now deduce two relations connecting polynomials $g_{k,l}(x)$ with such polynomials of a lower index. We consider the difference $P_{k,l}(n) - P_{k,l-1}(n)$. The first term is equal to the number of partitions of n into no more than k terms not exceeding l, $n = a_1 + \cdots + a_j$, $j \leq k$, $a_i \leq l$. And the second term is the same except for the condition of not exceeding $l - 1$. Obviously, the difference is equal to the number of partitions $n = a_1 + \cdots + a_j$, where again $j \leq k$, $a_i \leq l$, and at least one of the $a_1, \ldots, a_j$ is equal to l, for example, $a_1 = l$. Subtracting this term, we obtain the partition $n - l = a_2 + \cdots + a_j$, where the number of terms now does not exceed $k - 1$ and each term still does not exceed l. We thus obtain a one-to-one correspondence between the considered $P_{k,l}(n) - P_{k,l-1}(n)$ partitions of the number n and the $P_{k-1,l}(n - l)$ partitions of the number $n - l$. Therefore,

$$P_{k,l}(n) - P_{k,l-1}(n) = P_{k-1,l}(n - l). \tag{2}$$

By definition, the number $P_{k-1,l}(n - l)$ is equal to the coefficient of x^{n-l} in the polynomial $g_{k-1,l}(x)$ and therefore to the coefficient of x^n in the polynomial $g_{k-1,l}(x)x^l$. Relation (2) now establishes an equality of the coefficients of x^n in the polynomials $g_{k,l} - g_{k,l-1}$ and $g_{k-1,l}(x)x^l$. Because the equality holds for any n, we obtain the relation

$$g_{k,l}(x) = g_{k,l-1}(x) + g_{k-1,l}(x)x^l. \tag{3}$$

The second relation is deduced completely analogously. We consider the difference $P_{k,l}(n) - P_{k-1,l}(n)$. The first term is equal to the number of partitions of n into no more than k terms not exceeding l. The second is similar, but the number of terms in the partition does not exceed $k-1$. This means the difference is the number of partitions $n = a_1 + \cdots + a_k$ into exactly k terms, each of which does not exceed l. We now substract 1 from each term, and if the term was already 1, we omit it. As a result, we obtain a partition $n - k = b_1 + \cdots + b_j$, where $j \leq k$ and $b_i \leq l - 1$. Obviously, the difference $P_{k,l}(n) - P_{k-1,l}(n)$ is equal to the number of such partitions. In other words, we have proved the relation

$$P_{k,l}(n) - P_{k-1,l}(n) = P_{k,l-1}(n - k).$$

As before, from this, we obtain the relation

$$g_{k,l}(x) = g_{k-1,l}(x) + g_{k,l-1}(x)x^k. \tag{4}$$

We can use relations (3) and (4) to find an explicit formula for the polynomial $g_{k,l}(x)$. It follows from them that

$$g_{k,l-1}(x) + g_{k-1,l}(x)x^l = g_{k-1,l}(x) + g_{k,l-1}(x)x^k,$$

whence

$$g_{k,l-1}(x)(1 - x^k) = g_{k-1,l}(x)(1 - x^l).$$

This means that

$$g_{k,l-1}(x) = g_{k-1,l}(x)\frac{1 - x^l}{1 - x^k}.$$

Replacing l in this relation with $l + 1$, we obtain

$$g_{k,l}(x) = g_{k-1,l+1}(x)\frac{1 - x^{l+1}}{1 - x^k}. \tag{5}$$

We can now apply relation (5) to the polynomial $g_{k-1,l+1}(x)$. As a result, we obtain

$$g_{k,l}(x) = g_{k-2,l+2}(x)\frac{1 - x^{l+1}}{1 - x^k}\frac{1 - x^{l+2}}{1 - x^{k-1}}.$$

This process can be repeated k times, and because $g_{0,l+k}(x) = 1$, we obtain

$$g_{k,l}(x) = \frac{(1 - x^{l+1})(1 - x^{l+2}) \cdots (1 - x^{l+k})}{(1 - x^k)(1 - x^{k-1}) \cdots (1 - x)}. \tag{6}$$

Formula (6) takes a more symmetric appearance if the numerator and denominator in the right-hand side are multiplied by $(1 - x)(1 - x^2) \cdots (1 - x^l)$. We let $h_m(x)$ denote the polynomial $(1-x)(1-x^2) \cdots (1 - x^m)$. Then formula (6) becomes

$$g_{k,l}(x) = \frac{h_{k+l}(x)}{h_k(x)h_l(x)}. \tag{7}$$

The expression in the right-hand side has a structure analogous to the binomial coefficient C_{k+l}^k, where $h_k(x)$ is analogous to $k!$. The polynomials $g_{k,l}(x)$ defined by equality (7) are called *Gauss polynomials*. As in the case of binomial coefficients, it is not immediately obvious that the rational fraction $h_{k+l}(x)/\bigl(h_k(x)h_l(x)\bigr)$ is a polynomial. This follows, of course, from the connection with partitions, that is, from its determination by formula (1) (see, however, Problem 3 at the end of this section).

We now show some properties of the Gauss polynomials $g_{k,l}(x)$, analogous to the known properties of binomial coefficients. From formula (7), we obtain the relation

$$g_{k,l}(x) = g_{l,k}(x), \tag{8}$$

analogous to the property of binomial coefficients $C_{k+l}^k = C_{k+l}^l$ (because the polynomials $g_{k,l}(x)$ are analogous to the binomial coefficients C_{k+l}^k). Relations (3) and (4) pass into each other if relation (8) is used. Setting $g_{k,l}(x) = g_{l,k}(x)$, we apply relation (3) to $g_{l,k}(x)$ and obtain $g_{l,k}(x) = g_{l,k-1}(x) + g_{l-1,k}(x)$. Again using (8), we then set $g_{l,k-1} = g_{k-1,l}$ and $g_{l-1,k} = g_{k,l-1}$. We thus obtain relation (4) from (3). Both relations are analogous to the equality $C_n^k = C_{n-1}^k + C_{n-1}^{k-1}$, formula (26) in Chap. 2 (we must replace n with $k + l$, and then C_{n-1}^k is obtained from C_n^k by replacing l with $l - 1$ and C_{n-1}^{k-1} by replacing k with $k - 1$). Finally, a direct connection (and not an analogy) follows from the relation

$$g_{k,l}(1) = C_{k+l}^k. \tag{9}$$

Directly substituting $x = 1$ in relation (6) is prohibited: the numerator and denominator both become zero. We divide the numerator and denominator by $(1 - x)^k$; more precisely, we divide each of the k factors in the numerator and denominator by $1 - x$. In general, the polynomial $1 - x^m$ is divisible by $1 - x$ for any m, and $(1 - x^m)/(1 - x) = 1 + x + \cdots + x^{m-1}$ (see formula (12) in Chap. 1). Therefore,

$$\frac{1 - x^m}{1 - x}(1) = m.$$

Dividing each factor in the numerator and denominator in formula (6) by $1 - x$ and then setting $x = 1$, we obtain

$$g_{k,l}(1) = \frac{(l + k) \cdots (l + 2)(l + 1)}{1 \cdot 2 \cdots k}$$

(we reverse the order of the factors in the numerator and denominator). This shows that $g_{k,l}(1) = C_{k+l}^k$.

Finally, we use the fact that $g_{k,l}(x)$ is a polynomial, not a number, to show an important property of the Gauss polynomials that does not have an analogue for binomial coefficients. Namely, we prove that the polynomial $g_{k,l}(x)$ *is reciprocal* for any k and l. We recall (see Sec. 10) that a polynomial $f(x) = a_0 + a_1 x + \cdots + a_n x^n$ of degree n is said to be reciprocal if its coefficients that are equally distant from its beginning and end are equal to each other, that is, $a_k = a_{n-k}$ for $k = 0, 1, \ldots, n$. A polynomial $f(x)$ of degree n is reciprocal if and only if $x^n f(1/x) = f(x)$ (this is also proved in Sec. 10). The polynomial $g_{k,l}(x)$ has the degree kl. This follows from representation (1) and the fact that $P_{k,l}(kl) \geq 1$ because at least one partition of kl into k terms equal to l exists: $kl = l + l + \cdots + l$ (k terms). This can also be deduced from representation (6) by calculating the degrees of the numerator and the denominator and subtracting the second from the first. It is therefore sufficient to verify the relation $x^{kl} g_{k,l}(1/x) = g_{k,l}(x)$. This immediately follows from representation (6). We note that the equality

$$\left(1 - \frac{1}{x^m}\right) = (-1)x^{-m}(1 - x^m)$$

holds for any m. Therefore, such a relation holds for any factor in the numerator and denominator in the right-hand side of formula (6). Because the number of factors in the numerator is exactly the same as the number of factors in the denominator (it is equal to k), the multiplier (-1) cancels altogether. We factor out x^{-m} from each factor $1 - x^m$ of degree m. We obtain $g_{k,l}(1/x) = x^{-N} g_{k,l}(x)$, where N is the difference between the degrees of the numerator and the denominator. But this difference is equal to the degree of the polynomial $g_{k,l}(x)$, that is, kl. Therefore, $N = kl$, $g_{k,l}(1/x) = x^{-kl} g_{k,l}(x)$, and hence $x^{kl} g_{k,l}(1/x) = g_{k,l}(x)$. This means that the polynomial $g_{k,l}(x)$ is reciprocal.

Listing the properties of the Gauss polynomials $g_{k,l}(x)$ leads to the corresponding properties of partitions, more precisely, of the numbers $P_{k,l}(n)$, if we use definition (1) to pass to the Gauss polynomial coefficients. Thus, relation (8) gives the equality

$$P_{k,l}(n) = P_{l,k}(n), \tag{10}$$

that is, *the number of partitions of n into no more than k terms not exceeding l is equal to the number of its partitions into no more than l terms not exceeding k.* The reciprocity of the polynomial $g_{k,l}(x)$ implies that

$$P_{k,l}(n) = P_{k,l}(kl - n). \tag{11}$$

Relation (9) implies that for given numbers k and l, the sum of all numbers $P_{k,l}(n)$ for $n = 0, 1, \ldots, kl$ is equal to C_{k+l}^k, that is,

$$P_{k,l}(0) + P_{k,l}(1) + \cdots + P_{k,l}(kl) = C_{k+l}^k. \tag{12}$$

Of course, such simple properties of partitions can be proved without using the generating function $g_{k,l}(x)$ (see Problems 4, 5, and 6). However, it is simpler to discover them using the generating function.

In conclusion, we recall that in addition to reciprocity, we considered one other property of polynomials in Sec. 10—unimodality. For a reciprocal polynomial $a_0 + a_1 x + \cdots + a_N x^N$, unimodality implies that $a_i \leq a_{i+1}$ for $i+1 \leq N/2$. It then follows from reciprocity that $a_j \geq a_{j+1}$ for $j \geq N/2$. It turns out that the Gauss polynomials $g_{k,l}(x)$ have the unimodality property. By definition, this means that

$$P_{k,l}(n) \leq P_{k,l}(n+1) \quad \text{for } n+1 \leq \frac{kl}{2}.$$

For a long time, the proof of this assertion was based on a connection with a completely different branch of algebra. Namely, the difference $P_{k,l}(n+1) - P_{k,l}(n)$ for $n+1 \leq kl/2$ turned out to coincide with the number of elements of a certain set connected with a completely different concept. A purely combinatorial proof was found only about ten years ago, and it is rather more complicated. It would be interesting to find a simple proof of these simple inequalities.

As a *second example*, we consider some already known properties of natural numbers that can be deduced particularly elegantly using generating functions. This is connected with the possibility of writing natural numbers in a system of representation with a given base.

We begin with the binary system of representation. From any natural number n, we can factor out the largest power of 2 that divides it and represent the number in the form $n = 2^k m$, where k is a natural number or zero and m is an odd natural number. This means that m has the form $2r + 1$, and n can therefore be represented as $n = 2^k + 2^{k+1} r$. The same reasoning can now be applied to r. Continuing this process, we eventually represent n as a sum of *different* powers of 2:

$$n = 2^{k_1} + 2^{k_2} + \cdots + 2^{k_m}, \quad \text{where } k_1 < k_2 < \cdots < k_m.$$

In other words, we represent n in the form

$$n = a_0 + a_1 2 + a_2 2^2 + \cdots + a_N 2^N, \tag{13}$$

where the coefficients $a_0, a_1, \ldots, a_N$ can take the value 0 or 1. Eliminating all terms for which $a_i = 0$, we return to the representation of n

as the sum of different powers of 2. Representation (13) is called the *binary representation* of the number n or its representaion in the *binary system of representation*. We prove that representation (13) is unique for a given n.

Let $n = b_0 + b_1 2 + b_2 2^2 + \cdots + b_M 2^M$ be a different representation. Then $a_0 = b_0$ because $a_0 = b_0 = 1$ if n is odd and $a_0 = b_0 = 0$ if n is even. We can therefore set $b_0 = a_0$ in the second representation and obtain

$$\frac{n - a_0}{2} = a_1 + a_2 2 + \cdots + a_n 2^{N-1},$$

$$\frac{n - a_0}{2} = b_1 + b_2 2 + \cdots + b_m 2^{M-1}.$$

Because $(n - a_0)/2 \leq n/2 < n$, we obtain two different representations for the number $(n - a_0)/2$, which is less than n. Applying induction, we can assume that our assertion has been proved for $(n - a_0)/2$, and this means that $a_1 = b_1$, $a_2 = b_2$, and so on. (Incidentally, the reader has probably already used this reasoning to solve Problem 5 in Sec. 1.)

For a given value of the number N, we obtain the largest number n in representation (13) when all the numbers a_i take their largest value, that is, when all $a_i = 1$. Then $n = 1 + 2 + 2^2 + \cdots + 2^N = (2^{N+1} - 1)/(2 - 1) = 2^{N+1} - 1$. Therefore, for a given value N, all numbers less than 2^{N+1} can be written in form (13), and obviously only these numbers.

On the other hand, we consider the product

$$(1 + x)(1 + x^2)(1 + x^4)(1 + x^8) \cdots (1 + x^{2^N}). \tag{14}$$

Opening the parentheses, we must take one term from each pair of parentheses, that is, either 1 or x^{2^i} from $(1 + x^{2^i})$. As a result, we obtain a term x^n, where n is a sum of different powers of 2, that is, a sum of different numbers 2^i for certain $i \leq N$. As we saw, we thus obtain any number $n \leq 2^{N+1} - 1$ and, moreover, each such number exactly once. That is, opening the parentheses in product (14), we obtain all terms x^n with $n \leq 2^{N+1} - 1$ and with the coefficient 1. In other words, the statement that any number $n \leq 2^{N+1} - 1$ has a unique binary representation yields the identity

$$(1 + x)(1 + x^2)(1 + x^4)(1 + x^8) \cdots (1 + x^{2^N}) =$$
$$1 + x + x^2 + x^3 + \cdots + x^{2^{N+1}-1}. \tag{15}$$

You can easily verify that all our reasoning can be taken in the reverse order, that is, the existence of a unique binary representation for all numbers $n \leq 2^{N+1} - 1$ follows from identity (15).

But how can we verify equality (15) directly and thus obtain another proof of the existence and uniqueness of the binary representation? For this, it is sufficient to apply the formula

$$1 + x + x^2 + x^3 + \cdots + x^{2^{N+1}-1} = \frac{1 - x^{2^{N+1}}}{1 - x},$$

which we have encountered many times, to the right-hand side of equality (15). This means that to prove identity (15), it is sufficient to verify the identity

$$(1-x)(1+x)(1+x^2)(1+x^4)(1+x^8)\cdots(1+x^{2^N}) = 1 - x^{2^{N+1}}.$$

But this is obvious! Multiplying the first two factors, we obtain $1 - x^2$. Multiplying $1 - x^2$ by $1 + x^2$, we obtain $1 - x^4$, and so on, until we multiply $1 - x^{2^N}$ by $1 + x^{2^N}$ and obtain $1 - x^{2^{2N+1}}$.

We now consider the completely analogous case for the decimal system of representation. We divide an arbitrary natural number n by 10 with a remainder: $n = 10n_1 + a_0$, where $0 \le a_0 \le 9$. We then divide n_1 by 10 with a remainder: $n_1 = 10n_2 + a_1$, where $0 \le a_1 \le 9$. Substituting, we obtain $n = 10^2 n_2 + 10a_1 + a_0$. Continuing this process, we finally obtain

$$n = 10^k a_k + 10^{k-1} a_{k-1} + \cdots + 10a_1 + a_0,$$

where $0 \le a_i \le 9$ for all a_i, for some value k. This is our customary decimal representation of the number n. It is unique. Indeed, writing this formula in the form

$$n = 10(10^{k-1} a_k + 10^{k-2} a_{k-1} + \cdots + a_1) + a_0 \quad \text{or} \quad n = 10m + a_0,$$

where $m = 10^{k-1} a_k + 10^{k-2} a_{k-1} + \cdots + a_1$, we see that a_0 is the remainder from dividing n by 10. But division with a remainder is unique (see Theorem 4 in Chap. 1). Therefore, if there exist different decimal representations of n, then at least a_0 must be the same in all of them—it is the remainder from dividing n by 10. If a different decimal representation $n = 10^l b_l + 10^{l-1} b_{l-1} + \cdots + 10b_1 + b_0$ exists, where $0 \le b_i \le 9$ for all b_i, then we can therefore state that $a_0 = b_0$. Hence,

$$10^k a_k + 10^{k-1} a_{k-1} + \cdots + 10a_1 = 10^l b_l + 10^{l-1} b_{l-1} + \cdots + 10b_1.$$

Dividing by 10, we obtain

$$10^{k-1} a_k + 10^{k-2} a_{k-1} + \cdots + a_1 = 10^{l-1} b_l + 10^{l-2} b_{l-1} + \cdots + b_1,$$

that is, we have two decimal representations for the number $m = (n - a_0)/10$. Because $m \le n/10 < n$, we can use the induction assumption that m has a unique decimal representation, and this means that $a_1 = b_1$, $a_2 = b_2$, and so on.

It is clear that the number n with the decimal representation $n = 10^k a_k + 10^{k-1} a_{k-1} + \cdots + 10 a_1 + a_0$ does not exceed the number $9(10^k + 10^{k-1} + \cdots + 10 + 1)$ (because all $a_i \leq 9$), and this number is equal to $9(10^{k+1} - 1)/(10 - 1) = 10^{k+1} - 1$. Therefore, decimal representations of the form $10^k a_k + 10^{k-1} a_{k-1} + \cdots + 10 a_1 + a_0$ with a given k include all numbers not exceeding $10^{k+1} - 1$ (that is, less than 10^{k+1}) and only these numbers.

We consider the product

$$(1 + x + x^2 + \cdots + x^9)(1 + x^{10} + x^{20} + \cdots + x^{90}) \cdots$$
$$(1 + x^{10^k} + x^{2 \cdot 10^k} + \cdots + x^{9 \cdot 10^k}).$$

Opening parentheses, we must take a factor x^{a_0} from the first pair of parentheses, where a_0 takes one of the values $0, 1, \ldots, 9$, then take a factor $x^{10 a_1}$ from the second pair of parentheses, where a_1 takes one of the same possible values, and so on for each pair of parentheses. As a result, we obtain the term $x^{a_0 + 10 a_1 + \cdots + 10^k a_k}$, and, as we saw, this is an arbitrary term x^n, where n is any number not exceeding $10^{k+1} - 1$. And again, this is a statement of the existence of the decimal representation. Moreover, each such term is obtained exactly one time, that is, with the coefficient 1. Thus, the identity

$$(1 + x + x^2 + \cdots + x^9)(1 + x^{10} + x^{20} + \cdots + x^{90})$$
$$\cdots (1 + x^{10^k} + x^{2 \cdot 10^k} + \cdots + x^{9 \cdot 10^k})$$
$$= 1 + x + x^2 + \cdots + x^{10^{k+1} - 1} \quad (16)$$

follows from the existence and uniqueness of the decimal representation in the form $n = 10^k a_k + 10^{k-1} a_{k-1} + \cdots + 10 a_1 + a_0$ for all numbers $n \leq 10^{k+1} - 1$.

As in the case of the binary representation, all our arguments can be used in the reverse order, and this means that the existence and uniqueness of the decimal representations follow from identity (16). We now prove identity (16) directly and thus obtain another proof of the existence and uniqueness of the decimal representations. This is very simply done. We again apply the formula

$$1 + x + x^2 + \cdots + x^{10^{k+1} - 1} = \frac{x^{10^{k+1}} - 1}{x - 1}$$

to the right-hand side of equality (16). We transform each expression in parentheses in the left-hand side analogously:

$$1 + x + x^2 + \cdots + x^9 = \frac{x^{10} - 1}{x - 1},$$

$$1 + x^{10} + x^{20} + \cdots + x^{90} = \frac{x^{100} - 1}{x^{10} - 1},$$

$$\vdots$$

$$1 + x^{10^k} + x^{2 \cdot 10^k} + \cdots + x^{9 \cdot 10^k} = \frac{x^{10^{k+1}} - 1}{x^{10^k} - 1}.$$

Relation (16) then becomes

$$\frac{x^{10} - 1}{x - 1} \frac{x^{100} - 1}{x^{10} - 1} \cdots \frac{x^{10^{k+1}} - 1}{x^{10^k} - 1} = \frac{x^{10^{k+1}} - 1}{x - 1}.$$

This is perfectly obvious: the numerator of each fraction in the left-hand side cancels with the denominator of the next fraction leaving only $x^{10^{k+1}} - 1$ (the numerator of the last fraction) divided by $x - 1$ (the denominator of the first fraction). And this is what we have in the right-hand side.

A system of representation with an arbitrary base can be treated in literally the same way.

Problems:

1. Find an explicit form of the polynomial $g_{k,1}(x)$ from the definition and from formula (6).
2. Find an explicit form of the polynomial $g_{k,2}(x)$. (This is more difficult than Problem 1.)
3. Let the rational expressions $g_{k,l}(x)$ be *defined* by formula (7). Prove that relations (3) and (4) hold for them, and use this to prove that they are polynomials (without using formula (1) and the connection with partitions).
4. Prove that $P_{k,l}(n) = P_{l,k}(n)$ without using the properties of Gauss polynomials. *Hint:* The partition $n = a_1 + \cdots + a_k$, $a_1 \geq a_2 \geq \cdots \geq a_j$, can be represented as a table of points with a_1 points in the first row, a_2 points in the second row, and so on. For example, $13 = 7 + 3 + 1 + 1 + 1$ is represented as

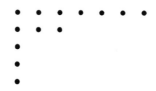

With each such table, associate a "transposed" table whose rows are the columns of the first table. For example, the following "transposed" table is associated with the preceding table:

$$
\begin{array}{ccccc}
\bullet & \bullet & \bullet & \bullet & \bullet \\
\bullet & \bullet & & & \\
\bullet & \bullet & & & \\
\bullet & & & & \\
\bullet & & & & \\
\bullet & & & & \\
\bullet & & & &
\end{array}
$$

5. Prove that $P_{k,l}(n) = P_{k,l}(kl - n)$ without using the properties of Gauss polynomials. *Hint*: The partition $n = a_1 + \cdots + a_j$, $j \le k$, $a_i \le l$, can be associated with the partition $kl - n = (l - a_1) + (l - a_2) + \cdots + (l - a_j) + l + \cdots + l$, where the term $l - a_i$ is omitted if it is equal to zero and the number of terms l is equal to $k - j$.

6. Prove that $P_{l,k}(0) + P_{l,k}(1) + \cdots + P_{l,k}(kl) = C_{k+l}^k$ without using the properties of Gauss polynomials. *Hint*: The partition $a_1 + \cdots + a_j$, $j \le k$, $a_i \le l$, of a number not exceeding kl can be associated with the subset $\{a_1 + 1, a_2 + 2, \ldots, a_j + j\}$ of the set $\{1, 2, \ldots, k + l\}$.

7. Prove that any the weight of any object measurable in an integer number of kilograms less than 2^n can be determined with a pan balance if we have n weights of $1, 2, 4, \ldots, 2^{n-1}$ kilograms (the object is placed on one pan and a combination of weights on the other).

8. Placing weights on both pans of the balance is allowed during weighing. Prove that the weight of any object measurable in an integer number of kilograms less than $(3^n - 1)/2$ can be determined if we have n weights of $1, 3, 9, \ldots, 3^{n-1}$ kilograms. What identity corresponds to this assertion? Prove this identity directly. *Hint*: Prove the existence and uniqueness of a ternary representation in the form $m = a_0 + a_1 3 + \cdots + a_{n-1} 3^{n-1}$, where $a_0, a_1, \ldots, a_{n-1}$ take the values $1, 0,$ or -1, for all integers m between $-(3^n - 1)/2$ and $+(3^n - 1)/2$.

22. Power Series

In the preceding section, we saw that it is often convenient in investigating the properties of a finite sequence $a_0, \ldots, a_n$ to examine the polynomial $f(x) = a_0 + a_1 x + \cdots + a_n x^n$—the generating function of the sequence. But what to do if the sequence is infinite, for example, if it is the sequence of natural numbers or of Bernoulli numbers? Without thinking much, given an infinite sequence $a_0, a_1, \ldots, a_n, \ldots$, we can easily write

$$
f(x) = a_0 + a_1 x + \cdots + a_n x^n + \ldots . \tag{17}
$$

But what is the meaning of the expression in the right-hand side?

We return to the case of finite sequences and polynomials. To deduce properties of different sequences in Sec. 21, we used identities relating polynomials. In order to do this, we did not need to answer the general question "what is a polynomial" and only needed to know when we consider two polynomials equal and how to perform operations with them.

We clarify these same questions regarding the expression in formula (17) and then show how we can use this to obtain a number of unexpected results concerning properties of infinite sequences.

The expression in formula (17) is called a *power series*. The coefficient a_0 is called its *free term*. How should we understand the equality of power series? We had two concepts of equality for polynomials, which, as was shown in Chap. 2, are equivalent. One concept is that after combining like terms, the coefficients of identical powers of x are identical. The other concept is that the two polynomials take identical values for identical values of the unknown x. Applying the second concept to power series would require explaining what the *value* of the power series given by formula (17) is for some value $x = \alpha$. And this would require explaining what the sum of an infinite number of terms $a_n x^n$ is. Such an explanation could be given, relying upon the material in Chap. 5, but it would not be applicable to all cases. But this is a more complex level of understanding. However, if we stop with the first concept of equality, then no question arise. We consider two power series $f(x) = a_0 + a_1 x + \cdots + a_n x^n + \cdots$ and $g(x) = b_0 + b_1 x + \cdots + b_n x^n + \cdots$ equal if $a_0 = b_0$, $a_1 = b_1$, and, in general, $a_n = b_n$ for all n. If the representation of a polynomial in powers of x is analogous to the representation of natural numbers in the decimal system of representation (we pointed out this analogy in Sec. 7), then the power series is analogous to infinite decimal fractions representing real numbers. Newton pointed out this analogy many times.

We now discuss operations on power series. We define them exactly the same way as for polynomials—with combining like terms and opening parentheses. The sum of the power series

$$f(x) = a_0 + a_1 x + a_2 x^2 + \cdots + a_n x^n + \cdots$$

and

$$g(x) = b_0 + b_1 x + b_2 x^2 + \cdots + b_n x^n + \cdots$$

is defined as the power series

$$(a_0 + b_0) + (a_1 + b_1)x + (a_2 + b_2)x^2 + \cdots + (a_n + b_n)x^n + \cdots.$$

We define the product of the same two power series, opening parentheses in the expression

$$(a_0 + a_1 x + a_2 x^2 + \cdots + a_n x^n + \cdots)(b_0 + b_1 x + b_2 x^2 + \cdots + b_n x^n + \cdots)$$

and combining like terms. In other words, we must bring together like terms contained among the terms $a_n b_m x^{n+m}$. Therefore, the coefficient of x^l in the new power series is the sum $a_0 b_l + a_1 b_{l-1} + \cdots + a_l b_0$. We note that this is a *finite* sum for a given l, that is, among the terms $a_n b_m x^{n+m}$ with $n + m = l$ are only a *finite* number of like terms. Therefore, we can multiply two power series and always obtain a completely determined result.

We thus define the operations of addition and multiplication of arbitrary power series. They are determined by exactly the same formulas as for polynomials. Moreover, these operations can be determined using the operations on polynomials. For this, we call the polynomial $a_0 + a_1 x + \cdots + a_n x^n$, obtained by dropping all terms with powers of x higher than n from power series (17), the *nth partial sum of this series*. We note that to calculate the terms with powers of x not exceeding n in the sum or the product of two power series, we need only know such terms (with powers of x not exceeding n) in those two series. Therefore, to find the nth partial sum of the sum or the product of two series, we can perform the operation (addition or multiplication) on the nth partial sums of those two series and drop any terms with powers of x higher than n from the resulting polynomial. Because operations on power series thus reduce to operations on polynomials, the operations on power series have the same properties: commutative, associative, and distributive. In other words, the following relations (analogous to Axioms I and II for real numbers in Sec. 14) hold for power series $f(x)$, $g(x)$, and $h(x)$:

$$f(x) + g(x) = g(x) + f(x),$$
$$\big(f(x) + g(x)\big) + h(x) = f(x) + \big(g(x) + h(x)\big),$$
$$f(x)g(x) = g(x)f(x),$$
$$\big(f(x)g(x)\big)h(x) = f(x)\big(g(x)h(x)\big),$$
$$\big(f(x) + g(x)\big)h(x) = f(x)h(x) + g(x)h(x).$$

All this long explanation was needed to be able to operate with power series just as with polynomials. Namely, such was the attitude of mathematicians in the 18th century, especially Euler, looking at power series as polynomials whose degree is infinite for some reason, but which are none the worse for it. For this same reason, we placed this chapter under the topic of polynomials ("infinite-degree polynomials").

We can now pass to the properties of power series. We see that some operations on power series are performable, while they are impossible within the framework of polynomials.

Theorem 49. *Any power series $f(x) = a_0 + a_1 x + \cdots$ with a nonzero free term a_0 has an inverse power series $f(x)^{-1}$.*

To prove the theorem, we must find a power series $g(x) = b_0 + b_1 x + \cdots$ such that $f(x)g(x) = 1$. Then $g(x)$ will be equal to $f(x)^{-1}$. Multiplying the power series $f(x)$ and $g(x)$ by the rule given above, we obtain the power series $a_0 b_0 + (a_0 b_1 + a_1 b_0)x + \cdots + (a_0 b_n + a_1 b_{n-1} + \cdots + a_n b_0)x^n + \ldots$. For this power series to be equal to 1, it is necessary that $a_0 b_0 = 1$ and the remaining coefficients all be equal to zero. Hence, we obtain the equality $a_0 b_0 = 1$ and $b_0 = a_0^{-1}$; a_0^{-1} exists because $a_0 \neq 0$ by the condition of the theorem. The next equation (for the coefficient of x) gives $a_0 b_1 + a_1 b_0 = 0$ and hence $b_1 = -a_0^{-1} b_0 a_1 = -a_0^{-2} a_1$. We thus in turn determine each following equation for the coefficients $b_2, b_3, \ldots, 1$. We assume that we have considered the coefficients of $1, x, x^2, \ldots, x^{n-1}$ and determined $b_0, b_1, b_2, \ldots, b_{n-1}$. Equating the coefficient of x^n in $f(x)g(x)$ to zero gives the equation

$$a_0 b_n + a_1 b_{n-1} + \cdots + a_n b_0 = 0$$

and hence $b_n = -a_0^{-1}(a_1 b_{n-1} + \cdots + a_n b_0)$. Because $b_0, b_1, \ldots, b_{n-1}$ have already been determined, this gives the value for b_n, and the theorem is thus proved.   □

In particular, we can see that any polynomial $f(x) = a_0 + a_1 x + \cdots + a_n x^n$ with a nonzero free term a_0 has an inverse power series $f(x)^{-1}$. Therefore, every rational expression $g(x)/f(x)$, where $f(x)$ and $g(x)$ are polynomials and the free term of $f(x)$ differs from zero, can be represented in the form of a power series.

We verify our conclusion with a simple example. The polynomial $1 - x$ should have an inverse power series $(1 - x)^{-1}$. We prove that this series coincides with the series $1 + x + x^2 + x^3 + \ldots$, all of whose coefficients are equal to 1. We must prove that $(1 - x)(1 + x + x^2 + x^3 + \cdots) = 1$. The left-hand side of the equality is equal to $1 + x + x^2 + x^3 + \cdots - x(1 + x + x^2 + x^3 + \cdots)$. We can see that all terms indeed cancel here except for 1. The result obtained can be written in the form

$$\frac{1}{1 - x} = 1 + x + x^2 + x^3 + \cdots + x^n + \ldots. \tag{18}$$

This relation is analogous to relation (2) in Sec. 15 but has a completely different sense. Replacing x in (18) with $-x$, we obtain

$$\frac{1}{1 + x} = 1 - x + x^2 - x^3 + \cdots + (-1)^n x^n + \ldots. \tag{19}$$

We recall our discussion in Sec. 6. There, with any given sequence $a_0, a_1, \ldots$, we associated two other sequences: $Sa = (b_0, b_1, \ldots)$ and $\Delta a = (c_0, c_1, \ldots)$, where $b_0 = a_0$, $b_1 = a_0 + a_1$, $b_2 = a_0 + a_1 + a_2, \ldots$, and $c_0 = a_0$, $c_1 = a_1 - a_0$, $c_2 = a_2 - a_1, \ldots$. We call the power series $f(x) = a_0 + a_1 x + a_2 x^2 + \cdots$ the *generating function of the sequence a* by analogy with finite sequences in Sec. 21.

How can we find the generating functions for the power series Sa and Δa? Let $s(x)$ denote the power series $1 + x + x^2 + x^3 + \cdots + x^n + \ldots$. The coefficients of the power series $s(x)f(x)$ are just a_0, $a_0 + a_1$, $a_0 + a_1 + a_2, \ldots$, that is, the series is the generating function of the sequence Sa. It is even more obvious that the coefficients of the power series $(1 - x)f(x)$ are equal to a_0, $a_1 - a_0$, $a_2 - a_1, \ldots$, that is, the series is the generating function of the sequence Δa. Because $s(x) = (1 - x)^{-1}$, the operations of multiplying a power series by $s(x)$ and by $1 - x$ are mutually inverse. This makes the proved property in Sec. 6—"the operations S and Δ on sequences are mutually inverse"—obvious at a glance.

We now turn to another operation on power series.

Theorem 50. *If a power series $f(x) = a_0 + a_1 x + a_2 x^2 + \cdots$ has a nonzero free term a_0 and the kth root of it can be extracted, then the kth root of the entire series $f(x)$ can be extracted in the form of a power series. This power series is uniquely determined by its free term, for which any value $\sqrt[k]{a_0}$ can be taken.*

The theorem asserts that for any b_0 such that $b_0^k = a_0$, there exists a power series $b_0 + b_1 x + b_2 x^2 + \cdots$ such that

$$a_0 + a_1 x + a_2 x^2 + \cdots = (b_0 + b_1 x + b_2 x^2 + \cdots)^k. \qquad (20)$$

We prove it by consecutively determining the coefficients $b_0, b_1, b_2, \ldots$ such that the terms of degree $0, 1, 2, \ldots$ in relation (20) coincide. Equating the terms of degree 0, we obtain the condition $b_0^k = a_0$ for b_0. The existence of such a number b_0 is guaranteed by the conditions of the theorem. We note that because $a_0 \neq 0$, we have $b_0 \neq 0$.

We compare terms of degree 1. In the right-hand side, we can discard all terms of a degree higher than 1 ($b_2 x^2$, etc.) because products containing them as a factor cannot be of degree 1. Therefore, the term of degree 1 is the same as in $(b_0 + b_1 x)^k$. We can use the binomial formula to expand this expression, and we see that the first-degree term has the form $k b_0^{k-1} b_1 x$. The equality of first-degree terms in relation (20) means that $a_1 = k b_0^{k-1} b_1$. Because we already determined b_0 and it is nonzero, we obtain $b_1 = b_0^{-(k-1)} a_1 / k$. For these values of b_0 and b_1, the zeroth- and first-degree terms in the left- and right-hand sides of relation (20) coincide.

We can obviously continue thus. We assume that the coefficients $b_0, b_1, \ldots, b_n$ have been determined such that the terms of degrees

$0, 1, \ldots, n$ in relation (20) coincide. We prove that it is possible to choose b_{n+1} such that the terms of degree $n + 1$ in relation (20) coincide. We let $u(x)$ denote the nth partial sum of the series $f(x)$, that is, the polynomial $b_0 + b_1 x + b_2 x^2 + \cdots + b_n x^n$, and let $v(x)$ denote the power series $b_{n+2} x^{n+2} + \ldots$. The right-hand side of relation (20) then takes the form $(u(x) + b_{n+1} x^{n+1} + v(x))^k$. The power series $v(x)$ contains only terms of a degree higher than $n + 1$; terms of degree $n + 1$ cannot arise from multiplications with it. We can therefore discard $v(x)$: terms of degree $n + 1$ in the right-hand side of relation (20) are the same as in the polynomial $(u(x) + b_{n+1} x^{n+1})^k$. Using the binomial formula to expand this expression, we see that terms of degree $n + 1$ can arise only from the terms $u(x)^k + k u(x)^{k-1} b_{n+1} x^{n+1}$. The term of degree $n + 1$ in the polynomial $u(x)^k$ depends only on the coefficients in the polynomial $u(x)$, which we already know according to the induction assumption. We let $F(b_0, b_1, \ldots, b_n) x^{n+1}$ denote this term. The term of degree $n + 1$ in the polynomial $k u(x)^{k-1} b_{n+1} x^{n+1}$ can arise only from the free term of $u(x)$ and consequently has the form $k b_0^{k-1} b_{n+1} x^{n+1}$. Therefore, the term of degree $n + 1$ in the right-hand side of relation (20) has the form $\big(F(b_0, b_1, \ldots, b_n) + k b_0^{k-1} b_{n+1} \big) x^{n+1}$. The equality of terms of degree $n + 1$ in relation (20) implies that

$$a_{n+1} = F(b_0, b_1, \ldots, b_n) + k b_0^{k-1} b_{n+1}.$$

Setting

$$b_{n+1} = \frac{1}{k} b_0^{-(k-1)} \big(a_{n+1} - F(b_0, b_1, \ldots, b_n) \big)$$

satisfies this relation. Thus consecutively determining the coefficients b_i, we can satisfy equality (20). The theorem is proved. $\qquad\square$

For example, if $a_0 > 0$, then the power series $f(x) = a_0 + a_1 x + a_2 x^2 + \cdots$, according to Theorem 50, has a unique kth root power series $\sqrt[k]{f(x)}$ with a positive free term; it can be written as $f(x)^{1/k}$. Raising it to an arbitrary natural power m, we obtain the power series $f(x)^{m/k}$, that is, $f(x)^\alpha$, where α is any positive rational number. Applying Theorem 49, we can write $f(x)^{-\alpha}$ as a power series, that is, the power series $f(x)^\alpha$ exists for any rational—positive or negative—value α. This raises some interesting questions. For example, how can we write the series $(1 + x)^\alpha$ in an explicit form for any rational number α? We are speaking, of course, of extending the binomial formula to fractional (perhaps, rational) exponents. In Sec. 6, we used properties of the derivative of a polynomial to deduce the binomial formula for integer exponents. To argue analogously in our case, we must introduce the concept of the derivative of a power series.

We have a completely explicit formula for the derivative of a polynomial, formula (15) in Chap. 2. It is perfectly applicable to power series. We proceed thus, defining the derivative of the power series $f(x) = a_0 + a_1 x + a_2 x^2 + \cdots + a_n x^n + \cdots$ as the power series

$$f'(x) = a_1 + a_2 x + \cdots + a_n x^{n-1} + \ldots. \tag{21}$$

To make this definition less formal, we explain it as follows. The power series $f(x) = a_0 + a_1 x + a_2 x^2 + \cdots + a_n x^n + \cdots$ is equal to the sum of a polynomial of degree not higher than n, $p(x) = a_0 + a_1 x + a_2 x^2 + \cdots + a_n x^n$ (the nth partial sum) and the series $u(x) = a_{n+1} x^{n+1} + \ldots$, which contains only terms of degree higher than n. Formula (21) then indicates that the terms of a degree lower than n are the same as in the polynomial $p'(x)$. In other words, $f'(x) = p'(x) + v(x)$, where $v(x)$ is a power series containing only terms of a degree not lower than n. Summarizing, the $(n-1)$th partial sum of the series $f'(x)$ is equal to the derivative of the nth partial sum of the series $f(x)$. This rule defines the terms of degree lower than n in the derivative. Because it holds for any n, it uniquely defines the derivative.

Using this rule, we can easily demonstrate that the properties of the derivative proved for polynomials in Sec. 5 also hold for power series. We speak of the relations

$$\begin{aligned}
(f_1 + f_2)' &= f_1' + f_2', \\
(f_1 + \cdots + f_n)' &= f_1' + \cdots + f_n', \\
(f_1 f_2)' &= f_1' f_2 + f_1 f_2', \\
(f_1 \cdots f_k)' &= f_1' f_2 \cdots f_k + \cdots + f_1 f_2 \cdots f_k', \\
(f^k)' &= k f^{k-1} f'.
\end{aligned} \tag{22}$$

As an example, we show how to deduce the properties of the derivative for products. We represent each of the power series f_1 and f_2 as the sum of its nth partial sum and a series containing only terms of degree higher than n: $f_1 = p_1 + u_1$ and $f_2 = p_2 + u_2$. Then $f_1 f_2 = p_1 p_2 + (p_1 u_2 + p_2 u_1 + u_1 u_2) = p_1 p_2 + v$, where v contains only terms of degree higher than n. The nth partial sum p of the series $f_1 f_2$ is therefore obtained from $p_1 p_2$ by discarding terms of a degree higher than n, that is, $p = p_1 p_2 - w$, where w is a polynomial containing all terms of a degree higher than n in $p_1 p_2$. In view of the rule formulated above, the $(n-1)$th partial sum of the series $(f_1 f_2)'$ is equal to $p' = (p_1 p_2)' + w' = p_1' p_2 + p_1 p_2' + w'$ (here, we use formulas for the derivatives of polynomials in Sec. 5). The $(n-1)$th partial sum of the series $(f_1 f_2)'$ is thus obtained from the polynomials $p_1' p_2 + p_1 p_2'$ by discarding terms of a degree higher than $n - 1$. On the other hand, $f' = p'_1 + u'_1$, $f'_2 = p'_2 + u'_2$, and $f'_1 f_2 + f_1 f'_2 = p'_1 p_2 + p_1 p'_2 + \varphi$, where $\varphi = p'_1 u_2 + p_2 u'_1 + u'_1 u_2 + p'_2 u_1 + p_1 u'_2 + u_1 u'_2$ and contains only terms of a degree not less than n (even $2n - 3$). Therefore, the $(n-1)$th partial sum of the series $f'_1 f_2 + f_1 f'_2$ is also obtained from the polynomial $p'_1 p_2 + p_1 p'_2$ by discarding terms of a degree higher than $n - 1$, that is, it coincides with the $(n-1)$th partial sum of the series $(f_1 f_2)'$. Because this is true for any n, we have $(f_1 f_2)' = f'_1 f_2 + f_1 f'_2$. Other formulas in (22) concerning derivatives of products are obtained from this using induction just as in the case of polynomials. The formulas for the derivatives of sums are obvious—the reader can easily verify this.

We are now prepared to give a meaning for $(1 + x)^\alpha$, where α is a rational number, as a power series and to deduce the formula for that series. For simplicity, we consider only the case where α is positive (see Problem 7). Let $\alpha = p/q$, where p and q are natural numbers. Clearly, $(1 + x)^p$ has the free term 1, that is, its free term is nonzero and has a qth root, namely, 1. According to Theorem 50, there exists a power series $f(x) = 1 + a_1 x + \cdots$ such that

$$f(x)^q = (1 + x)^p. \tag{23}$$

We consider the derivatives of both sides of equality (23). Using the deduced properties of derivatives of power series given in (22) and of polynomials (formula (20) in Chap. 2), we obtain

$$q f'(x) f(x)^{q-1} = p(1 + x)^{p-1}.$$

We multiply both sides of this equality by $(1 + x)f(x)$ and obtain

$$q f'(x) f(x)^q (1 + x) = p(1 + x)^p f(x).$$

Using equality (23), we can cancel the factors $f(x)^q$ in the left-hand side and $(1 + x)^p$ in the right-hand side. Recalling that $p/q = \alpha$, we obtain

$$f'(x)(1 + x) = \alpha f(x). \tag{24}$$

Let $f(x) = 1 + a_1 x + \cdots + a_n x^n + \ldots$. We equate the coefficients of x^{n-1} in both sides of equality (24). According to the definition $f'(x) = a_1 + a_2 x + \cdots + n a_n x^{n-1} + \ldots$, we thus obtain

$$n a_n + (n - 1) a_{n-1} = \alpha a_{n-1}.$$

Hence,

$$a_n = \frac{\alpha - n + 1}{n} a_{n-1}.$$

Applying this formula r times, we obtain

$$a_n = \frac{(\alpha - n + 1)(\alpha - n + 2) \cdots (\alpha - n + r)}{n(n-1)\cdots(n-r+1)} a_{n-r}.$$

Because $a_0 = 1$, we obtain

$$a_n = \frac{\alpha(\alpha - 1) \cdots (\alpha - n + 1)}{n!}$$

for $r = n$. In other words,

$$(1+x)^\alpha = 1 + \alpha x + \frac{\alpha(\alpha - 1)}{2} x^2 + \cdots +$$
$$\frac{\alpha(\alpha - 1) \cdots (\alpha - n + 1)}{n!} x^n + \dots . \qquad (25)$$

We note that if α is equal to a natural number m, then all the coefficients of this series beginning with the $(m+1)$th coefficient become zero, and we obtain the usual binomial formula. Formula (25) also holds for negative α (see Problem 7). It is fair to call formula (25) the Newton binomial formula because it is exactly the series that Newton deduced (even for any real number α, although we do not explain what this means here). The formula with natural exponents was known earlier, to Pascal, for example.

Concluding this section, we consider an application of generalized binomial formula (25) to the determination of a sequence of numbers that is important for several questions, the so-called *Catalan numbers*. They are connected with different problems concerning partitions into parts. For example, we assume that we are concerned with calculating the product of n numbers $a_1, a_2, \dots, a_n$ taken in a definite order and want to find their product using a sequence of multiplications of two numbers each time. For this, we must place parentheses in the product $a_1 a_2 \cdots a_n$ such that each pair of parentheses contains the product of exactly two definite numbers (we consider such a product a definite number). The number of ways of placing such parentheses is called the Catalan number c_n. We set $c_1 = 1$. It is obvious that $c_2 = 1$. For the product of three factors $a_1 a_2 a_3$, there are two ways to place the parentheses: $(a_1 a_2)a_3$ and $a_1(a_2 a_3)$. Therefore, $c_3 = 2$. For $n = 4$, the possible placements are $((a_1 a_2)a_3)a_4$, $(a_1(a_2 a_3))a_4$, $(a_1 a_2)(a_3 a_4)$, $a_1((a_2 a_3)a_4)$, and $a_1(a_2(a_3 a_4))$, and therefore $c_4 = 5$.

The Catalan numbers are connected by an important relation. The last product we find in calculating $a_1 a_2 \cdots a_n$ determines a placement of parentheses $(a_1 \cdots a_k)(a_{k+1} \cdots a_n)$. Within each pair of parentheses, we can place parentheses in any way, that is, c_k ways in the first pair and c_{n-k} in the second pair. In all, there are $c_k c_{n-k}$ possible placements of parentheses with the given final placement. The total number of all placements is equal to the sum of these numbers for $k = 1, 2, \ldots, n - 1$. In other words, for $n \geq 2$, we have the relation

$$c_n = c_1 c_{n-1} + c_2 c_{n-2} + \cdots + c_{n-1} c_1. \qquad (26)$$

The right-hand side of relation (26) is reminiscent of the formula for the coefficient in the product of two power series and suggests considering the power series

$$f(x) = c_1 x + c_2 x^2 + \cdots + c_n x^n + \cdots$$

(the generating function for the Catalan numbers). The right-hand side in relation (26) is equal to the coefficient of x^n in the series $(f(x))^2$. Relation (26) suggests that the coefficients of the series $f(x)$ and $(f(x))^2$ are equal for all terms of degree 2 and higher. But $f(x)$ has a term x of degree 1, and $(f(x))^2$ does not have such a term. Therefore,

$$(f(x))^2 = f(x) - x.$$

We see that the series $f(x)$ satisfies the quadratic equation $y^2 - y + x = 0$. Therefore, the series $f(x)$ can be calculated explicitly:

$$f(x) = \frac{1}{2}\left(1 - \sqrt{1 - 4x}\right).$$

We place a minus sign before the square root because we consider that the series $\sqrt{1 - 4x}$ has the free term 1 and $f(x)$ has the free term zero.
According to formula (25)

$$\sqrt{1 - 4x} = 1 + \frac{1}{2}(-4x) + \frac{(1/2)(1/2 - 1)}{2}(-4x)^2$$
$$+ \cdots + \frac{(1/2)(1/2 - 1)\cdots(1/2 - n + 1)}{n!}(-4x)^n + \cdots.$$

Hence,
$$c_n = -\frac{(1/2)(1/2 - 1)\cdots(1/2 - n + 1)}{n!}(-4)^n.$$

We can simplify this formula further:

$$c_n = -\frac{1}{2}\frac{(-1)(-3)\cdots(-2n + 3)}{n!}(-2)^n$$
$$= \frac{1 \cdot 3 \cdot 5 \cdots (2n - 3)}{n!}2^{n-1}.$$

We multiply the numerator and denominator of the last expression by $(n-1)!$ and combine each factor in $1 \cdot 2 \cdots (n-1)$ with the factor 2. We obtain the product of even natural numbers not exceeding $2n-2$. In the numerator, there is the product of odd numbers less than $2n-2$. Together, their product yields $(2n-2)!$. As a result, we obtain

$$c_n = \frac{(2n-2)!}{n!(n-1)!}.$$

Because $C_{2n-2}^{n-1} = (2n-2)!/((n-1)!(n-1)!)$, we can write this formula as

$$c_n = \frac{1}{n} C_{2n-2}^{n-1}.$$

Problems:

1. Find the coefficients of the power series $1/(1-x)^2$ by directly squaring the series $1/(1-x)$.
2. Find a formula for the coefficients of the power series $1/(1-x)^n$. *Hint:* Use induction on n and the connection between multiplying a series by $1/(1-x)$ and applying the operation S to its coefficients.
3. Find the coefficients of the power series $1/((1-ax)(1-bx))$.
4. Prove that the series $1 + x + x^2/2! + \cdots + x^n/n! + \cdots$ and $1 - x + x^2/2! - \cdots + (-1)^n x^n/n! + \ldots$ are inverse to each other.
5. Find a formula for the derivative of $1/(1-x)^n$. Use it to determine the coefficients of the power series $1/(1-x)^n$ (induction on n).
6. Find all power series $f(x)$ for which $f'(x) = f(x)$.
7. Prove that formula (25) also holds for negative α. Prove that for negative integers α, the result agrees with the results in Problems 2 and 5. *Hint:* Set $\alpha = -p/q$, where p and q are natural numbers, set $f(x) = (1+x)^\alpha$, and use the relation $f(x)^q (1+x)^p = 1$.
8. How many ways can a convex polygon with $n+1$ sides be divided into triangles by diagonals that do not intersect within the polygon? Prove that this number is equal to the Catalan number c_n.
9. Let $f(x)$ be a polynomial of degree n. Prove that the coefficient of x^k in the power series $f(x)/(1-x)$ is equal to $f(1)$ if $k > n$.
10. Set $f_n(x) = x + 2^n x^2 + 3^n x^3 + \ldots$. Prove that $f_n(x) = u_n(x)/(1-x)^{n+1}$, where $u_n(x)$ is a polynomial of degree $n+1$ satisfying the relation $u_{n+1}(x) = x(1-x)u_n'(x) + (n+1)x u_n(x)$. *Hint:* Prove that $x f_n'(x) = f_{n+1}(x)$. Find $f_0(x)$ and use induction.
11. Prove the relation
$$n^n - C_n^1 (n-1)^n + C_n^2 (n-2)^n - \cdots + (-1)^{n-1} C_n^{n-1} \cdot 1 = n!.$$

   Hint: Use the results in Problems 9 and 10. Prove that under the conditions in Problem 10, $u_n(1) = n!$.

23. Partitio Numerorum

Euler gave the Latin name *partitio numerorum* to the branch of mathematics in which power series are used to study the partition of natural

numbers into terms. As an introduction to Sec. 21, we gave examples of
problems of partitioning numbers in which using polynomials sufficed.
In the general case, it is necessary to use sums of power series.

Let $f_n(x)$, $n = 0, 1, 2, \ldots$, be an infinite sequence of power series
and, moreover, such that each series $f_n(x)$ begins with a power of x
that increases without limit as n increases. In other words, for each
exponent N, the term ax^N is nonzero in only a finite number of series
$f_n(x)$. Then, to calculate the coefficient of x^N in the infinite sum $f_0(x) +$
$f_1(x) + \cdots + f_n(x) + \ldots$, we need only add a finite number of series:
$f_0(x) + f_1(x) + \cdots + f_m(x)$. And every Nth partial sum of the obtained
series coincides with the Nth partial sum of the finite sum of the series
$f_0(x) + f_1(x) + \cdots + f_m(x)$. Because of this, the actual calculation of an
infinite sum (that is, its partial sums) always reduces to calculating the
partial sums of some finite sum of series. Those rules deduced in Sec. 22
for sums of a finite number of series remain valid for infinite sums (if
the series $f_n(x)$ satisfies the requirements formulated above). In fact,
only after such an explanation do we have the right to say that a power
series $f(x)$ is the sum of its terms—in this case, $f_n(x) = a_n x^n$.

The same comments also relate to infinite products of the form

$$(1 + f_0(x))(1 + f_1(x))(1 + f_2(x)) \cdots (1 + f_n(x)) \ldots, \qquad (27)$$

where the power series $f_n(x)$ satisfy the same condition: $f_n(x)$ begins
with a power of x that increases without limit as n increases. Then, for
each exponent N, the series $f_k(x)$ beginning from some number $m + 1$,
that is, for $k > m$, does not contain terms of degree N. Therefore, terms
of a fixed degree N in product (27) can only be obtained from a finite
product $(1 + f_0(x)) \cdots (1 + f_m(x))$.

Using these observations, we can now show the generating functions
for the number of partitions of one or another type. Thus, the number
of partitions into terms not exceeding m has the generating function

$$(1 + x + x^2 + x^3 + \cdots)(1 + x^2 + x^4 + x^6 + \cdots) \cdots (1 + x^m + x^{2m} + x^{3m} + \cdots).$$

Indeed, in such a partition of the number n, the term 1 occurs a_1 times,
the term 2 occurs a_2 times, $\ldots$, the term m occurs a_m times: $n =$
$1 \cdot a_1 + 2 \cdot a_2 + \cdots + m \cdot a_m$ (some a_i may be equal to zero). To this
partition corresponds the term obtained by multiplying the term x^{a_1} in
the first pair of parentheses, the term x^{2a_2} in the second, $\ldots$, and the
term x^{ma_m} in the mth. The coefficient of x^n is equal to the total number
of all partitions into terms not exceeding m, that is, the coefficient of x^n
in the infinite product written above. Taking formula (18) into account,
we can write this series in the form

$$\frac{1}{(1-x)(1-x^2)\cdots(1-x^m)}. \tag{28}$$

Completely analogously, the number of partitions of the number n into arbitrary terms has the generating function

$$\frac{1}{(1-x)(1-x^2)\cdots(1-x^m)\cdots}. \tag{29}$$

The number of partitions into odd terms has the generating function

$$\frac{1}{(1-x)(1-x^3)(1-x^5)\cdots(1-x^{2m+1})\cdots}, \tag{30}$$

and into even terms, the generating function

$$\frac{1}{(1-x^2)(1-x^4)\cdots(1-x^{2m})\cdots}.$$

If we are interested only in partitions into different terms, then the generating function is

$$(1+x)(1+x^2)(1+x^3)\cdots(1+x^m)\cdots. \tag{31}$$

Here, only those partitions are allowed in which 1 occurs a_1 times, 2 occurs a_2 times, $\ldots$, and m occurs a_m times and, moreover, each a_i must be either zero or one. But each factor in product (31) in fact contains exactly the two terms x_i^a with $a_i = 0$ or 1.

These formulas immediately yield a number of applications.

Theorem 51. *The number of partitions of the number n into different terms is equal to the number of partitions into (possibly identical) odd terms.*

For example, the number 6 has three partitions into different terms, $6 = 1 + 5 = 1 + 2 + 3 = 2 + 4$, and three partitions into odd terms, $6 = 1 + 5 = 1 + 1 + 1 + 3 = 3 + 3$.

In the language of generating functions, the theorem implies that power series (30) and (31) coincide. To prove this, we write series (31) in the form

$$(1+x)(1+x^2)(1+x^3)\cdots = \frac{1-x^2}{1-x}\frac{1-x^4}{1-x^2}\frac{1-x^6}{1-x^3}\cdots.$$

The numerator contains the product of all factors $1 - x^{2n}$, and the denominator contains the product of all factors $1 - x^m$. The factors in the denominator with an even m together make up the expression in the numerator. After they are canceled, only the factors $1 - x^m$ with an odd m remain in the denominator, that is, we have series (30).   □

The next property concerns numbers of all partitions of the natural number n into arbitrary natural terms. The number of all partitions is denoted by $p(n)$. As stated above, the series $1 + p(1)x + p(2)x^2 + \cdots + p(n)x^n + \cdots$ is given by formula (29).

Theorem 52. *The inequality*

$$p(n) - 2p(n - 1) + p(n - 2) \geq 0$$

holds for $n \geq 2$.

This inequality implies that

$$p(n - 1) \leq \frac{p(n) + p(n - 2)}{2}.$$

In other words, if the points with the coordinates $(n, p(n))$, $n = 1, 2, \ldots$, are graphed on the plane, then each point after the first is not above the line joining its two nearest neighbors (Fig. 41). That is, if we pound a nail at each point $(n, p(n))$ and stretch a string on them, then we obtain an infinite convex polygon.

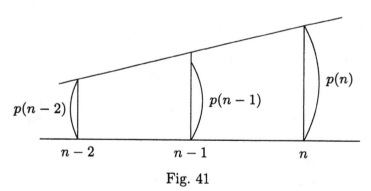

Fig. 41

The first ten values in the series $p(n)$ are $p(1) = 1$, $p(2) = 2$, $p(3) = 3$, $p(4) = 5$, $p(5) = 7$, $p(6) = 11$, $p(7) = 15$, $p(8) = 22$, $p(9) = 30$, and $p(10) = 42$. You can verify the theorem experimentally with these values. The convexity of the resulting polygon is connected with the very rapid increase of the numbers $p(n)$: already $p(50) = 204226$.

We preface the proof of the theorem with one comment. For each sequence $a = (a_0, a_1, a_2, \ldots)$, we defined a sequence $\Delta a = (a_0, a_1 - a_0, a_2 - a_1, \ldots)$ in Sec. 6. We apply this operation once more and obtain the sequence $\Delta\Delta a = (a_0, a_1 - 2a_0, a_2 - 2a_1 + a_0, \ldots)$. The term b_n

in this sequence has the form $a_n - 2a_{n-1} + a_{n-2}$ for $n \geq 2$, just the expression in the formulation of Theorem 52. On the other hand, if $f(x) = a_0 + a_1 x + \cdots$ is the generating function of the sequence a, then, as we saw in Sec. 22, the generating function of the sequence Δa is the series $(1 - x)f(x)$. Therefore, the sequence $\Delta \Delta a$ has the generating function $(1 - x)^2 f(x)$. We obtain the identity

$$(1 - x)^2 f(x) = a_0 + (a_1 - 2a_0)x + (a_2 - 2a_1 + a_0)x^2$$
$$+ \cdots + (a_n - 2a_{n-1} + a_{n-2})x^n + \cdots$$

if $f(x) = a_0 + a_1 x + a_2 x^2 + \ldots$.

We can now turn to the proof of the theorem. In view of the preceding comment, the theorem is equivalent to the statement that the coefficients of the series $(1 - x)^2 \big(p(0) + p(1)x + p(2)x^2 + \cdots + p(n)x^n + \cdots \big)$ are nonnegative beginning with the coefficient of x^2. Because

$$p(0) + p(1)x + p(2)x^2 + \cdots + p(n)x^n + \cdots$$
$$= (1 - x)^{-1}(1 - x^2)^{-1}(1 - x^3)^{-1} \ldots,$$

we must show that the coefficients of the series

$$(1 - x)(1 - x^2)^{-1}(1 - x^3)^{-1} \cdots$$

are nonnegative beginning with the coefficient of x^2.

We set $g(x) = (1 - x^3)^{-1}(1 - x^4)^{-1} \ldots$. The series we are interested in has the form $(1 - x)(1 - x^2)^{-1}g(x)$. Because $(1 - x)(1 - x^2)^{-1} = (1 - x)/\big((1 - x)(1 + x)\big) = (1 + x)^{-1}$, it is equal to $(1 + x)^{-1}g(x)$. We easily notice that $g(x)$ is also the generating function of a very simple number of partitions. Namely, reasoning literally the same as when considering the series (28), (29), and (30), we can verify that $g(x)$ is the generating function for the number of partitions into terms not less than 3. Letting $q(n)$ denote the number of such partitions of the natural number n, we obtain $g(x) = 1 + q(1)x + q(2)x^2 + \ldots$. Because $(1 + x)^{-1} = 1 - x + x^2 - x^3 + \ldots$, the coefficient of x^n in the series $(1 + x)^{-1}g(x)$ is equal to $q(n) - q(n - 1) + q(n - 2) - \cdots + (-1)^n$ (we need only recall the rule for multiplying power series: each term of the first is multiplied by every term of the second). It thus remains to prove the inequality

$$q(n) - q(n - 1) + q(n - 2) - \cdots + (-1)^n \geq 0. \qquad (32)$$

Inequality (32) follows from the obvious inequality $q(n) \geq q(n - 1)$. Indeed, increasing the largest term in some partition of the number $n - 1$ by 1, we obtain a partition of the number n. If the first partition consisted of terms greater than 2, then so does the second. In view of this inequality, sum (32) can be segregated into $n/2$ nonnegative sums of two terms $q(n-2k) - q(n-2k-1)$ (for odd n) and one more term $1 > 0$ (for even n). This proves the theorem. $\square$

Up to now, we obtained assertions about the numbers of partitions almost from nothing—from multiplying power series. Euler found a more refined method for calculating the coefficients of certain products using so-called functional equations that they satisfy. We illustrate this method with one example.

We consider a question concerning the partition of a natural number into a given number of *different* terms. For this, Euler proposed introducing a new unknown z and the series $G(x,z) = (1+z)(1+xz)(1+x^2z)(1+x^3z)\ldots$. Expanding it in powers of z, we obtain the equality

$$G(x,z) = 1 + u_1(x)z + u_2(x)z^2 + \cdots + u_m(x)z^m + \ldots, \qquad (33)$$

where $u_i(x)$ is a power series in the unknown x. Here, the term $u_m(x)z^m$ consists of terms obtained by multiplying m terms x^iz with different exponents i from the product $G(x,z)$. Consequently, the coefficient of x^nz^m is equal to the number of partitions of n into m different terms. In other words, $u_m(x)$ is the generating function for such partitions.

If we replace z with xz in $G(x,z)$, then we obtain all the factors in $G(x,z)$ except the first. Therefore,

$$G(x,z) = (1+z)G(x,xz). \qquad (34)$$

This is the functional equation for the series $G(x,z)$. On the other hand, substituting xz for z in series (33), we see that

$$G(x,xz) = 1 + u_1(x)xz + u_2(x)x^2z^2 + \cdots + u_m(x)x^mz^m \ldots .$$

Multiplying this expression for $G(x,xz)$ by $1+z$ and substituting in relation (34), we obtain

$$u_m(x) = u_m(x)x^m + u_{m-1}(x)x^{m-1}.$$

Hence,

$$u_m(x) = \frac{x^{m-1}}{1-x^m}u_{m-1}(x). \qquad (35)$$

Applying this relation to $u_{m-1}(x)$ and substituting in equality (35), we obtain

$$u_m(x) = \frac{x^{(m-1)+(m-2)}}{(1-x^m)(1-x^{m-1})}u_{m-2}(x).$$

Proceeding thus m times and taking $u_0(x) = 1$ into account, we obtain the expression for $u_m(x)$:

$$u_m(x) = \frac{x^{(m-1)+(m-2)+\cdots+1}}{(1-x^m)(1-x^{m-1})\cdots(1-x)}$$
$$= \frac{x^{m(m-1)/2}}{(1-x)(1-x^2)\cdots(1-x^m)}. \qquad (36)$$

But we already encountered the series $1/\big((1-x)(1-x^2)\cdots(1-x^m)\big)$ (see formula (28)). It is the generating function for the number of partitions into terms not exceeding m. Formula (36) shows that *the number of partitions of the number n into m different terms is equal to the number of partitions of the number $n - m(m-1)/2$ into terms not exceeding m.*

In connection with generating function (29), Euler consider the logically simpler product

$$(1-x)(1-x^2)(1-x^3)\dots. \qquad (37)$$

This is a very interesting expression. In Sec. 21, we noted an analogy between Gauss polynomials $g_{k,l}(x)$ and binomial coefficients. The analogy was based on formula (7), in which the polynomial $h_m(x)$ was analogous to the number $m!$. We recall that

$$h_m(x) = (1-x)(1-x^2)\cdots(1-x^m).$$

From this standpoint, product (37) is analogous to "infinity factorial." For numbers, such a combination of words is senseless; it does not have a meaning even for polynomials. But we can reach a totally rigorous definition of the expression if we use power series.

Euler expanded this expression in a series up to the term x^{51} and obtained

$$(1-x)(1-x^2)(1-x^3)\cdots = 1 - x - x^2 + x^5 + x^7 - x^{12} - x^{15}$$
$$+ x^{22} + x^{26} - x^{35} - x^{40} + x^{51} + \dots.$$

He was amazed by the regularity observed here: all coefficients are equal to 0, +1, or −1. Moreover, the exponents of terms with nonzero coefficients form a sequence that was familiar to Euler: these are numbers of the form $n(3n+1)/2$ for $n = -1, 1, -2, 2, -3, 3, -4, 4, -5, 5, -6$. These numbers generally provoked interest then in connection with "figure numbers," already known in antiquity. Namely, a triangular number is the number of points in a regular triangle if there are $n+1$ points at

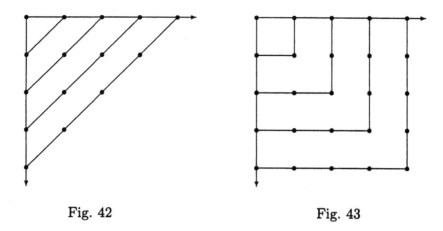

Fig. 42   Fig. 43

a distance n from a vertex of the triangle (Fig. 42). These numbers are
$1, 3, 6, 10, \ldots, n(n+1)/2, \ldots$.

A square number is the number of points in a square if these are
placed at equal distances and the number of points on a side is equal to
n (Fig. 43). In other words, it is simply a perfect square n^2.

Pentagonal numbers are obtained from a regular pentagon by placing
a point at each vertex and then similarly enlarging it, retaining one
vertex A and the direction of the sides coming from it. On each side
that does not intersect the vertex A, we place two points at the first
enlargement, three points at the next, and so on. A pentagonal number
is the number of points obtained after n such operations (Fig. 44).

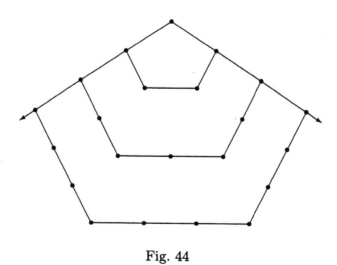

Fig. 44

Consequently, a pentagonal number is equal to the sum of an arithmetic progression:

$$V = 1 + 4 + 7 + \cdots + (3n - 2)$$
$$= 1 + (1 + 3) + (1 + 2 \cdot 3) + \cdots + (1 + (n - 1)3)$$
$$= \underbrace{1 + \cdots + 1}_{n \text{ times}} + (1 + 2 + \cdots + (n - 1))3$$
$$= n + \frac{n(n - 1)}{2} \cdot 3 = \frac{3n^2 - n}{2}.$$

By analogy with these numbers, numbers of the same form with a negative $n = -m$, that is, the numbers $(3m^2 + m)/2$, are also called pentagonal numbers.

Euler proposed an expression for the product $(1-x)(1-x^2)(1-x^3) \cdots$ in the form of a series consisting of the terms

$$(-1)^n x^{n(3n-1)/2} + (-1)^n x^{n(3n+1)/2}$$

for $n = 0, 1, 2, \ldots$. He called this a "noteworthy observation that I, however, am unable to prove with geometric rigor." We would now call it a hypothesis. Euler stated this hypothesis in 1741. He found a proof of it nine years later in 1750. In view of its connection with pentagonal numbers, it is called the Euler pentagon theorem. Its proof is somewhat more complicated than the reasoning we have so far given, and we relegate the proof to a supplement.

The Euler pentagon theorem yields new properties of numbers of partitions. First, the product $(1 - x)(1 - x^2)(1 - x^3) \cdots$ is also a generating function. Namely, completely analogously to the expansion of product (31), each term x^n is obtained from some partition of n into different terms. But this term now enters with a plus sign if the number of terms is even and with a minus sign if the number of terms is odd. The coefficient of x^n is therefore equal to the difference between the number of partitions of n into an even number and an odd number of different terms. The Euler pentagon theorem therefore has the following formulation.

The number of partitions of a natural number n into different even terms is equal to the number of partitions of it into different odd terms if n is not a pentagonal number and the difference between those two numbers is equal to $(-1)^m$ if n has the form $m(3m - 1)/2$ for a positive or negative integer m.

Another consequence of the pentagon theorem is as follows. We recall that product (29) coincides with the series $1 + p(1)x + p(2)x^2 + \cdots$ and its inverse product (37) is the sum of terms of the form $(-1)^n x^{n(3n\pm1)/2}$ according to the Euler theorem. We multiply these two series according to the usual rule ("a term of one is multiplied by a term of the other and then like terms are combined") and then write that the product is equal to one, that is, a series with all zero coefficients except for the free term. We write the coefficient of x^n (for $n > 0$) in the product and equate it to zero. We obtain the relation

$$p(n) - p(n-1) - p(n-2) + p(n-5) + \cdots = 0.$$

This sum contains the terms $(-1)^m \big(p(n-m_1) + p(n-m_2)\big)$, where $m_1 = m(3m-1)/2$ and $m_2 = m(3m+1)/2$. Only values of m_1 and m_2 not exceeding n are considered, and $p(0)$ is consider to be one. The relation obtained expresses $p(n)$ in terms of the values $p(n')$ with $n' < n$ and gives a convenient method for consecutively calculating the numbers $p(n)$. For example,

$$p(10) = p(9) + p(8) - p(5) - p(3),$$
$$p(9) = p(8) + p(7) - p(4) - p(2),$$
$$p(8) = p(7) + p(6) - p(3) - p(1),$$
$$p(7) = p(6) + p(5) - p(2) - 1,$$
$$p(6) = p(5) + p(4) - p(1),$$
$$p(5) = p(4) + p(3) - 1,$$
$$p(4) = p(3) + p(2),$$
$$p(3) = p(2) + p(1),$$
$$p(2) = p(1) + 1,$$
$$p(1) = 1.$$

From this, moving up from the bottom, we obtain $p(1) = 1$, $p(2) = 2$, $p(3) = 3$, $p(4) = 5$, $p(5) = 7$, $p(6) = 11$, $p(7) = 15$, $p(8) = 22$, $p(9) = 30$, and $p(10) = 42$.

Problems:

1. The monetary unit "kopeck" and coins with the value 1, 5, 10, and 50 kopecks were recently reintroduced. Let a_n be the number of ways the sum of n kopecks can be produced using such coins. Prove that the power series $1 + a_1 x + a_2 x^2 + \cdots$ is equal to $1/p(x)$, where $p(x)$ is a polynomial, and find this polynomial.
2. Solve the problem similar to Problem 1 where a_n is the number of ways the sum n can be formed from terms equal to given numbers $k_1, \ldots, k_r$.
3. Prove that if partitions differing in the order of their terms are considered different partitions in Problem 2, then the number of partitions of n into m terms is equal to the coefficient of x^n in $(x^{k_1} + \cdots + x^{k_r})^m$. Solve Problem 2 under these new conditions.
4. Prove that every natural number n can be represented as a sum of natural numbers in 2^{n-1} ways if partitions differing in the order of their terms are considered different partitions.
5. How many different monomials of degree m with the unknowns $x_1, \ldots, x_n$ are there? *Hint:* Represent the series that is the sum of all monomials $x_1^{r_1} \cdots x_n^{r_n}$ in the form $1/p(x_1, \ldots, x_n)$, where $p(x_1, \ldots, x_n)$ is a polynomial, and then set $x_1 = \cdots = x_n = y$.

6. We set $F_m = 1/((1-x)(1-x^2)\cdots(1-x^m))$. The equality $F_m = F_{m-1} + x^m F_m$ follows from the obvious relation $(1-x^m)F_m = F_{m-1}$. Deduce from this that the number of partitions of the number n into terms $1,\ldots,m$ is equal to the sum of the number of such partitions of the number $n-m$ and the number of partitions of n into terms $1, 2, \ldots, m-1$. Note that the number of partitions of n into terms $1,\ldots,m$ is equal to $P_{n,m}(n)$ in the notation in Sec. 21. Therefore, the relation obtained above is a consequence of equality (2) (for what values of k, l, and n?). But it is now obtained without any reasoning about partitions, purely following from properties of power series.

7. The German mathematician Node attracted Euler's attention to the theory of partitions. In his letter, Node wrote, "How can the number of partitions be determined for sufficiently large numbers n? For example, what is the number of partitions of 50 into terms not exceeding 7 equal to? Into 7 different terms?" Euler answered Node in two weeks, pointing out the connection of his question with power series and showing how Node's questions could be answered using this method. Euler published his results half a year later. In particular, he deduced the relation given in Problem 6 and thus found the sequence of numbers of partitions beginning with the smallest n and m all the way to $n = 69$ and $m = 11$. Try to reconstruct Euler's line of reasoning and set up a table of the numbers of partitions of $1, 2, \ldots, 49, 50$ into terms not excedding $1, 2, 3, \ldots, 7$. The table should have the form of a rectangle with the numbers $m = 1, \ldots, 7$ placed horizontally at the top and the numbers $n = 1, \ldots, 50$ place vertically at the left. At the intersection of the nth row and the mth column, place the number of partitions of n into terms not exceeding m. In this way, prove that the number of partitions of 50 into terms not exceeding 6 is equal to 18 138 and the number of partitions of 50 into terms not exceeding 7 is equal to 522. *Hint:* The assertion in italics on page 249 is useful here.

8. Prove that the number of partitions of n in which only odd terms can occur more than once is equal to the number of partitions in which each term occurs no more than three times.

9. Represent the product $(1 + xz)(1 + x^2 z)(1 + x^4 z)(1 + x^8 z)\cdots$ in the form of a series $1 + u_i(x)z + u_2(x)z^2 + \ldots$, and find $u_k(x)$.

10. Represent the power series $(1 - xz)^{-1}(1 - x^2 z)^{-1}(1 - x^3 z)^{-1}\cdots$ in the form $1 + v_1(x)z + v_2(x)z^2 + \ldots$, and find $v_k(x)$. Interpret the result obtained as a relation between numbers of partitions analogously to the interpretation of formula (36).

Supplement 1: The Euler Pentagon Theorem

We give two proofs of the Euler pentagon theorem here. There now exist several different proofs of it. We first give the proof belonging to Euler himself. It is a remarkable example of pure algebra. It does not use anything except opening parentheses and grouping terms, while the operations are combined so intricately that Euler himself found the proof almost ten years after he stated the theorem in the form of a hypothesis.

The idea of the proof is very natural. We expand the product

$$(1 - x)(1 - x^2)(1 - x^3)\cdots(1 - x^n)\cdots \qquad (38)$$

step by step, representing it at each step as the sum of a polynomial of degree N and an expression divisible by x^{N+1}. Moreover, N increases at each step. We thus calculate the partial sums of the expansion of product (38) in a power series. We begin with the finite product

$$(1 - a_1)(1 - a_2) \cdots (1 - a_n).$$

Opening the last pair of parentheses, we write it as

$$(1 - a_1)(1 - a_2) \cdots (1 - a_n) = (1 - a_1)(1 - a_2) \cdots (1 - a_{n-1})$$
$$- a_n(1 - a_1)(1 - a_2) \cdots (1 - a_{n-1}).$$

We can now apply the same technique to the first term, which becomes the sum of the product of $n-2$ factors $(1 - a_1)(1 - a_2) \cdots (1 - a_{n-2})$ and the term $-a_{n-1}(1 - a_1)(1 - a_2) \cdots (1 - a_{n-2})$. We then similarly transform the term $(1 - a_1)(1 - a_2) \cdots (1 - a_{n-2})$. We perform $n-1$ such transformations until we obtain the term $1 - a_1$. As a result, we obtain the identity

$$(1 - a_1)(1 - a_2) \cdots (1 - a_n)$$
$$= 1 - a_1 - a_2(1 - a_1) - a_3(1 - a_1)(1 - a_2)$$
$$- \cdots - a_n(1 - a_1)(1 - a_2) \cdots (1 - a_{n-1}). \qquad (39)$$

Identity (39) can be applied to an infinite product $(1 - u_1)(1 - u_2) \cdots (1 - u_n) \ldots$, where the $u_i(x)$ are power series beginning with ever higher powers of x. We apply this reasoning to the case where $u_n = x^n$, in which the "series" begins with the term x^n (and ends with it). We obtain the identity

$$(1 - u_1)(1 - u_2) \cdots (1 - u_n) \cdots$$
$$= 1 - u_1 - u_2(1 - u_1) - u_3(1 - u_1)(1 - u_2)$$
$$- \cdots - u_n(1 - u_1)(1 - u_2) \cdots (1 - u_{n-1}) - \ldots . \qquad (40)$$

Indeed, if we consider terms of a degree not exceeding n, then all factors in the right-hand side beginning with the $(n+1)$th and all terms in the left-hand side beginning with the $(n+1)$th can be discarded because they do not contain terms of a degree not exceeding n. But then we obtain identity (39) with $a_i = u_i$. That is, the terms of a degree not exceeding n in the left-hand and right-hand sides of equality (40) coincide. Because this is true for any n, equality (40) is valid.

Substituting $u_i = x^i$ in relation (40), we obtain the equality

$$(1 - x^1)(1 - x^2) \cdots (1 - x^n) \cdots$$
$$= 1 - x^1 - x^2(1 - x^1) - x^3(1 - x^1)(1 - x^2)$$
$$- \cdots - x^n(1 - x^1)(1 - x^2) \cdots (1 - x^{n-1}) - \ldots . \qquad (41)$$

This is the first step in our chain of transformations. Letting P_0 denote the product $(1 - x) \cdots (1 - x^n) \ldots$, we factor out x^2 from every term in equality (41) except the first two and set

$$P_1 = 1 - x + x(1 - x)(1 - x^2)$$
$$+ \cdots + x^m(1 - x) \cdots (1 - x^{m+1}) + \ldots .$$

Equality (41) then takes the form

$$P_0 = 1 - x - x^2 P_1. \tag{42}$$

We now transform P_1. We write P_1 in the form

$$P_1 = Q_0 + Q_1 + \cdots + Q_k + \cdots ,$$

where

$$Q_k = x^k(1 - x)(1 - x^2) \cdots (1 - x^{k+1}).$$

We open the first pair of parentheses in the product Q_k. We obtain an equality that we can write as

$$Q_k = A_k - B_k,$$

where

$$A_k = x^k(1 - x^2)(1 - x^3) \cdots (1 - x^{k+1}), \qquad A_0 = 1,$$
$$B_k = x^{k+1}(1 - x^2)(1 - x^3) \cdots (1 - x^{k+1}), \qquad B_0 = x.$$

The expression P_1 can be written in the form

$$P_1 = A_0 - B_0 + A_1 - B_1 + \cdots + A_k - B_k + \ldots . \tag{43}$$

We now note the expression $A_k - B_{k-1}$ can be written more simply for $k \geq 2$:

$$A_k - B_{k-1} = x^k(1 - x^2) \cdots (1 - x^{k+1}) - x^k(1 - x^2) \cdots (1 - x^k)$$
$$= -x^{2k+1}C_{k-2},$$

where

$$C_k = (1 - x^2) \cdots (1 - x^{k+2}), \quad k \geq 2, \qquad C_0 = 1 - x^2,$$

and this implies that

$$C_{k-2} = (1 - x^2)(1 - x^3) \cdots (1 - x^k).$$

Writing expansion (43) in the form

$$P_1 = A_0 - B_0 + A_1 + (-B_1 + A_2) \\ + (-B_2 + A_3) + \cdots + (-B_{k-1} + A_k) + \ldots,$$

we obtain the representation

$$P_1 = 1 - x + x(1 - x^2) - x^5 C_0 - x^7 C_1 - \cdots - x^{2k+5} C_k - \ldots.$$

We can write it differently, combining like terms at the beginning,

$$P_1 = 1 - x^3 - x^5 P_2, \qquad (44)$$

where

$$P_2 = C_0 + x^2 C_1 + \cdots + x^{2k} C_k + \ldots.$$

In the expanded form, we have

$$P_2 = 1 - x^2 + x^2(1 - x^2)(1 - x^3) + \cdots \\ + x^{2k}(1 - x^2)(1 - x^3) \cdots (1 - x^{k+2}) + \ldots.$$

With this, in fact, the essence of the proof is finished. Equalities (42) and (44) give the beginning of the inductive expansion of product (38) with the segregation of partial sums of ever higher degree from it. We need only formulate the process of passing from the nth step to the $(n+1)$th and write the result explicitly.

We set

$$P_n = 1 - x^n + x^n(1 - x^n)(1 - x^{n+1}) \\ + x^{2n}(1 - x^n)(1 - x^{n+1})(1 - x^{n+2}) + \cdots \\ + x^{kn}(1 - x^n)(1 - x^{n+1}) \cdots (1 - x^{n+k}) + \ldots.$$

We transform this expression as we did P_1. We set

$$Q_k = x^{nk}(1 - x^n)(1 - x^{n+1}) \cdots (1 - x^{n+k}).$$

Then $P_n = Q_0 + Q_1 + \cdots + Q_k + \ldots$. We open the first pair of parentheses in the product Q_k:

$$Q_k = A_k - B_k, \\ A_k = x^{nk}(1 - x^{n+1}) \cdots (1 - x^{n+k}), \\ B_k = x^{n(k+1)}(1 - x^{n+1}) \cdots (1 - x^{n+k}).$$

We consider the difference $A_k - B_{k-1}$ for $n \geq 2$:

$$A_k - B_{k-1} = x^{nk}(1 - x^{n+1})\cdots(1 - x^{n+k-1})(-x^{n+k})$$
$$= -x^{nk+n+k}(1 - x^{n+1})\cdots(1 - x^{n+k-1})$$
$$= -x^{nk+n+k}C_{k-2},$$

where
$$C_k = (1 - x^{n+1})\cdots(1 - x^{n+1+k}), \quad k \geq 0.$$

The exponent of the highest power of x that divides $A_k - B_{k-1}$, $nk+n+k$, can be written in the form

$$nk + n + k = (n + 1)(k - 2) + 3n + 2.$$

Therefore, we can write P_n in the form

$$P_n = A_0 - B_0 + A_1 + (-B_1 + A_2) + \cdots + (-B_{k-1} + A_k) + \cdots$$
$$= A_0 - B_0 + A_1 - x^{3n+2}(C_0 + x^{n+1}C_1 + x^{2(n+1)}C_2 + \cdots).$$

The sum

$$C_0 + x^{n+1}C_1 + x^{2(n+1)}C_2 + \cdots$$
$$= 1 - x^{n+1} + x^{n+1}(1 - x^{n+1})(1 - x^{n+2})$$
$$+ x^{2(n+1)}(1 - x^{n+1})(1 - x^{n+2})(1 - x^{n+3}) + \cdots$$
$$+ x^{k(n+1)}(1 - x^{n+1})\cdots(1 - x^{n+1+k}) + \cdots$$

coincides with P_{n+1} by definition. Because

$$A_0 - B_0 + A_1 = 1 - x^n + x^n - x^{2n+1} = 1 - x^{2n+1},$$

we consequently obtain the relation

$$P_n = 1 - x^{2n+1} - x^{3n+2}P_{n+1}. \tag{45}$$

Our process for expanding product (38) in a series is described completely. It remains to see just what is obtained as a result. We use the same formula to express P_{n-1} in terms of P_n and substitute expression (45) for P_n. We obtain

$$P_{n-1} = 1 - x^{2n-1} - x^{3n-1}(1 - x^{2n+1} - x^{3n+2}P_{n+1}).$$

We similarly express P_{n-2} in terms of P_{n-1} and substitute the expression just obtained for P_{n-1}. We obtain

$$P_{n-2} = 1 - x^{2n-3} - x^{3n-4}(1 - x^{2n-1} - x^{3n-1}(1 - 2^{2n+1} - x^{3n+2}P_{n+1})).$$

We thus pass through n steps to P_0 and obtain an expression in which pairs of terms $1 - x^{2n+1}$ enter with alternating signs and, moreover, the expression $1 - x^{2n+1}$ has the sign of $(-1)^n$. Furthermore, each such expression is multiplied by a certain power of x, namely, the factor x^{3n-1} arises in passing from P_n to P_{n-1}, the factor x^{3n-4} arises in passing from P_{n-1} to P_{n-2}, and so on. As a result, the sum $1 - x^{2n+1}$ enters P_0 multiplied by $x^{2+5+\cdots+(3n-1)}$. The exponent here is the sum of the arithmetic progression

$$2 + (2 + 3) + \cdots + (2 + 3(n - 1)) = 2n + 3(1 + 2 + \cdots + (n - 1))$$

$$= 2n + 3\frac{n(n - 1)}{2} = \frac{3n^2 + n}{2}.$$

Product (38) is therefore equal to the sum of the terms

$$(-1)^n x^{(3n^2+n)/2}(1 - x^{2n+1}) = (-1)^n x^{(3n^2+n)/2}$$
$$+ (-1)^{n+1} x^{(3n^2+n)/2+2n+1}.$$

We see that $(3n^2 + n)/2$ is the nth pentagonal number. Moreover,

$$\frac{3n^2 + n}{2} + 2n + 1 = \frac{3n^2 + 5n + 2}{2} = \frac{3(n + 1)^2 - (n + 1)}{2}.$$

This is the pentagonal number corresponding to the negative value $-(n + 1)$. Therefore, product (38) is equal to the sum of the terms $(-1)^n x^{(3n^2+n)/2}$ for $n = 0, -1, 1, -2, 2, \ldots$. And this is the assertion of the Euler pentagon theorem.   □

We note that we could omit the deduction of the formula for P_1 in terms of P_2 (that is, formula (44)) and only deduce formula (45) because formula (44) is obtained from (45) for $n = 1$. In effect, we gave the same arguments two times to make the logic of our transformations clearer.

We now give a second proof of the pentagon theorem. It is based on an identity found in the 19th century by Gauss and Jacobi. The matter concerns calculating the infinite product

$$(1 + xz)(1 + xz^{-1})(1 + x^3z)(1 + x^3z^{-1})$$
$$\cdots (1 - x^{2n-1}z)(1 - x^{2n-1}z^{-1})\ldots, \tag{46}$$

where the exponent of x for z and z^{-1} ranges all odd numbers. This is a more complicated expression than what we considered earlier because it contains not only positive also negative powers of z. We verify that such an expression has a perfectly determined meaning just the same as for infinite products composed of power series. If we consider the first n pairs of factors in product (46), then we obtain the expression

$$(1 + xz)(1 + xz^{-1})(1 + x^3z)(1 + x^3z^{-1})$$
$$\cdots (1 - x^{2n-1}z)(1 - x^{2n-1}z^{-1}), \qquad (47)$$

which is an ordinary algebraic fraction. Opening all parentheses, we obtain terms of the form $x^m z^r$, where m takes positive values and r takes both positive and negative values. If we consider the next pair of factors in product (46), then we add only terms containing powers of x with exponents greater than $2n$ when opening parentheses. Therefore, the coefficient of z^r is a power series in x, and to calculate its terms of a degree not higher than $2n$, it is sufficient to consider finite product (47). After expansion, infinite product (46) thus becomes the sum of expressions $A_r(x)z^r$, where $A_r(x)$ is a power series in x and r takes any integer value. But in view of the symmetry of expression (46) with respect to z amd z^{-1}, the expression obtained by expansion is also symmetric; therefore, the coeeficient $A_r(x)$ of z^r, $r > 0$, is equal to the coefficent $A_{-r}(x)$ of z^{-r}. As a result, product (46) can be written as

$$A_0(x) + A_1(x)(z + z^{-1}) + \cdots + A_r(x)(z^r + z^{-r}) + \ldots . \qquad (48)$$

Our task is to calculate the power series $A_0(x), A_1(x), \ldots$, which we do in two stages.

The first stage is completely parallel to the resoning we used in Sec. 23. We let $F(z)$ denote product (46) and replace z in it with x^2z. Each pair of factors $1+x^{2k-1}z$ and $1+x^{2k-1}z^{-1}$ yields similar factors $1+x^{2k+1}z$ and $1 + x^{2k-3}z^{-1}$ when z is replaced with x^2z. The factors in product (46) thus change places but are otherwise identical except that, first, the factor $1+xz$ is missing (all factors $1+x^{2k+1}z$ have an exponent $2k+1 \geq 3$) and, second, the original factor $1+xz^{-1}$ yields the new factor $1+x^{-1}z^{-1}$, which was absent in the original product. All this can be written in the form of one formula,

$$F(x^2z)\frac{1 + xz}{1 + x^{-1}z^{-1}} = F(z).$$

But, obviously, $(1+xz)/(1+x^{-1}z^{-1}) = xz$, and we can write this formula in the form

$$F(x^2z)xz = F(z). \qquad (49)$$

We now recall that we can consider the product $F(z)$ expanded in form (48), and we apply relation (49) to that representation. We obtain

$$\bigl(A_0(x) + A_1(x)(x^2 z + x^{-2} z^{-1}) + \cdots$$
$$+ A_r(x)(x^{2r} z + x^{-2r} z^{-r}) + \cdots\bigr)xz$$
$$= A_0(x) + A_1(x)(z + z^{-1}) + \cdots + A_r(x)(z^r + z^{-r}) + \ldots .$$

We equate terms containing z^r. In the left-hand side, they are obtained from the term containing z^{r-1} after multiplication by xz, that is, from the term $A_{r-1}(x)x^{2(r-1)}z^{r-1}$ after multiplication by xz. In the right-hand side, they are obtained from the term containing z^r, that is, from $A_r(x)z^r$. As a result, we obtain

$$A_{r-1}(x)x^{2r-1} = A_r(x).$$

We see that all the series $A_r(x)$ are expressed in terms of one another. In particular,

$$A_r(x) = x^{2r-1}A_{r-1}(x) = x^{2r-1+2r-3}A_{r-2}(x)$$
$$= \cdots = x^{2r-1+2r-3+\cdots+1}A_0(x). \tag{50}$$

The exponent of x is the sum of the first r odd numbers. We already encountered an analogous sum in the first proof of the pentagonal theorem:

$$1 + (1+2) + (1+2\cdot2) + \cdots$$
$$+ (1+2(r-1)) = r + 2(1+2+\cdots+r-1)$$
$$= r + 2\frac{r(r-1)}{2}$$
$$= r + r(r-1) = r^2.$$

In short, the sum of the first r odd natural numbers is equal to r^2 (see Fig. 43). We can rewrite relation (50):

$$A_r(x) = x^{r^2}A_0(x).$$

We could verify that considering terms with negative powers of z leads to exactly the same relation, but we omit this here.

We see that $A_0(x)$ can be factored out of the entire expression (48) and a very elegant expression of our product (46) is obtained:

$$(1 + xz)(1 + xz^{-1})(1 + x^3 z)(1 + x^3 z^{-1})$$
$$\cdots (1 + x^{2r-1} z)(1 + x^{2r-1} z^{-1}) \cdots$$
$$= A_0(x)\bigl(1 + x(z + z^{-1}) + \cdots + x^{r^2}(z^r + z^{-r}) + \cdots\bigr), \tag{51}$$

but the factor $A_0(x)$ in it remains undetermined.

The reasoning used here completely follows the method used in Sec. 23 to calculate product (32). However, we had an already known (free) term there, and all the other terms were expressed in terms of it. In our case, we do not have such a term, and it becomes the factor $A_0(x)$, which we must still determine, in formula (51).

We now pass to the second stage of the proof—calculation of the series $A_0(x)$. We recall that to find the terms of this series of a degree not exceeding $2n$, it is sufficient to consider finite product (47). We can give some coefficients in it explicitly. For example, the coefficient of z^n is obtained if the terms $x^{2r-1}z$ are taken from pairs of parentheses enclosing expressions of the type $1+x^{2r-1}z$ and the term 1 is taken from pairs of parentheses containing expressions of the type $1+x^{2r-1}z^{-1}$. As a result, we obtain the term $x^{1+3+\cdots+(2n-1)}z^n = x^{n^2}z^n$. Applying the same technique as in the first stage of the proof, we can express all the others, $A_0(x)$ in particular, in terms of it. This is the plan of the proof.

Let $f(z)$ denote product (47) (for some fixed n). We again replace z with x^2z. We now obtain a greater change than previously with the same replacement in product (46). As before, the factor $1+xz$ is now missing at the beginning of $f(x^2z)$, and the new factor $1 + x^{-1}z^{-1}$ appears. But in addition, a change now occurs at the end of product (48): the new factor $1 + x^{2n-1}x^2z = 1 + x^{2n+1}z$ now appears, and the factor $1 + x^{2n-1}z_{-1}$ is missing (after substituting x^2z for z, the exponent of x decreases). The remaining factors in $f(x^2z)$ and $f(z)$ are identical. We again obtain a relation between them, only more complicated:

$$f(x^2z)\frac{1+xz}{1+x^{-1}z^{-1}}\frac{1+x^{2n-1}z^{-1}}{1+x^{2n+1}z} = f(z). \tag{52}$$

As we saw, $(1 + xz)/(1 + x^{-1}z^{-1}) = xz$, we have $xz(1 + x^{2n-1}z^{-1}) = zx + x^{2n}$, and relation (52) becomes

$$f(x^2z)(zx + x^{2n}) = f(z)(1 + x^{2n+1}z). \tag{53}$$

We now represent $f(z)$ expanded in powers of z and z^{-1}:

$$f(z) = a_0(x) + a_1(x)(z + z^{-1}) + \cdots + a_n(x)(z^n + z^{-n}). \tag{54}$$

We substitute this expression in relation (53):

$$\begin{aligned}
&\big(a_0(x) + a_1(x)(x^2z + x^{-2}z^{-1}) \\
&\quad + \cdots + a_n(x)(x^{2n}z^n + x^{-2n}z^{-n})\big)(xz + x^{2n}) \\
&= \big(a_0(x) + a_1(x)(z + z^{-1}) \\
&\quad + \cdots + a_n(x)(z^n + z^{-n})\big)(1 + x^{2n+1}z).
\end{aligned}$$

We equate the coefficients of z^r in both sides of the equality (considering $r \geq 1$). In the left-hand side, such a term is obtained from terms containing z^{r-1} when multiplied by xz and containing z^r when multiplied by x^{2n}. In the right-hand side, such a term is obtained from terms containing z^{r-1} when multiplied by $x^{2n+1}z$ and containing z^r when multiplied by 1. As a result, we obtain the relation

$$a_{r-1}(x)x^{2r-1} + a_r(x)x^{2r+2n} = a_{r-1}(x)x^{2n+1} + a_r(x).$$

Transferring terms with $a_{r-1}(x)$ and $a_r(x)$ to different sides, we obtain

$$a_{r-1}(x)x^{2r-1}(1 - x^{2n-2r+2}) = a_r(x)(1 - x^{2n+2r}). \qquad (55)$$

Relation (55) allows expressing each coefficient $a_r(x)$ in terms of others. For example,

$$a_r(x) = \frac{a_{r-1}(x)x^{2r-1}(1 - x^{2n-2r+2})}{1 - x^{2n+2r}}.$$

Replacing r with $r - 1$ in relation (55), we similarly express $a_{r-1}(x)$. Substituting this expression, we obtain

$$a_r(x) = \frac{a_{r-2}(x)x^{2r-1+2r-3}(1 - x^{2n-2r+2})(1 - x^{2n-2r+4})}{(1 - x^{2n+2r})(1 - x^{2n+2r-2})}.$$

Performing this process r times, we obtain a power of x equal to $x^{(2r-1)+(2r-3)+\cdots+1}$ in the numerator, that is, as we saw, x^{r^2}. Finally, we obtain

$$a_r(x) = a_0(x)x^{r^2}\frac{(1 - x^{2n-2r+2})(1 - x^{2n-2r+4})\cdots(1 - x^{2n})}{(1 - x^{2n+2r})(1 - x^{2n+2r-2})\cdots(1 - x^{2n+2})}.$$

Because we know the coefficient $a_n(x)$ (it is equal to x^{n^2}), we can set $r = n$ in this relation and obtain

$$x^{n^2} = a_0(x)x^{n^2}\frac{(1 - x^2)(1 - x^4)\cdots(1 - x^{2n})}{(1 - x^{2n+2})\cdots(1 - x^{4n})}.$$

This can be rewritten as

$$a_0(x) = \frac{(1 - x^{2n+2})\cdots(1 - x^{4n})}{(1 - x^2)(1 - x^4)\cdots(1 - x^{2n})}. \qquad (56)$$

We recall the conclusion we obtained: to find the coefficients of the powers of x of a degree not higher than $2n$ in product (46), it is sufficient to find the same terms in finite product (47). In particular, this relates

to terms in $A_0(x)$: in the exponents not exceeding $2n$, they coincide
with the corresponding terms in $a_0(x)$. But all the powers of x in the
numerator in formula (56) have an exponent exceeding $2n$. In calculating
terms of a degree not exceeding $2n$, they can therefore be discarded. We
see that the terms of a degree not exceeding $2n$ in the series $A_0(x)$ are
the same as in the series

$$\frac{1}{(1 - x^2)(1 - x^4) \cdots (1 - x^{2n})}.$$

Our result is valid for any n. This proves that the equality

$$A_0(x) = \frac{1}{(1 - x^2)(1 - x^4) \cdots (1 - x^{2n}) \cdots}$$

holds, where all binomials $1 - x^{2n}$ with natural numbers n are multiplied.
Combined with formula (51), this completely determines product (46).
Multiplying by the denominator and grouping the factors differently (we
discussed the lawfulness of this operation in Secs. 15 and 16), we obtain
the relation

$$(1 + xz)(1 + xz^{-1})(1 - x^2)$$
$$\cdots (1 + x^{2n-1}z)(1 + x^{2n-1}z^{-1})(1 - x^{2n}) \cdots$$
$$= 1 + x(z + z^{-1}) + x^4(z^2 + z^{-2})$$
$$+ \cdots + x^{n^2}(z^n + z^{-n}) + \ldots, \qquad (57)$$

which is quite elegant in itself. In the left-hand side, factors occur in
goups of three: $(1 + x^{2n-1}z)(1 + x^{2n-1}z^{-1})(1 - x^{2n})$ for $n = 1, 2, \ldots$.

The pentagon theorem is a consequence of identity (57). Indeed, we
set $x = y^3$ and $z = -y$ in it. Then in the left-hand side, we have

$$1 + x^{2n-1}z = 1 - y^{6n-2},$$
$$1 + x^{2n-1}z^{-1} = 1 - y^{6n-4},$$
$$1 - x^{2n} = 1 - y^{6n},$$

that is, the left-hand side of identity (57) contains all products of $1 - y^n$
with even natural numbers n. In the right-hand side, the term $x^{n^2}z^n$
yields $(-1)^n y^{3n^2+n}$, and the term $x^{n^2}z^{-n}$ yields $(-1)^n y^{3n^2-n}$. We obtain
only terms with even powers of y in both the left-hand and right-hand
series in the identity. We can therefore set $y^2 = t$. As a result, the
left-hand side contains all factors $1 - t^n$ with natural number n, and the
right-hand side contains the sum of terms $(-1)^n t^{(3n^2 \pm n)/2}$. This equality
is the pentagon theorem. $\qquad \square$

Problems:

1. Find the representation in the form of a power series for the product
$$(1-x)^2(1-x^2)(1-x^3)^2(1-x^4)\cdots(1-x^{2n+1})^2(1-x^{2n+2})\ldots.$$

2. Find the representation in the form of a power series for the product
$$(1+x)^2(1-x^2)(1+x^3)^2(1-x^4)\cdots(1+x^{2n+1})^2(1-x^{2n+2})\ldots.$$

3. Prove the identity
$$(1-x^2)(1+x)(1-x^4)\cdots(1+x^n)(1-x^{2n+2})\cdots$$
$$= 1+x+x^3+x^6+\cdots+x^{n(n+1)/2}+\ldots.$$

4. Prove the identity
$$\frac{(1-x^2)(1-x^4)(1-x^6)\cdots}{(1-x)(1-x^3)(1-x^5)\cdots} = 1+x+x^3+x^6+\cdots+x^{n(n+1)/2}+\ldots.$$

Supplement 2: Generating Function for Bernoulli Numbers

We consider one remarkable power series,

$$e(x) = 1 + \frac{x}{1!} + \frac{x^2}{2!} + \cdots + \frac{x^n}{n!} + \ldots. \tag{58}$$

If we prove that a number can be substituted in place of x, then we thus obtain an important function, namely, we can prove that $e(x) = e^x$, where e is the base of the natural logarithms. But we stay within the realm of the purely algebraic theory of power series. All the same, we show that the series $e(x)$ retains certain properties of an exponential function.

We introduce a new unknown y and then consider the series $e(y)$ and $e(x+y)$. We prove the identity

$$e(x+y) = e(x)e(y). \tag{59}$$

Indeed, we substitute $x+y$ in place of x in formula (58). The term of degree n has the form $(x+y)^n/n!$. We expand $(x+y)^n$ according to the binomial formula and use the expression we found for binomial coefficients, formula (25) in Sec. 6:

$$(x+y)^n = x^n + \frac{n!}{1!(n-1)!}x^{n-1}y + \frac{n!}{2!(n-2)!}x^{n-2}y^2 + \cdots + y^n.$$

Therefore,

$$\frac{(x+y)^n}{n!} = \frac{x^n}{n!} + \frac{x^{n-1}}{(n-1)!}\frac{y}{1!} + \frac{x^{n-2}}{(n-2)!}\frac{y^2}{2!} + \cdots + \frac{y^n}{n!}.$$

This is the sum of expressions of the form

$$\frac{x^k}{k!}\frac{y^{n-k}}{(n-k)!}$$

for $k = n, n-1, \ldots, 0$, that is, the sum of the products of terms of degree k in the series $e(x)$ and terms of the degree $n-k$ in this series $e(y)$. But exactly such are the terms of degree n in the series $e(x)e(y)$. This proves identity (59).

In essence, formula (59) is equivalent to the binomial formula and contains all the binomial formulas for all values of n.

Thanks to this remarkable property of the series $e(x)$, it is convenient to use it to construct new types of generating functions. Let a be a sequence $\alpha_0, \alpha_1, \alpha_2, \ldots, \alpha_n, \ldots$. We introduce a series that is denoted by $e(ax)$ (a is a sequence!),

$$e(ax) = \alpha_0 + \frac{\alpha_1 x}{1!} + \frac{\alpha_2 x^2}{2!} + \frac{\alpha_3 x^3}{3!} + \cdots + \frac{\alpha_n x^n}{n!} + \cdots.$$

This is also called the *factorial generating function* of the sequence a. If we defined addition of sequences by element, that is, for $a = (\alpha_0, \alpha_1, \ldots, \alpha_n, \ldots)$ and $b = (\beta_0, \beta_1, \ldots, \beta_n, \ldots)$, we set $a + b = (\alpha_0 + \beta_0, \alpha_1 + \beta_1, \ldots, \alpha_n + \beta_n, \ldots)$, then obviously

$$e((a+b)x) = e(ax) + e(bx). \tag{60}$$

We apply the notation introduced in the supplement to Chap. 2 to power series. Namely, if $f(t, x) = f_0(t) + f_1(t)x + \cdots + f_n(x)x^n + \cdots$ is a power series whose coefficients are polynomials and if a is a sequence, then we set

$$f(a, x) = f_0(a) + f_1(a)x + \cdots + f_n(a)x^n + \cdots.$$

The meaning of the expression $f(a)$, where $f(t)$ is a polynomial, was defined in the supplement to Chap. 2: if $f(t) = a_0 + a_1(t) + \cdots + a_m t^m$ and $a = (\alpha_0, \alpha_1, \ldots, \alpha_n, \ldots)$, then $f(a) = a_0 + a_1\alpha_1 + \cdots + a_m\alpha_m$. In this notation, the factorial generating function for a sequence a is written as $e(ax)$. It is easy to see that the analogue of relation (59) holds:

$$e((\alpha + a)x) = e(\alpha x)e(ax). \tag{61}$$

Here, α is a number, $a = (\alpha_0, \alpha_1, \ldots, \alpha_n, \ldots)$ is a sequence, and $\alpha + a$ denotes the sequence $a' = (\alpha_0 + \alpha, \ldots, \alpha_n + \alpha, \ldots)$. The proof is exactly the same as for relation (59). The term of degree n in the left-hand side is equal to $(\alpha + a)^m/m!$ by definition, that is, to the sum of terms

$$\frac{1}{m!}\frac{m!}{k!(m-k)!}\alpha^k a_{n-k}x^n = \frac{1}{k!}\alpha^k x^k \frac{1}{(n-k)!}a_{n-k}x^{n-k}.$$

But this is the product of the term of degree k in $e(\alpha x)$ and the term of degree $n - k$ in $e(ax)$. By the definition of multiplication of power series, the sum of all such terms is the term of degree n in the product $e(\alpha x)e(ax)$. This proves equality (61).

We now use the identities introduce to find the factorial generating function for the sequence of Bernoulli numbers $B = (B_0, B_1, \ldots, B_n, \ldots)$. We recall that the Bernoulli numbers are defined using the relation

$$(B+1)^m - B_m = m, \quad m = 1, 2, \ldots . \tag{62}$$

We consider the factorial generating function for the three sequences in relation (62). These are the sequence $(1 + B)^m$, the sequence B_m, and the sequence (in the right-hand side) with the form $0, 1, \ldots, n, \ldots$, that is, the sequence of all nonnegative integers. The last sequence is denoted by N. In view of property (60), we can write all relations (62) in the form

$$e\big((1 + B)x\big) - e(Bx) = e(Nx), \tag{63}$$

and in view of property (62), $e\big((1 + B)x\big) = e(x)e(Bx)$. It remains to find the series $e(Nx)$. Its term of degree n is equal to

$$\frac{n}{n!}x^n = \frac{1}{(n-1)!}x^{n-1}x.$$

Therefore, $e(Nx) = xe(x)$, and relation (63) becomes

$$e(Bx)(e(x) - 1) = xe(x).$$

Hence,

$$e(Bx) = \frac{xe(x)}{e(x) - 1}. \tag{64}$$

This is the form of the factorial generating function for the Bernoulli numbers. We note that the series $e(x) - 1$ in the denominator has the free term 0. We can therefore factor out x from this series, which then cancels with the x in the numerator, and the remaining power series has a free term equal to 1 and therefore has an inverse according to Theorem 49 (in Sec. 22). It is very simple to deduce all properties of the Bernoulli numbers from this form of the generating function. For example, we prove that all Bernoulli numbers with an odd index greater than 1 are equal to zero (see Problem 3 in the supplement to Chap. 2). As we know, $B_1 = 1/2$. This follows easily from formula (64). Consequently, our assertion implies that the series $e(Bx) - x/2$ contains only terms with even powers of x. If we replace x with $-x$ in the power series $f(x)$, then terms with even powers of x do not change, and terms with odd powers of x change their signs. That a series $f(x)$ contains only terms with even powers of x is equivalent to the series does not change under the substitution of $-x$ for x, that is, $f(-x) = f(x)$.

Therefore, we must verify that the series $e(Bx) - x/2$ does not change when x is replaced with $-x$. Using expression (64) for the series $e(Bx)$, we find that our assertion is equivalent to the identity

$$\frac{xe(x)}{e(x) - 1} - \frac{x}{2} = \frac{-xe(-x)}{e(-x) - 1} + \frac{x}{2}.$$

We divide both sides by x and subtract $1/2$ from both sides. We let u denote $e(x)$. According to identity (59), $e(-x) = u^{-1}$. Our equality now becomes

$$\frac{u}{u - 1} - 1 = -\frac{u^{-1}}{u^{-1} - 1}.$$

This is obvious—we put the left-hand side over the common denominator and multiply the numerator and denominator in the right-hand side by u.

We use this new approach to deduce the connection between Bernoulli numbers and sums of powers of consecutive natural numbers, $S_m(n) = 1^m + 2^m + \cdots + n^m$. In the supplement to Chap. 2, we proved the formula

$$S_m(n) = \frac{1}{m + 1}\left((B + n)^{m+1} - B_{m+1}\right),$$

which can be written with m replaced with $m - 1$:

$$S_{m-1}(n) = \frac{1}{m}\left((B + n)^m - B_m\right). \tag{65}$$

We deduce this formula again.

For this, we consider the factorial generating function for the sequence $(B + n)^m - B_m$ (for a fixed n). In view of property (60), it can be rewritten in the form $e\left((B + n)x\right) - e(Bx)$. According to identity (61), this series is equal to $e(Bx)e(nx) - e(Bx)(e(nx) - 1)$. It easily follows from property (60) (by induction on n) that $e(nx) = e(x)^n$. Substituting this expression and expression (64) for $e(Bx)$, we write our series in the form

$$xe(x)\frac{e(x)^n - 1}{e(x) - 1}.$$

In view of identity (12) in Chap. 1, we have

$$\frac{e(x)^n - 1}{e(x) - 1} = 1 + e(x) + \cdots + e(x)^{n-1}.$$

Therefore,

$$xe(x)\frac{e(x)^n - 1}{e(x) - 1} = x\big(e(x) + \cdots + e(x)^n\big).$$

Again replacing $e(x)^k$ with $e(kx)$ for all $k = 1, 2, \ldots, n$, we obtain

$$xe(x)\frac{e(x)^n - 1}{e(x) - 1} = x\big(e(x) + e(2x) + \cdots + e(nx)\big).$$

We find the coefficient of x^m in the series in the right-hand side. It is equal to the coefficient of x^{m-1} in the series $e(x) + e(2x) + \cdots + e(nx)$. In $e(kx)$, the coefficient of x^{m-1} is equal to $k^{m-1}/(m-1)!$, and in the entire sum, it is equal to

$$\frac{1}{(m-1)!} + \frac{2^{m-1}}{(m-1)!} + \cdots + \frac{n^{m-1}}{(m-1)!} = \frac{S_{m-1}(n)}{(m-1)!}.$$

We set $\alpha_m = (B + n)^m - B_m$ and let a denote the entire sequence $\alpha_0, \alpha_1, \ldots, \alpha_n, \ldots$. We have proved that the coefficient of x^m in the series $e(ax)$ is equal to $S_{m-1}(n)/(m-1)!$. By definition, it is equal to $\alpha_m/m!$. Therefore,

$$\frac{\alpha_m}{m!} = \frac{S_{m-1}(n)}{(m-1)!},$$

whence follows equality (65).

Problems:

1. Define the sequence B'_n, where $B'_1 = -1/2$ and $B'_n = B_n$ for $n \geq 2$. Prove that the factorial generating function for the sequence B'_n has the form $e(B't) = t/(e^t - 1)$. For the sequence B'_n, prove the relation $(B' + 1)^m = B'_m$ for $m \geq 2$.
2. Verify the relation

$$e\left(\left(B - \frac{1}{2}\right)x\right) = 2e\left(B\frac{x}{2}\right) - e(Bx).$$

3. Prove that the Bernoulli polynomial $B_m(x)$ for even m has the root $x = -1/2$. Hint: Use Problems 2 and 3 in the supplement to Chap. 2.

4. Prove that infinite sum (58) exists not only as a power series but also in the sense of the definition given in Sec. 15, where any real number is taken for x. *Hint:* Replace x with $|x|$, and reduce the assertion to the case where $x > 0$. For $x > 0$, prove that

$$\frac{x^n}{n!} + \cdots + \frac{x^{n+k}}{(n+k)!} \leq \frac{x^n}{n!}(1 + \alpha + \cdots + \alpha^k),$$

where $\alpha = x/n$, and then apply formula (2) in Chap. 5 for $n > x$.

Dates of Lives of Mathematicians Mentioned in the Text

The section in the text in which the mathematician is mentioned is shown on the right in the following table. The symbol S13 denotes the supplement that follows Sec. 13.

B.C.

Pythagoras	c580–c500	1
Theodorus of Cyrene	c470–c399	2
Theaetetus	c414–369	2
Euclid	c465–c300	2, 3, 11
Archimedes	287?–212	6
Eratosthenes	c276?–195?	13

A.D.

Leonardo da Pisa (Fibonacci)	c1170–1240?	3
Francois Viète	1540–1603	8
Galileo Galilei	1564–1642	18
René Descartes	1596–1650	4
Pierre de Fermat	1601–1665	6
Blaise Pascal	1623–1662	6, 22
Isaac Barrow	1630–1677	6
Isaac Newton	1642–1727	6, 22
Etienne Rolle	1652–1719	17
Jakob Bernoulli	1654–1705	6, S6, 10, S10, 15, 20, 22, S23b
Leonhard Euler	1707–1783	3, S6, 9, 11, 12, 13, S13, 15, 22, 23, S23a
Etienne Bézout	1730–1783	4
Pierre Laplace	1749–1827	10
Carl Friedrich Gauss	1777–1855	2, S13, 21, 23, S23a
Bernard Bolzano	1781–1848	17

Jacques Sturm	1803–1855	S17
Carl Jacobi	1804–1851	S23a
Eugéne Catalan	1814–1894	22
Pafnutii L'vovich Chebyshev	1821–1894	S10, S13, S20
Bernhard Riemann	1826–1866	S13
Richard Dedekind	1831–1916	7, 18, 19
Georg Cantor	1845–1918	19
Henri Poincaré	1854–1912	2

Index

Universitext

Aksoy, A.; Khamsi, M. A.: Methods in Fixed Point Theory

Alevras, D.; Padberg M. W.: Linear Optimization and Extensions

Andersson, M.: Topics in Complex Analysis

Aoki, M.: State Space Modeling of Time Series

Aupetit, B.: A Primer on Spectral Theory

Bachem, A.; Kern, W.: Linear Programming Duality

Bachmann, G.; Narici, L.; Beckenstein, E.: Fourier and Wavelet Analysis

Badescu, L.: Algebraic Surfaces

Balakrishnan, R.; Ranganathan, K.: A Textbook of Graph Theory

Balser, W.: Formal Power Series and Linear Systems of Meromorphic Ordinary Differential Equations

Bapat, R.B.: Linear Algebra and Linear Models

Benedetti, R.; Petronio, C.: Lectures on Hyperbolic Geometry

Berberian, S. K.: Fundamentals of Real Analysis

Berger, M.: Geometry I, and II

Bliedtner, J.; Hansen, W.: Potential Theory

Blowey, J. F.; Coleman, J. P.; Craig, A. W. (Eds.): Theory and Numerics of Differential Equations

Börger, E.; Grädel, E.; Gurevich, Y.: The Classical Decision Problem

Böttcher, A; Silbermann, B.: Introduction to Large Truncated Toeplitz Matrices

Boltyanski, V.; Martini, H.; Soltan, P. S.: Excursions into Combinatorial Geometry

Boltyanskii, V. G.; Efremovich, V. A.: Intuitive Combinatorial Topology

Booss, B.; Bleecker, D. D.: Topology and Analysis

Borkar, V. S.: Probability Theory

Carleson, L.; Gamelin, T. W.: Complex Dynamics

Cecil, T. E.: Lie Sphere Geometry: With Applications of Submanifolds

Chae, S. B.: Lebesgue Integration

Chandrasekharan, K.: Classical Fourier Transform

Charlap, L. S.: Bieberbach Groups and Flat Manifolds

Chern, S.: Complex Manifolds without Potential Theory

Chorin, A. J.; Marsden, J. E.: Mathematical Introduction to Fluid Mechanics

Cohn, H.: A Classical Invitation to Algebraic Numbers and Class Fields

Curtis, M. L.: Abstract Linear Algebra

Curtis, M. L.: Matrix Groups

Cyganowski, S.; Kloeden, P.; Ombach, J.: From Elementary Probability to Stochastic Differential Equations with MAPLE

Dalen, D. van: Logic and Structure

Das, A.: The Special Theory of Relativity: A Mathematical Exposition

Debarre, O.: Higher-Dimensional Algebraic Geometry

Deitmar, A.: A First Course in Harmonic Analysis

Demazure, M.: Bifurcations and Catastrophes

Devlin, K. J.: Fundamentals of Contemporary Set Theory

DiBenedetto, E.: Degenerate Parabolic Equations

Diener, F.; Diener, M.(Eds.): Nonstandard Analysis in Practice

Dimca, A.: Singularities and Topology of Hypersurfaces

DoCarmo, M. P.: Differential Forms and Applications

Duistermaat, J. J.; Kolk, J. A. C.: Lie Groups

Edwards, R. E.: A Formal Background to Higher Mathematics Ia, and Ib

Marcus, D. A.: Number Fields

Martinez, A.: An Introduction to Semiclassical and Microlocal Analysis

Matsuki, K.: Introduction to the Mori Program

Mc Carthy, P. J.: Introduction to Arithmetical Functions

Meyer, R. M.: Essential Mathematics for Applied Field

Meyer-Nieberg, P.: Banach Lattices

Mines, R.; Richman, F.; Ruitenburg, W.: A Course in Constructive Algebra

Moise, E. E.: Introductory Problem Courses in Analysis and Topology

Montesinos-Amilibia, J. M.: Classical Tessellations and Three Manifolds

Morris, P.: Introduction to Game Theory

Nikulin, V. V.; Shafarevich, I. R.: Geometries and Groups

Oden, J. J.; Reddy, J. N.: Variational Methods in Theoretical Mechanics

Øksendal, B.: Stochastic Differential Equations

Poizat, B.: A Course in Model Theory

Polster, B.: A Geometrical Picture Book

Porter, J. R.; Woods, R. G.: Extensions and Absolutes of Hausdorff Spaces

Radjavi, H.; Rosenthal, P.: Simultaneous Triangularization

Ramsay, A.; Richtmeyer, R. D.: Introduction to Hyperbolic Geometry

Rees, E. G.: Notes on Geometry

Reisel, R. B.: Elementary Theory of Metric Spaces

Rey, W. J. J.: Introduction to Robust and Quasi-Robust Statistical Methods

Ribenboim, P.: Classical Theory of Algebraic Numbers

Rickart, C. E.: Natural Function Algebras

Rotman, J. J.: Galois Theory

Rubel, L. A.: Entire and Meromorphic Functions

Rybakowski, K. P.: The Homotopy Index and Partial Differential Equations

Sagan, H.: Space-Filling Curves

Samelson, H.: Notes on Lie Algebras

Schiff, J. L.: Normal Families

Sengupta, J. K.: Optimal Decisions under Uncertainty

Séroul, R.: Programming for Mathematicians

Seydel, R.: Tools for Computational Finance

Shapiro, J. H.: Composition Operators and Classical Function Theory

Simonnet, M.: Measures and Probabilities

Smith, K. E.; Kahanpää, L.; Kekäläinen, P.; Traves, W.: An Invitation to Algebraic Geometry

Smith, K. T.: Power Series from a Computational Point of View

Smoryński, C.: Logical Number Theory I. An Introduction

Stichtenoth, H.: Algebraic Function Fields and Codes

Stillwell, J.: Geometry of Surfaces

Stroock, D. W.: An Introduction to the Theory of Large Deviations

Sunder, V. S.: An Invitation to von Neumann Algebras

Tamme, G.: Introduction to Étale Cohomology

Tondeur, P.: Foliations on Riemannian Manifolds

Verhulst, F.: Nonlinear Differential Equations and Dynamical Systems

Wong, M. W.: Weyl Transforms

Zaanen, A.C.: Continuity, Integration and Fourier Theory

Zhang, F.: Matrix Theory

Zong, C.: Sphere Packings

Zong, C.: Strange Phenomena in Convex and Discrete Geometry

Printed in Italy by Legoprint S.p.A., Lavis (Trento)

CPSIA information can be obtained
at www.ICGtesting.com
Printed in the USA
LVHW021510090921
697458LV00004B/173

9 783540 422532